PROTEIN EVOLUTION

Protein Evolution

Second edition

László Patthy
Institute of Enzymology
Biological Research Center
Hungarian Academy of Sciences
Budapest
Hungary

Blackwell
Publishing

© 1999, 2008 by Blackwell Publishing Ltd

BLACKWELL PUBLISHING
350 Main Street, Malden, MA 02148-5020, USA
9600 Garsington Road, Oxford OX4 2DQ, UK
550 Swanston Street, Carlton, Victoria 3053, Australia

First edition published 1999 by Blackwell Publishing Ltd
Second edition published 2008

1 2008

Library of Congress Cataloging-in-Publication Data

Patthy, László.
    Protein evolution / László Patthy. — 2nd ed.
        p. ; cm.
    Includes bibliographical references and index.
    ISBN 978-1-4051-5166-5 (pbk. : alk. paper) 1. Proteins—Evolution. I. Title.
    [DNLM: 1. Proteins—physiology. 2. Evolution, Molecular. 3. Protein Conformation.
QU55 P3185p 2007]
QP551.P354 2007
572'.6—dc22

                                                        2007018803

A catalogue record for this title is available from the British Library.

Set in 9.5/13pt Meridien
by Graphicraft Limited, Hong Kong
Printed and bound in Singapore
by Utopia Press Pte Ltd

For further information on
Blackwell Publishing, visit our website at
www.blackwellpublishing.com

Contents

Preface to the first edition

'Nothing in biology makes sense except in the light of evolution'
(Theodosius Dobzhansky)

Progress with the Human Genome Project, the sequencing of the genomes of a number of model organisms such as those of prototypical bacteria (*Mycoplasma genitalium, Mycoplasma pneumoniae, Haemophilus influenzae, Escherichia coli, Borrelia burgdorferi, Helicobacter pylori, Aquifex aeolicus, Mycobacterium tuberculosis*), cyanobacteria (*Synechocystis sp.*), Archaea (*Methanoccoccus jannaschii, Archaeoglobus fulgidus, Methanobacterium thermoautotrophicum*), unicellular fungi (*Saccharomyces cerevisiae*), plants (*Arabidopsis thaliana*), nematodes (*Caenorhabditis elegans*), arthropods (*Drosophila melanogaster*), and rodents (*Mus musculus*) has led to rapid development of the science called genomics.

The flood of information on gene and protein sequences has revolutionized molecular and evolutionary biology and calls for more efficient computational tools for the elucidation of the functions of the newly identified genes and proteins. Reliable prediction of the function of a novel gene/protein requires a complex analysis of genomic and protein sequence information, including sensitive homology searches (assigning the novel protein to a family), detailed evolutionary analyses (identifying the closest relatives, prediction of the most probable function of the last common ancestor), analysis of regulatory features of the gene, and predictions of the tissue distribution, subcellular localization and structure–function aspects of the protein. In genome-sequencing projects the same type of complex analyses have to be carried out on a massive scale, in an automated way. As a result, there is now a spectacular burst of activity in the field of bioinformatics, to extract functional information from genomic and sequence data. Functional genomics absolutely requires the development of novel algorithms and software that fully exploit the principles governing the evolution of protein structure and function.

The major goal of this book is to provide a concise summary of recent advances in our understanding of the principles of the evolution of protein structure and function. The majority of the text will be dedicated to areas at the frontiers of protein evolution research; the well-established aspects will be summarized more briefly. The book will briefly survey techniques that are currently available for the analysis of the evolutionary history of proteins (homology searches and analysis of molecular phylogenies based on comparisons of nucleic acid sequences and exon–intron organization of protein-coding genes, comparison of amino acid sequences and three-dimensional

structures of proteins), as well as for the prediction of the structure, function and structure–function aspects of proteins. The emphasis will be on the
conceptual basis of these techniques, what their limitations are and how much
information can be obtained. Although the danger of misinterpretation is real
(false claims for 'homology', and errors in the prediction of the structure and
function of proteins) evolutionary analyses offer very useful tools for science
and biotechnology. These techniques can provide valuable hints for experimental design years ahead of direct functional or crystallographic information on novel proteins, and may facilitate the identification of genes that may
have biotechnological or medical importance.

The book aims to provide the background necessary for the reader to
make an informed judgement of the value, reliability and limitations of the
techniques used for evolutionary analyses. The book is written for beginners
in protein evolution, functional genomics and bioinformatics, and contains
instructive examples that illustrate how biological information may be dis-
tilled from sequence data.

Preface to the second edition

Just as the first edition, the new version of this book is also intended for beginners in protein evolution, genome evolution and molecular bio-informatics to provide a concise summary of the principles of the evolution of protein structure and function. I tried to write the book in a readable form at a level accessible to undergraduates and graduate students as well as to newcomers to the field.

The manuscript of the first edition of this book was completed in 1998, at a time when the genomics revolution (and the associated revolution in bioinformatics) was just starting to unfold. As a result of the rapid increase in the number of completely sequenced genomes and the development of novel tools of bioinformatics some of the chapters of the first edition aged very rapidly. This is especially true for Chapter 8 (Protein Evolution by Assembly from Modules) and Chapter 9 (Genome Evolution and Protein Evolution), therefore in the second edition these chapters had to be modified and expanded significantly. The updated Chapter 9 now covers the genomes of key model organisms sequenced since the completion of the first edition and also discusses important novel issues of the evolution of the genomes of organelles and viruses. The chapters also dedicate more space to the issue of lateral gene transfer and alternative splicing.

The other chapters of the book, summarizing well-established aspects of proteins, protein-coding genes and general principles of protein evolution were updated to focus on new developments in these areas, giving key references that may guide readers in the most recent primary literature.

Unlike the first version of this book, the second edition includes references to the most widely used databases and bioinformatic tools that are currently available for the analysis of the evolutionary history of proteins, as well as for the prediction of the structure, function and structure–function aspects of proteins and protein-coding genes. (The web sites were selected based on their current popularity, which seems to guarantee that they would survive during the lifetime of the second edition.)

Many colleagues helped in finding errors in the first edition and gave valuable suggestions for improvement. I wish to thank László Bányai and Mária Trexler for carefully reviewing the entire manuscript and for their help in the preparation of the figures.

László Patthy
October 2006, Budapest

Acknowledgements

I wish to thank Hedvig Tordai, Mária Trexler and László Bányai for carefully reviewing the entire manuscript and for their valuable suggestions.

Introduction

Protein evolution is a discipline whose participants are interested in understanding the principles and driving forces whereby new protein-coding genes, new protein structures and new biological functions emerge as a consequence of changes in the genetic material during the course of evolution. Since *de novo* formation of protein-coding genes has a vanishingly low probability, these studies usually involve the comparison of proteins (protein-coding genes) derived from a common ancestor. Using information on extant protein-coding genes, attempts are made to reconstruct the ancestral gene/protein and the actual events that led to the formation of its descendants. Aside from its theoretical interest, understanding the rules that govern the changes in protein sequence, structure and function has obvious practical importance for functional genomics and drug discovery. These rules may form the basis for the development of more sensitive homology search protocols, may permit more reliable prediction of the function of novel genes and may guide protein-engineering studies.

Chapter 1: Protein-coding genes

The genetic material of most organisms is double-stranded DNA, except that some viruses use single-stranded DNA or RNA. In DNA the information is coded as sequences of the purine and pyrimidine nucleotides: adenine (A), guanine (G), cytosine (C) and thymine (T); RNA has the same nucleotides except that thymine is replaced by uracil (U). The nucleotide sequences of the two strands of DNA are complementary, with characteristic purine/ pyrimidine pairings: A pairs with T, G pairs with C.

Specific regions of the genetic material that encode some type of essential function are called **genes**; the regions between genes are usually referred to as **intergenic regions**. In its broadest sense, a gene is any type of sequence of genomic DNA or RNA that is indispensable for a specific biological function, irrespective of whether that function requires the transcription or translation of the gene. In this sense we may speak of nontranscribed **regulatory genes**, transcribed **RNA genes** and translated **protein-coding genes**.

Nontranscribed regulatory genes include regions that define the sites for initiation and termination of DNA replication, genes that serve as sites for the attachment to chromosomes of the segregation machinery during meiosis and mitosis, etc. A special group of nontranscribed regulatory genes are those DNA segments that serve as sites for the attachment of proteins regulating the expression of genes. Although in many cases they may be part of transcribed genes, sometimes they are located outside the boundaries of such genes. For example, enhancer sequences can stimulate transcription of genes that are thousands of bases away. The basis of such remote control is that the binding of regulatory proteins induces the formation of DNA loops, thereby bringing sites distant in linear sequence into proximity. Recent studies indicate that enhancer elements can also act in *trans*, i.e. they can interact with promoter(s) on other chromosomes (Lomvardas et al., 2006; Savarese & Grosschedl, 2006).

In a more conventional, narrower sense, the term gene is used to refer to **structural genes**, which are transcribed. The **noncoding RNA genes**, which are only transcribed, may yield stable RNA products such as **ribosomal RNA** (rRNA), **transfer RNA** (tRNA), **small nuclear RNA** (snRNA), **antisense RNA**, **ribozyme RNA** etc. **Small nucleolar RNAs** (snoRNA) are involved in directing the processing and modification of eukaryotic rRNAs in the nucleolus. **MicroRNAs** (miRNAs) represent an increasingly important class of noncoding RNA genes whose products are ~22 nt sequences that play important roles in the regulation of translation and degradation of **messenger RNAs** (mRNAs) through base pairing to partially complementary

1

sites in the untranslated regions (UTRs) of the message. MicroRNAs, together with **small interfering RNAs** (siRNAs), **piwi-interacting RNAs** (piRNAs), **repeat-associated small interfering RNAs** (rasiRNAs) and other types of **nontranslated RNAs** (ntRNAs) fulfil important roles in genome control (Bickel & Morris, 2006; Carthew, 2006; Lau et al., 2006; Vagin et al., 2006).

In the case of the protein-coding genes the RNA transcript is subsequently translated into protein. In this book we will concentrate on this latter category of genes.

1.1 Structure of protein-coding genes

The schematic organization of a eukaryotic protein-coding gene is shown in Fig. 1.1. Such genes may be first subdivided into **transcribed** and **non-transcribed parts**. The region upstream from the transcribed region (**5′ flanking region**) contains most of the signals that regulate the initiation of the transcription (gene expression) process. Since these regulatory sequences may promote the transcription process, they are usually referred to as promoters, and the corresponding region is usually called the **promoter region**. The promoter region for RNA polymerase II is located on the 5′ side of the **initiation site for transcription** and usually contains the so-called **TATA box** (**Hogness box**), which is closest to the start site, 19–27 base pairs upstream of the start point of transcription. The **CAAT box** is farther upstream, and there are one or more copies of the **GC box**, consisting of the sequence GGGCGG or its variants. The CAAT and GC boxes control the initial binding of the RNA polymerase, while the TATA box controls the start point of

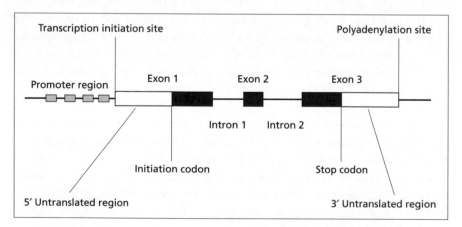

Fig. 1.1 Schematic structure of a eukaryotic protein-coding gene. The large rectangles represent exons, the small boxes represent regulatory elements in the promoter region. The translated regions of exons are shown in black. The thin lines separating exons represent the introns.

transcription. The positions of the CAAT box and GC boxes vary from one promoter to another.

The 3′ part of the gene contains signals for the **termination** of the transcription process and **poly(A) addition**.

The transcribed part of a typical eukaryotic gene consists of **exons** and **introns**. Introns, or intervening sequences, are those transcribed sequences that are removed later from the **primary transcript**, during the processing of the **pre-mRNA** (pre-messenger RNA) molecules to **mRNA** (messenger RNA). Genomic sequences that remain in the mature mRNA following **splicing** are referred to as **exons**, irrespective of whether they are translated or not. Exons or parts of exons that are translated are collectively referred to as **protein-coding regions**. Those regions of mRNA that are not translated are called the **5′** and **3′ untranslated regions**, depending on whether they are upstream or downstream of the translated region.

Protein-coding genes of bacteria differ from those of eukaryotes primarily in that the polypeptide chain is encoded by a single continuous segment, which is usually not interrupted by introns. Promoters of bacteria contain a −10 sequence and a −35 sequence, so called because they are located 10 bp and 35 bp upstream of the initiation site of transcription. The −10 sequence, also known as the **Pribnow box**, has the sequence TATAAT or its variant, while the −35 sequence has a variant of the sequence TTGACA. In prokaryotes several protein-coding genes may be arranged consecutively to form a unit of gene expression, called an **operon**. They are transcribed into one molecule of mRNA but are subsequently translated into different proteins. Such a unit ensures coordinated expression of the genes that belong to the same operon.

1.2 Transcription

In eukaryotes, transcriptionally active and inactive regions of DNA have distinguishable features of **chromatin** structure. Transcriptionally inactive chromatin generally exists in a highly condensed conformation, whereas transcriptionally active DNA adopts a more open conformation thanks to an extensive acetylation of the four types of histones (see section 4.4.2). During interphase most of the chromatin is much less tightly packed than in the mitotic chromosomes. Such diffusely staining material, called **euchromatin**, contains transcriptionally active regions interspersed with nontranscribed DNA sequences. Some regions of chromatin contain fibres that are highly condensed throughout the cell cycle and form dark-staining **heterochromatin**, which is essentially transcriptionally inactive. Facultative heterochromatin can be active or inactive, as in the case of mammalian X chromosome inactivation. Constitutive heterochromatin is always in the inactive state, such as the repetitive DNA sequences found around the **centromeres** of chromosomes.

Genes encoding proteins essential for general cell functions are transcribed and expressed in all cell types; they are usually called **housekeeping genes**. The expression of other genes may be largely restricted to a specific cell type, a specific physiological state, or a specific developmental stage. The regulation of the expression of genes is determined by the interaction of **regulatory proteins** with **control elements**.

Expression of a protein-coding gene starts with the transcription of the nucleotide sequence into a complementary RNA molecule. Eukaryotic RNA polymerases cannot initiate transcription by themselves: short sequence elements in the immediate vicinity of a gene act as recognition signals for **transcription factors** that bind to the DNA, guiding and activating the polymerase. For example, GC boxes bind the transcription factor Sp 1, the CAAT box binds CTF (CAAT-binding transcription factor), the heat-shock transcription factor (HSTF) binds to unique sequences present in the promoter of heat-shock genes, etc. Such short **control elements** are often clustered upstream of the coding sequence of a gene in the promoter region. After a number of transcription factors bind to the promoter region, an RNA polymerase binds to the **transcription factor complex** and is activated to initiate the synthesis of RNA. Most regulatory proteins bind to the promoter region, but others act at other regions, which can be upstream, downstream, or in the middle of a gene. **Enhancers** are frequently located at a considerable distance from the transcriptional start site and can enhance the transcriptional activity of specific eukaryotic genes by binding gene regulatory proteins. In this case the DNA between the promoter and enhancer loops out, allowing the proteins bound to the enhancer to interact with the transcription factors or with the RNA polymerase. The binding of regulatory proteins to the DNA molecule can also have a negative effect on transcription of a gene (**silencers**). A variety of mechanisms are used for various genes, often involving the binding of several different regulatory proteins to adjacent parts of the promoter region of the gene.

Regulation of the transcription of a protein-coding gene is critical for the control of cell metabolism, cell differentiation and development of organisms, therefore it is of great importance for evolutionary biology. Changes in transcription regulation have a great impact on the evolution of proteins, since a mutation in a control region may bring about a sudden change in the function of a protein (e.g. by changing the time or place of its expression) even if the structure of the protein is practically unchanged (see section 7.1.2). The **expression profile** is an important parameter of the biological function of a protein.

In the presence of appropriate transcription factors and regulatory proteins the gene is transcribed by RNA polymerase which copies one strand of the DNA into its complementary RNA sequence. Transcription is initiated at a site defined by specific sequences in the promoter region of the gene. DNA is copied in a 5' to 3' direction by assembling nucleoside 5'-triphosphates complementary to the template strand. In eukaryotic cells, the 5'-triphosphate group of the primary transcript is modified to produce a **5'-cap structure** by coupling it to the 5' position of a methylated guanine base. This modification plays a significant role in the stability and translation of the mRNA.

Transcription proceeds until the process reaches signals that terminate transcription. In prokaryotes intrinsic termination may be triggered by terminator sequences. The terminator sequences are usually sequences (palindromic sequences, inverted repeats) that form stable stem-loop structures, thereby stopping RNA polymerase and causing its dissociation from the DNA template.

Much less is known about the eukaryotic transcription termination signals of protein-coding genes, but it is clear that termination is intimately associated with post-transcriptional modification steps occurring at the 3' ends of the primary transcripts (Zhao, Hyman & Moore, 1999). In the case of most eukaryotic protein-coding genes the 3' end of the initial transcript is trimmed, and a **polyadenylate (poly(A)) tail** is added. The first step of polyadenylation is cleavage of the primary transcript by a specific endonuclease that recognizes the **polyadenylation signal** sequence, which has the consensus sequence AAUAAA. Two other weaker motifs are also important for recognition of the cleavage site: a GU-rich stretch that is downstream of the cleavage site and a U-rich auxiliary sequence that is upstream of the of AAUAAA motif. The sequence surrounding this cleavage site is not highly conserved, but it is most frequently preceded by the CA dinucleotide. The selection of the cleavage site appears to be determined by its distance from the AAUAAA motif: it is usually located 15–30 nucleotides downstream of the AAUAAA element. After cleavage by the endonuclease, **poly(A) polymerase** adds about 250 A residues to the 3' end of the transcript. The efficiency of transcription termination generally correlates with the strength of the polyadenylation signal as well as the presence of as yet poorly defined termination signals downstream of the poly(A) site.

It should be noted that an important class of mRNAs, those encoding replication-dependent histones, do not have a poly(A) tail. Nevertheless, the formation of the 3' ends of histone mRNAs also involves an endonucleolytic cleavage. In this case, however, the sequences defining the cleavage site include a highly conserved stem-loop structure upstream of the cleavage site and a purine-rich sequence (histone downstream element) 3' to the cleavage site. The histone downstream element forms a specific complex with U7 snRNP (**small nuclear ribonucleoprotein**) and this complex defines the exact position of the cleavage site.

Although the role of the poly(A) tail is not fully understood at present, several functions have been suggested: it may facilitate transport of the mRNA to the cytoplasm; it may control the lifetime of mRNA in the cytoplasm; or it may facilitate recognition of the mRNA by the ribosomal machinery.

The definition of the 3' end of mRNAs is not always unambiguous. There are numerous examples of **alternative polyadenylation** when there are

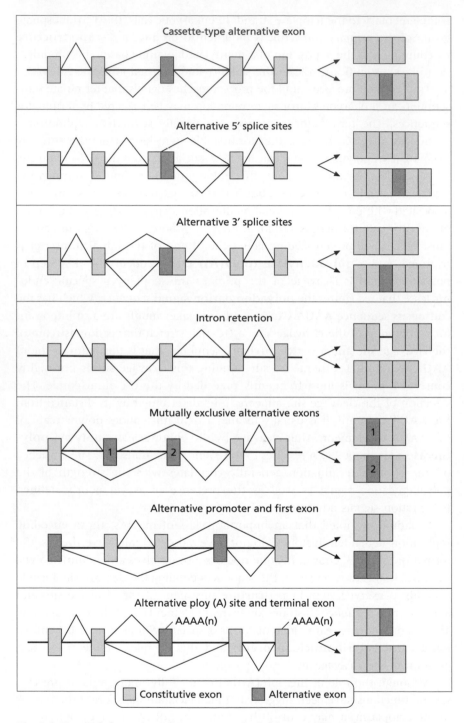

Fig. 1.2 Types of alternative splicing events. Shaded boxes indicate alternative exons, the thicker line indicates an intron retained. (Reprinted from Blencowe, B.J. (2006) Alternative splicing: new insights from global analyses. *Cell* **126**, 37–47. © 2006, with permission from Cell Press.)

multiple potential poly(A) sites in the 3′ untranslated region of a primary transcript. In these cases the efficiency with which a given poly(A) site is used depends on the strength of the polyadenylation signals. If alternative splicing (see below) affects regions that carry poly(A) signals then it will also influence the choice of polyadenylation sites.

In eukaryotic cells the primary transcript (pre-mRNA) of protein-coding genes is cut and **spliced** to remove the introns, producing mRNA. In the case of some genes the same set of introns is removed from all transcripts, producing a homogeneous population of mRNA molecules. In certain genes, however, there are alternative ways of splicing: **alternative splicing** may give rise to mRNA molecules with variant sequences (Blencowe, 2006). An important aspect of alternative splicing is that in such cases the distinction between exon and intron is obscured. Exons that are included in all mature mRNA isoforms are called **constitutive exons**, whereas exons that are present only in some alternatively spliced transcripts are called **alternative exons**.

The most common type of alternative splicing involves **cassette-type alternative exons**: an exon may be present in one population of mRNAs but may be absent from another population of mRNAs (**exon skipping**). Alternative choice of 5′ or 3′ splice sites within exon sequences are also frequent; other types of alternative splicing events include intron retention, mutually exclusive use of alternative exons, or differential selection of transcription initiation and 3′ end processing/termination sites (Fig. 1.2). If alternative splicing involves protein-coding regions, variant polypeptide sequences are produced, thereby generating a diversity of proteins from a single gene.

The introns of nuclear protein-coding genes (**pre-mRNA introns**) are spliced out of the pre-mRNA by complex ribonucleoprotein particles containing several proteins and small RNA molecules. These particles are called **spliceosomes**. The types of introns that are recognized and removed by these particles are also known as **spliceosomal introns**. Since there is now strong evidence that spliceosomal introns have played a major role in protein evolution, it is worth considering their special features in some detail (Figs 1.3 and 8.1).

Splicing begins with the cleavage of the phosphodiester bond between the upstream exon and the 5′ end of the intron (Fig. 1.3(a)). The attacking group in this reaction is the 2′-OH of an adenylate (A) residue, located about 20–50 bp upstream from the 3′ end of the intron. A 2′,5′-phosphodiester bond is formed between this A residue and the 5′-terminal phosphate of the intron. Since this adenylate residue remains joined to two other nucleotides by normal 3′,5′-phosphodiester bonds, a **branch** is generated at this site and a **lariat intermediate** is formed. In the next step of the splicing reaction the 3′-OH of the upstream exon attacks the phosphodiester bond linking the intron to the downstream exon, the exons become joined and the intron is released in lariat form.

Throughout this process, all participants of the reaction are held together by the spliceosome, which specifically recognizes the **5′ splice site (donor site)**, **3′ splice site (acceptor site)** and the **branch site** of mRNA precursors. The ribonucleic acid component of spliceosomes consists of a large variety of **small nuclear RNAs (snRNAs)** associated with specific proteins to form **snRNPs (small nuclear ribonucleoproteins)**. U1 snRNP recognizes the 5′ splice site sequence: U1 RNA contains a sequence that is complementary to the sequence at this splice site. U2 snRNP binds the branch site and a **poly-pyrimidine tract** in the vicinity of the 3′ end of the intron. The 3′ splice site is recognized by a particle containing U5 snRNP, as well as U4 and U6 snRNAs.

Of the entire length of introns the short consensus sequences containing the 5′ and 3′ splice sites, the branch site and the polypyrimidine tract appear to be most critical for splicing. Mutations in each of the three critical regions of spliceosomal introns lead to aberrant splicing. The importance of 5′ and 3′ splice sites and the branch site is reflected in the conservation of these sequences in introns as opposed to the lack of conservation in other parts of introns. In the vast majority of eukaryotic nuclear protein-coding genes – ranging from yeast to mammals – the spliceosomal introns begin with GT and end with AG (the **GT-AG rule**). It should be pointed out that there is a minor class of spliceosomal introns, where introns begin with AT and end in AC (**AT-AC introns**; Fig. 1.3(b)); these rare introns are spliced by a special kind of spliceosome containing a distinct set of snRNAs (Tarn & Steitz, 1997). Unlike the GT-AG spliceosome, the AT-AC spliceosome does not contain U1, U2, U4, or U6 snRNAs (both types contain U5 snRNA). In the case of AT-AC splicing, U11 and U12 snRNAs fulfil the roles of U1 and U2 snRNAs; the roles of U4 and U6 are performed by two highly diverged, novel variant snRNAs, U4atac and U6atac. The AT-AC and GT-AG introns and AT-AC and GT-AG spliceosomes thus illustrate the interdependence and **coevolution** of interacting components (Fig. 1.3). The fact that constituents of this minor spliceosome are also present in single-cell protists, plants and animals suggests an early evolutionary origin of this spliceosome in the eukaryotic lineage (Russell et al., 2006).

Outside these highly conserved regions, long stretches of DNA can be deleted from or inserted into introns without altering the site or efficiency of splicing. Moreover, chimeric introns created by fusion of the 5′ end of one intron and the 3′ end of a very different intron may be properly spliced provided that the splice sites and branch site are unaltered. This is also the case in one form of alternative splicing – exon skipping – when two differ-ent introns supply the donor and acceptor sites.

Nevertheless, studies employing site-directed mutagenesis of pre-mRNA reporters, binding of splicing factors and computational strategies have resulted in the identification of additional *cis*-acting sequence elements in exons and introns that are important for promoting splice site recognition (Blencowe, 2006). These sequences, referred to as **exonic splicing enhancers (ESEs)**, **exonic splicing silencers (ESSs)**, **intronic splicing enhancers (ISEs)** and

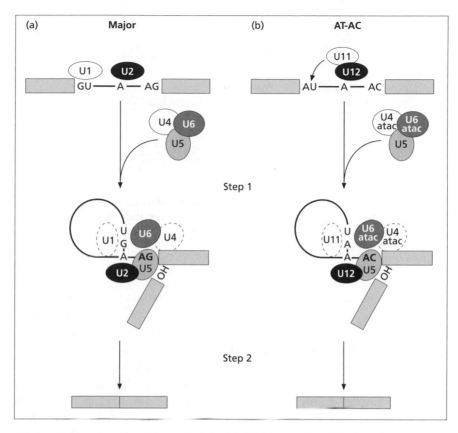

Fig. 1.3 Spliceosome assembly and splicing of GT–AG and AT–AC introns. The major spliceosome (a) involved in splicing of GT–AG introns and the minor spliceosome (b) responsible for the splicing of AT–AC introns contain unique snRNP components that identify the 5′ and 3′ splice sites characteristic of these two intron classes. (Reprinted from Tarn, W.Y. & Steitz, J.A. (1997) Pre-mRNA splicing: the discovery of a new spliceosome doubles the challenge. *Trends in Biochemical Sciences* **22**, 132–7. © 1997, with permission from Elsevier Science.) Reprinted from *Trends in Biochemical Sciences*, 22, Tarn, W.Y. & Steitz, J.A., Pre-mRNA splicing: the discovery of a new spliceosome doubles the challenge, 132–137, copyright 1997, with permission from Elsevier.

intronic splicing silencers (ISSs) also play important roles in the regulation of alternative splicing: silencers promote the skipping of an alternative exon, enhancers promote the inclusion of the alternative exon. Splicing enhancer and silencer sequences are usually short (5–10 nucleotide) sequences that are recognized by diverse RNA binding factors (including members of the serine/arginine-rich (SR) protein family) at the earliest stages of spliceosome formation. Cell- and tissue-specific expression of factors binding to specific splicing enhancer and silencer elements may regulate alternative splicing in a cell- and tissue-specific manner.

The number and size of introns varies from organism to organism and from gene to gene. Introns are rare and short in yeast, and relatively short

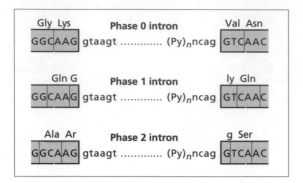

Fig. 1.4 Introns present in translated regions of protein-coding genes may split the reading frame between two codons (phase 0 intron), between the first and second nucleotide of a codon (phase 1 intron), or between the second and third nucleotide of a codon (phase 2 intron).

in the compact genome of *Caenorhabditis elegans*, but in vertebrates introns account for the major part of most protein-coding genes (see section 9.6.3). Some genes of vertebrates possess dozens of introns, some of which may be thousands of nucleotides in length. Others (e.g. the highly expressed replication-dependent histone genes) are devoid of introns.

Introns may occur in the 5′ and 3′ untranslated regions as well as in protein-coding regions of genes. In the latter case, introns may be located between two codons (**phase 0 introns**), may split codons between the first and second nucleotide (**phase 1 introns**), or between the second and third nucleotides (**phase 2 introns**) (Fig. 1.4).

1.3 Translation

Mature mRNA in eukaryotes migrates from the nucleus into the cytoplasm, where it is translated into polypeptide chains on ribosomes, which consist of rRNA molecules and a variety of ribosomal proteins. The nucleotide sequence of the mRNA molecule is read in a 5′ to 3′ direction by the ribosome, one **codon** – three nucleotides – at a time. Each codon codes for one specific amino acid, which is attached to a tRNA (transfer RNA) that carries the complementary anticodon. There are three possible **reading frames** by which trinucleotide codons can be read sequentially on an mRNA molecule. The actual reading frame used by the ribosome is determined by the position of the **initiating codon**.

Initiation of translation

Translation of the mRNA molecule is initiated by the formation of a complex that consists of mRNA, both subunits of ribosomes and initiation-factor proteins. The ribosome binds near the 5' end of the mRNA molecule where specific signals define the point to initiate translation. The start of synthesis of a polypeptide chain requires the presence of a methionyl tRNA, which interacts with the **initiating AUG codon** on the mRNA. In prokaryotic cells, the methionine attached to this tRNA is formylated on its amino group.

In prokaryotes, the initiating AUG codon on the mRNA is selected by the ribosome on the basis of a short upstream sequence – the **Shine–Dalgarno sequence** – in the mRNA. This sequence is complementary to a sequence near the 3' end of the small rRNA molecule.

In eukaryotic cells the recognition of the 5' end of the mRNA is the important first step in defining the initiating AUG codon. The 5' end of eukaryotic mRNAs has a special modified end (a methylated guanylate residue linked to the first ordinary nucleotide in a 5'–5' pyrophosphate linkage) that chemically 'caps' the RNA. Initiation of the translation of eukaryotic mRNAs involves recognition of this cap, followed by the location of a motif – the **Kozak consensus sequence** – surrounding the AUG initiation codon. The optimal Kozak consensus sequence has been identified as CCA/GCCAUGG. The small ribosomal subunit binds to the mRNA at the capped 5' end then scans the mRNA until it encounters the first AUG initiator codon. The significance of cap site recognition is illustrated by the fact that more than 90% of the AUGs used as initiation signals in eukaryotes are the first (the most 5') AUG in the molecule. If the sequence around the first AUG deviates significantly from the Kozak sequence the first AUG may be sometimes skipped and translation may be initiated at a downstream AUG, leading to **alternative translation initiation**.

Polypeptide elongation

The initiating Met-tRNA is bound to the initiating AUG codon of the mRNA at the so-called P (peptidyl-tRNA) site on the ribosome, where the growing peptide chain is held. Charged tRNA with an anticodon complementary to the next codon of the mRNA binds to the mRNA at an adjacent site on the ribosome – the A (aminoacyl-tRNA) site, which binds the aminoacyl-tRNAs that are to be added to the growing polypeptide chain. The initiating Met residue is transferred to the amino group of the second amino acid to generate the first peptide bond between the first and second amino acids, the uncharged tRNAMet is released from the P site, the dipeptidyl-tRNA is translocated to the P site, and the ribosome moves three nucleotides along the mRNA to the next codon. The process of translocations continues codon by codon until all the amino acids encoded by the mRNA have been added. In each translocation step, the ribosome with its attached peptidyl-tRNA moves three nucleotides closer to the 3' end of the mRNA. At high levels of gene expression (transcription) the rate of attachment may be limited by the availability of the appropriately charged tRNAs.

Termination

During translation of mRNA the polypeptide chain grows until a **termination codon** (UAA, UGA, or UAG in the standard genetic code) on the mRNA
is reached in the reading frame that was defined by the initiation codon.
These three codons are recognized by release-factor proteins, which cause
the polypeptide chain to be hydrolyzed from the last tRNA. The polypeptide
then dissociates from the ribosome, followed by the tRNA and the mRNA,
then the ribosome dissociates into two subunits.

Recent studies have characterized **mRNA surveillance** mechanisms that
ensure the rapid degradation of aberrant mRNAs that contain premature
termination codons as a result of mutations or errors during transcription or
splicing of RNA. **Nonsense-mediated mRNA decay** is thus an important
quality control mechanism inasmuch as it prevents the synthesis of truncated and potentially harmful proteins (Conti & Izaurralde, 2005). Although
nonsense-mediated mRNA decay has been shown to exist in all eukaryotes,
different mechanisms have evolved to discriminate natural stop codons
(of full-length proteins) from premature stop codons (of truncated proteins)
and to degrade the targeted mRNAs. In mammals the *cis*-acting signal is an
exon–exon boundary downstream of the premature termination codon: the
surveillance complex associates with spliced mRNAs through interactions with
components of the **exon–junction complex**, a multimeric protein complex
deposited by the spliceosome 20–24 nucleotides upstream of exon–exon
junctions. In yeast *cis*-acting downstream elements – independent of exon
boundaries – were shown to serve the recognition of premature termination
codons. Decay of aberrant mRNAs thus identified is initiated by decapping
and/or deadenylation followed by 5′ to 3′ as well as 3′ to 5′ exonucleolytic
digestion.

Aminoacyl-tRNA synthetases and tRNAs

Usually there is at least one class of tRNA molecule in every cell for each
type of amino acid. Amino acids are coupled to the appropriate tRNA molecules by appropriate aminoacyl-tRNA synthetases with high specificity;
these enzymes are very specific for both the amino acid and the tRNA that
they join together. This is very important for the fidelity of protein biosynthesis since all subsequent steps are determined only by the interaction
between the anticodon of the tRNA and the codon of the mRNA. The anticodon consists of three nucleotides that base pair with the complementary
three nucleotides of the corresponding codon on the mRNA. This base
pairing is always specific for at least two of the three sequential nucleotides,
but there is often 'wobble' in the interaction with the third. Thanks to this
'wobble', some tRNA molecules can recognize codons that differ in the third
nucleotide.

Table 1.1 The standard genetic code. **Bold** type indicates stop codons.

Codon	Amino acid	Codon	Amino acid	Codon	Amino acid	Codon	Amino acid
UUU	Phe	UCU	Ser	UAU	Tyr	UGU	Cys
UUC	Phe	UCC	Ser	UAC	Tyr	UGC	Cys
UUA	Leu	UCA	Ser	**UAA**	**Stop**	**UGA**	**Stop**
UUG	Leu	UCG	Ser	**UAG**	**Stop**	UGG	Trp
CUU	Leu	CCU	Pro	CAU	His	CGU	Arg
CUC	Leu	CCC	Pro	CAC	His	CGC	Arg
CUA	Leu	CCA	Pro	CAA	Gln	CGA	Arg
CUG	Leu	CCG	Pro	CAG	Gln	CGG	Arg
AUU	Ile	ACU	Thr	AAU	Asn	AGU	Ser
AUC	Ile	ACC	Thr	AAC	Asn	AGC	Ser
AUA	Ile	ACA	Thr	AAA	Lys	AGA	Arg
AUG	Met	ACG	Thr	AAG	Lys	AGG	Arg
GUU	Val	GCU	Ala	GAU	Asp	GGU	Gly
GUC	Val	GCC	Ala	GAC	Asp	GGC	Gly
GUA	Val	GCA	Ala	GAA	Glu	GGa	Gly
GUG	Val	GCG	Ala	GAG	Glu	GGG	Gly

Genetic code

Translation involves the sequential recognition of adjacent nonoverlapping triplets of nucleotides, called **codons**. The rules that define the correspondence between codons and amino acids is called the **genetic code**. The **standard genetic code** for nuclear protein-coding genes is nearly universal: with a few exceptions the translation of eukaryotic and prokaryotic genes is determined by the same set of rules. This standard genetic code is given in Table 1.1. Since codons are specified by three nucleotides, and since there are four different types of nucleotides, there are 64 possible codons. Three of these codons (UGA, UAA, UAG) act as signals for the termination of the translation process: these are called **stop codons** or **nonsense codons**. (The codons that define specific amino acids are called **sense codons**.) Since there are only 20 amino acids for 61 sense codons and most amino acids are encoded by more than one triplet, the code is highly **degenerate**. Leu, Arg and Ser are each encoded by six codons; Ala, Gly, Val, Pro, Thr each have four codons; Ile has three codons; Phe, Tyr, His, Asp, Asn, Glu, Gln, Lys, Cys are each encoded by two codons. Only Trp and Met residues are designated by single codons. Codons specifying the same amino acid are called **synonymous codons**; codons defining different amino acids are called **nonsynonymous codons**. Synonymous codons usually differ from each other at the third position only; these groups of codons are referred to as a **codon family** (e.g. Val, Ala, Gly). It should be pointed out that the third position of a codon corresponds to the position where 'wobble' may permit the same tRNA molecule to recognize different codons.

Correlation of tRNA abundance with codon usage

Although most amino acids are specified by more than one codon, the synonymous codons are not used with equal frequencies: some are preferentially employed in a given type of organism. In the case of many *Escherichia coli* genes a strong **codon-usage bias** has been observed: some synonymous codons are used frequently, while others are almost never observed. This is most striking in the case of the mRNAs for proteins that are synthesized in very large amounts (e.g. ribosomal proteins). The mRNAs of highly expressed proteins contain codons read by the most abundant isoacceptor tRNAs, whereas codons corresponding to minor tRNA species are practically absent. Similarly, in yeast the codons corresponding to the major tRNA isoacceptors are used with the highest frequency in highly expressed mRNAs. For example, for the abundant alcohol dehydrogenase isozyme I and glyceraldehyde phosphate dehydrogenase, more than 96% of the amino acids are coded by only 25 triplets, whose corresponding tRNAs are highly abundant.

In summary, the correlation between codon usage bias and expression level suggests that codons corresponding to rare tRNAs could slow down translation. In genes that are expressed at a high rate, codon choice is constrained by tRNA availability and seems to have evolved to optimize translational efficiency (Powell & Moriyama, 1997).

References

Bickel, K.S. & Morris, D.R. (2006) Silencing the transcriptome's dark matter: mechanisms for suppressing translation of intergenic transcripts. *Molecular Cell* **22**, 309–16.

Blencowe, B.J. (2006) Alternative splicing: new insights from global analyses. *Cell* **126**, 37–47.

Carthew, R.W. (2006) A new RNA dimension to genome control. *Science* **313**, 305–6.

Conti, E. & Izaurralde, E. (2005) Nonsense-mediated mRNA decay: molecular insights and mechanistic variations across species. *Current Opinion in Cell Biology* **17**, 316–25.

Lau, N.C., Seto, A.G., Kim, J., Kuramochi-Miyagawa, S., Nakano, T. et al. (2006) Characterization of the piRNA complex from rat testes. *Science* **313**, 363–7.

Lomvardas, S., Barnea, G., Pisapia, D.J. et al. (2006) Interchromosomal interactions and olfactory receptor choice. *Cell* **126**, 403–13.

Powell, J.R. & Moriyama, E.N. (1997) Evolution of codon usage bias in *Drosophila*. *Proceedings of the National Academy of Sciences of the USA* **94**, 7784–90.

Russell, A.G., Charette, M., Spencer, D.F. & Gray, M.W. (2006) An early evolutionary origin for the minor spliceosome. *Nature* **443**, 863–6.

Savarese, F. & Grosschedl, R. (2006) Blurring *cis* and *trans* in gene regulation. *Cell* **126**, 248–50.

Tarn, W.Y. & Steitz, J.A. (1997) Pre-mRNA splicing: the discovery of a new spliceosome doubles the challenge. *Trends in Biochemical Sciences* **22**, 132–7.

Vagin, V.V., Sigova, A., Li, C. et al. (2006) A distinct small RNA pathway silences selfish genetic elements in the germline. *Science* **313**, 320–4.

Zhao, J., Hyman, L. & Moore, C. (1999) Formation of mRNA 3′ ends in eukaryotes: mechanism, regulation and interrelationships with other steps in mRNA synthesis. *Microbiology and Molecular Biology Reviews* **63**, 405–45.

Useful internet resources

Noncoding, nontranslated RNA genes, untranslated regions of mRNAs

Rfam is a large collection of multiple sequence alignments of many common non-coding RNA families (http://www.sanger.ac.uk/Software/Rfam/index.shtml).

The **Noncoding RNA** database provides information on the sequences and functions of transcripts that do not code for proteins (http://biobases.ibch.poznan.pl/ncRNA/).

The **GtRNAdb** (Genomic tRNA database) contains tRNA identifications on complete or nearly complete genomes (http://lowelab.ucsc.edu/GtRNAdb/).

The **miRBase Sequence Database** is a searchable database of published miRNA sequences and annotation. The **miRBase Targets Database** is a new resource of predicted miRNA targets in animals (http://microrna.sanger.ac.uk/).

RNABase is an RNA structure database (http://www.rnabase.org/).

UTResource provides data and analysis tools for the functional classification of 5′ and 3′ UTRs of eukaryotic mRNAs (http://bighost.area.ba.cnr.it/BIG/UTRHome/).

Transcription factors, control elements, promoters, transcription start sites

TRANSFAC is a database on eukaryotic transcription factors, their genomic binding sites and DNA binding profiles (http://www.gene-regulation.com/pub/databases. html#transfac).

TESS (Transcription Element Search System) is a web tool for predicting transcription factor binding sites in DNA sequences. It can identify binding sites using site or consensus strings and positional weight matrices from the TRANSFAC, IMD and the CBIL-GibbsMat database (http://www.cbil.upenn.edu/tess/).

Enhancer Element Locator, or EEL, is a tool for locating distal gene enhancer elements in mammalian genomes by comparative genomics (http://www.cs.helsinki.fi/u/kpalin/EEL/).

Cister predicts regulatory regions in DNA sequences by searching for clusters of *cis*-elements (http://sullivan.bu.edu/~mfrith/cister.shtml).

EPD (Eukaryotic Promoter Database) is an annotated nonredundant collection of eukaryotic POL II promoters, for which the transcription start site has been determined experimentally (http://www.epd.isb-sib.ch/).

The **Promoter 2.0 Prediction Server** predicts transcription start sites of vertebrate Pol II promoters in DNA sequences (http://www.cbs.dtu.dk/services/promoter/).

PromoterInspector is a program that predicts eukaryotic Pol II promoter regions with high specificity in mammalian genomic sequences. The program focuses on the genomic context of promoters rather than their exact location (http://www.genomatix.de/online_help/help_gems/PromoterInspector_help.html).

WWW Promoter Scan predicts promoter regions based on scoring homologies with putative eukaryotic Pol II promoter sequences (http://thr.cit.nih.gov/molbio/proscan/).

Transcriptional Regulatory Element Database provides genome-wide promoter annotation for human, mouse and rat (http://rulai.cshl.edu/cgi-bin/TRED/tred.cgi?process=home).

Recognition of 3'-end cleavage and polyadenylation region

Polyadq decides whether a given AATAAA or ATTAAA hexamer is a true poly(A) signal (http://rulai.cshl.org/tools/polyadq/polyadq_form.html).

POLYAH identifies the 3'-processing sites of human mRNA precursors (http://www.softberry.com/berry.phtml?topic=polyah&group=programs&subgroup=promoter).

Splice site prediction

BGDP Splice Site Prediction by Neural Network predicts donor and acceptor sites using a separate neural network recognizer for each site (http://www.fruitfly.org/seq_tools/splice.html).

The **NetPlantGene** server produces neural network predictions of splice sites in *Arabidopsis thaliana* DNA (http://www.cbs.dtu.dk/services/NetPGene/).

The **NetGene2** server produces neural network predictions of splice sites in human, *C. elegans* and *A. thaliana* DNA (http://www.cbs.dtu.dk/services/NetGene2/).

Alternative splicing

The **Alternative Splicing Database (ASD) Project** aims to understand the mechanism of alternative splicing on a genome-wide scale by creating a database of alternative splice events and the resultant isoform splice patterns of genes from human and other model species (http://www.ebi.ac.uk/asd/).

The **Alternative Exon Database (AEDB)** is a manually generated database for human alternative exons and their properties – the data is gathered from literature where these exons have been experimentally verified (http://www.ebi.ac.uk/asd/aedb/index.html).

The **AltExtron Database** is a computer-generated high-quality data set of human transcript-confirmed constitutive and alternative exons and introns (http://www.ebi.ac.uk/asd/altextron/index.html).

Splicing enhancers, splicing silencers

The **ACESCAN2 Web Server** is an online tool for identifying candidate *cis*-elements (exonic splicing enhancers and silencers, intronic splicing enhancers) in alternative and constitutive splicing in mammalian exons (http://genes.mit.edu/acescan2/index.html).

The **FAS-ESS Web Server** is an online tool for the systematic identification and analysis of exonic splicing silencers (http://genes.mit.edu/fas-ess/).

ESEfinder is a web resource that identifies exonic splicing enhancers (http://rulai.cshl.edu/tools/ESE/index.html).

The **RESCUE-ESE Web Server** is an online tool for identifying candidate exonic splicing enhancers in vertebrate exons (http://genes.mit.edu/burgelab/rescue-ese/)

Exon–intron structure of protein-coding genes

The **ExInt™ database** is a relation database containing information on the exon–intron structure of protein-coding genes (http://sege.ntu.edu.sg/wester/exint/).

IE-Kb (Intron Exon Knowledge Base) is a relational database containing exon/intron gene structure data for eukaryotic genes derived from ExInt. IE-Kb allows ranking of exons and introns based on properties such as length, phase, composition etc. (http://sege.ntu.edu.sg/wester/iekb/)

Translation of nucleic acid sequences

Translate is a tool which allows the translation of a nucleotide (DNA/RNA) sequence to a protein sequence (http://au.expasy.org/tools/dna.html).

Transeq translates nucleic acid sequences to the corresponding peptide sequence. It can translate in any of the three forward or three reverse sense frames, or in all three forward or reverse frames, or in all six frames using either the standard or various nonstandard genetic codes (http://www.ebi.ac.uk/emboss/transeq/).

Prediction of translation start sites

The **NetStart Prediction Server** produces neural network predictions of translation start in vertebrate and *A. thaliana* nucleotide sequences (http://www.cbs.dtu.dk/services/NetStart/).

Genetic codes

The nonstandard genetic codes of various groups of organisms are found on NCBI's Taxonomy homepage (http://130.14.29.110/Taxonomy/Utils/wprintgc.cgi?mode=c).

Codon usage

The **Codon Usage Database** is an extended version of CUTG (Codon Usage Tabulated from GenBank). It contains information on the frequency of codon use in each organism (http://www.kazusa.or.jp/codon/).

Correspondence Analysis of Codon Usage. CodonW is a program designed to simplify correspondence analysis of codon and amino acid usage. It also calculates standard indices of codon usage (http://codonw.sourceforge.net/).

Chapter 2: Protein structure

2.1 The polypeptide backbone

The polypeptide backbone is a repeated structure consisting of the amide N, the Cα and the carbonyl C of the amino acyl residues:

$$
\begin{array}{c}
\quad\text{H}\quad\text{R}_1\quad\text{O}\quad\text{H}\quad\text{R}_2\quad\text{O}\quad\text{H}\quad\text{R}_3\quad\text{O} \\
\quad|\quad\;|\quad\;\|\quad\;|\quad\;|\quad\;\|\quad\;|\quad\;|\quad\;\| \\
-\text{N}-\text{C}-\text{C}-\text{N}-\text{C}-\text{C}-\text{N}-\text{C}-\text{C}- \\
\quad|\qquad\qquad|\qquad\qquad| \\
\quad\text{H}\qquad\quad\;\text{H}\qquad\quad\;\text{H}
\end{array}
$$

The peptide bonds connecting the amide N and the carbonyl C have partial double-bonded character, rotation of this bond is restricted and the amide N, the two α carbons and the carbonyl C have a strong tendency to be coplanar. Two configurations of such a planar peptide bond are possible, one in which the Cα atoms are *trans*, the other in which they are *cis*:

$$
\begin{array}{cc}
\begin{array}{c}
\text{O}\qquad\;\text{C}\alpha \\
\backslash\!\backslash\quad\;\; / \\
\text{C}-\text{N} \\
/\qquad\backslash \\
\text{C}\alpha\qquad\text{H}
\end{array}
&
\begin{array}{c}
\text{O}\qquad\;\text{H} \\
\backslash\!\backslash\quad\;\; / \\
\text{C}-\text{N} \\
/\qquad\backslash \\
\text{C}\alpha\qquad\text{C}\alpha
\end{array} \\
\textit{trans} & \textit{cis}
\end{array}
$$

The *trans* form is the energetically favoured conformation since in this form there are fewer repulsions. Consequently, peptide bonds of the backbone are nearly always of the *trans* isomer, unless the next residue is Pro, when the intrinsic stability of the *cis* isomer is less unfavourable. In folded proteins *cis* peptide bonds occur at about 5% of the bonds that precede Pro residues. Such *cis* peptide bonds are located primarily at tight bends of the polypeptide backbone.

2.2 The amino acids

Proteins are usually constructed of 20 different types of amino acids. Their structures, along with their three- and one-letter abbreviations are shown in Table 2.1. Nineteen of these amino acids have the general structure:

$$
\begin{array}{c}
\quad\text{H} \\
\quad| \\
\text{R}-\text{C}-\text{COOH} \\
\quad| \\
\quad\text{NH}_2
\end{array}
$$

In the case of proline the side chain is bonded to the nitrogen atom to give an imino acid. Except for glycine (where the side chain is a hydrogen

Table 2.1 The amino acids.

Trivial name	Three-letter code	One-letter code	Structural formula
Glycine	Gly	G	$H-CH-COOH$ with NH_2 below CH
Alanine	Ala	A	$CH_3-CH-COOH$ with NH_2 below CH
Cysteine	Cys	C	$CH_2-CH-COOH$ with SH, NH_2 below
Serine	Ser	S	$CH_2-CH-COOH$ with OH, NH_2 below
Threonine	Thr	T	$CH_3-CH-CH-COOH$ with OH, NH_2 below
Proline	Pro	P	ring structure with N–H and COOH
Valine	Val	V	$(H_3C)(H_3C)CH-CH-COOH$ with NH_2 below
Leucine	Leu	L	$(H_3C)(H_3C)CH-CH_2-CH-COOH$ with NH_2 below
Isoleucine	Ile	I	CH_3-CH_2, CH_3, $CH-CH-COOH$ with NH_2 below
Methionine	Met	M	$CH_2-CH_2-CH-COOH$ with $S-CH_3$, NH_2 below
Aspartic acid	Asp	D	$HOOC-CH_2-CH-COOH$ with NH_2 below
Asparagine	Asn	N	$H_2N-C(=O)-CH_2-CH-COOH$ with NH_2 below
Glutamic acid	Glu	E	$HOOC-CH_2-CH_2-CH-COOH$ with NH_2 below
Glutamine	Gln	Q	$H_2N-C(=O)-CH_2-CH_2-CH-COOH$ with NH_2 below

Continued on p. 20

Table 2.1 (*Continued*)

Trivial name	Three-letter code	One-letter code	Structural formula
Lysine	Lys	K	$CH_2-CH_2-CH_2-CH_2-CH-COOH$ with NH_2 and NH_2
Arginine	Arg	R	$H-N-CH_2-CH_2-CH_2-CH-COOH$ with $C=NH$, NH_2 and NH_2
Histidine	His	H	imidazole ring HN, N — $CH_2-CH-COOH$ with NH_2
Phenylalanine	Phe	F	phenyl ring — $CH_2-CH-COOH$ with NH_2
Tyrosine	Tyr	Y	HO — phenyl ring — $CH_2-CH-COOH$ with NH_2
Tryptophan	Trp	W	indole ring N, H — $CH_2-CH-COOH$ with NH_2

atom) the Cα atom is asymmetric; in the case of proteins it is always the L isomer.

The 20 amino acids differ in the chemical nature of their side chains, therefore they possess a variety of physicochemical properties. Here we will briefly emphasize only those special features of the amino acids that define their roles in proteins, affect their replaceability and are most pertinent to understanding substitution matrices (see section 3.3).

Glycine

Glycine is the simplest and smallest amino acid, with only a hydrogen atom side chain. Its most relevant feature is that the absence of a larger side chain gives the peptide backbones at Gly residues much greater conformational flexibility than at other residues.

Alanine

This amino acid residue has only a methyl ($-CH_3$) side chain. Relevant features are its small size (in this respect it is most similar to glycine) and the hydrophobic aliphatic side chain (in this respect it is more similar to aliphatic residues).

Valine, leucine, isoleucine and methionine

These amino acid residues have methylene ($-CH_2-$) and methyl ($-CH_3$) groups on their side chains. Their most important property is that their side chains are hydrophobic: they interact more favourably with each other and with other nonpolar residues (e.g. aromatic residues) than with water. The sulphur atom of the side chain of Met residues is relatively susceptible to oxidation by air and peroxides.

Proline

This cyclic imino acid has a compact side chain that is aliphatic like those of valine, leucine, isoleucine and methionine. Its most relevant feature, however, is that the cyclic five-membered ring imposes rigid constraints on rotation about its N–Cα bond of the backbone, and the peptide bonds preceding a Pro residue thus have a higher probability to adopt the *cis* configuration. Pro residues consequently have unique effects on the conformation of the polypeptide backbone.

Serine and threonine

The side chains of serine and threonine are small and aliphatic, but they also possess a polar hydroxyl group. These hydroxyl groups may serve as sites for post-translational modifications (e.g. phosphorylation by protein kinases, glycosylation, etc.).

Cysteine

Cysteine has a hydrophobic $-CH_2-$ group and a thiol group. The thiol group is the most reactive amino acid side chain and this reactivity is frequently exploited by enzymes. Thiols of cysteine residues form complexes of varying stability with a variety of metal ions (e.g. copper, iron, zinc, cobalt, molybdenum, manganese). This feature is the basis for the high-affinity binding of metal ions (e.g. by zinc-finger transcription factors).

The sulphur atom of cysteine residues can exist in a variety of oxidation states, but the disulphide is usually the end product in an oxidative milieu. **Disulphide bonds** between cysteine residues occur almost exclusively in extracytoplasmic proteins; two such residues linked by a disulphide bond are designated as **cystines**. Covalent disulphide bonds between cysteine residues are formed during folding of the proteins and endow the proteins with greater stability.

Aspartic acid and glutamic acid

The carboxyl groups of aspartic acid and glutamic acid side chains are ionized, negatively charged and very polar under physiological conditions. They can be effective chelators of metal ions. Although their side chains differ only by one methylene group, they differ markedly in their effects on the conformation and chemical reactivity of the peptide backbone, Asp–X peptide bonds being markedly more sensitive to cleavage in an acidic environment.

Asparagine and glutamine

The amide side chains are polar and can serve as hydrogen-bond donors and acceptors. The amide group of asparagine can deamidate to aspartic acid residue since its side chain readily interacts with the –NH group of the following residue to form a transient cyclic succinimidyl derivative. Since the absence of a side chain sterically favours succinimide formation, the deamidation reaction of asparagine occurs most rapidly if the following residue is a glycine. Succinimide formation and thus deamidation also depends on the conformation of the polypeptide backbone; 'programmed' deamidation may be important in the control of the unfolding and lifetime of some proteins.

Asparagine side chains are major sites for some post-translational modifications.

Lysine and arginine

The side chain of lysine is a hydrophobic chain with four methylene groups. The side chain has an ε-amino group that is usually ionized in an aqueous environment, and thus has a positive charge and is hydrophilic.

The side chain of arginine consists of three hydrophobic methylene groups and the strongly basic δ-guanidino group. In an aqueous environment, the guanidino group is ionized, positively charged over the entire pH range and is thus hydrophilic.

Lysine and arginine side chains are major sites for post-translational modifications.

Histidine

The imidazole side chain of histidine residues makes these residues extremely effective as catalysts: it is a tertiary amine that is intrinsically more nucleophilic than primary or secondary amines. It has a pK value near 7, so it is one of the strongest bases that can exist at neutral pH. Due to their special properties, histidine residues frequently participate in catalysis. With a pK value near 7, histidyl residues may or may not carry a positive charge at physiological pH (depending on their environment in folded proteins).

Imidazolyl side chains of proteins frequently participate in chelation of metal ions (e.g. zinc enzymes).

Phenylalanine, tyrosine and tryptophan

These bulky hydrophobic amino acid residues interact more favourably with each other and with other nonpolar amino acids (aliphatic amino acids) than with water. This 'hydrophobic interaction' is one of the main factors that stabilize the folded conformations of proteins: the aromatic residues frequently form the cores of protein folds. The aromatic ring of phenylalanine is unreactive, however, the hydroxyl group of the phenolic ring of tyrosine residues makes this aromatic ring relatively reactive. Tyrosyl side chains sometimes serve as sites of post-translational modifications (e.g. phosphorylation by tyrosine kinases, sulphation by tyrosylprotein sulphotransferase). A relevant property of tryptophan is that its indole side chain is the largest, bulkiest side chain found in proteins.

2.3 Covalent modifications of amino acid side chains

Various covalent modifications of proteins may occur either during or after assembly of the polypeptide chain. The side chains of many amino acids are covalently modified with concomitant modification of the biological activity of a protein. Most of these modifications are carried out with great specificity by enzymes, others may arise from nonenzymatic processes.

Recent studies have revealed that proteins are controlled by a vast and dynamic array of post-translational modifications, many of which create binding sites for specific protein-interaction domains. These protein-interaction domains read the state of the proteome and therefore couple post-translational modifications to cellular organization (Seet et al., 2006). For example, phosphotyrosine residues are specifically recognized by SH2 domains, bromodomains can interact specifically with acetylated lysine, recognition of methyllysines is achieved by the PHD finger domains (Dhalluin et al., 1999; Cosgrove, 2006). Here we will discuss some of those post-translational modifications that have clear biological (and evolutionary) significance.

2.3.1 Enzymatic modifications

Glycosylation

Attachment of carbohydrates is one of the most prevalent modifications of eukaryotic proteins, especially of secreted and membrane proteins.

With respect to their biological significance, the most relevant properties of carbohydrates are their highly variable structures, their hydrophilic nature and large size. Thanks to these properties, carbohydrates may be important for increasing protein solubility and may lengthen the biological life of a protein by protecting it from proteolysis. In some special cases (e.g. antifreeze glycoproteins; see section 6.1) they may interfere with the growth of ice crystals thereby having protective functions. Carbohydrate moieties of glycosylated proteins are often involved in mediating highly specific cell–cell or cell–matrix interactions (e.g. lectin-like cell-surface receptors of the selectin family; see Figs 8.7 and 8.10).

There are two types of glycosylation, called N-type or O-type depending on the atom of the protein to which the carbohydrate is attached. **N-glycosylation** occurs exclusively on the nitrogen atom of Asn side chains, whereas **O-glycosylation** occurs on the oxygen atoms of hydroxyls, primarily those of Ser and Thr residues of extracellular proteins or extracellular parts of transmembrane proteins. N-glycosylation occurs **cotranslationally** as the Asn residues enter the endoplasmic reticulum. The oligosaccharide chain is attached by **oligosaccharyl transferase** to an Asn residue that occurs in a characteristic sequence: -Asn-X-Ser-, -Asn-X-Thr-, or -Asn-X-Cys-, where X can be any residue except Pro.

O-glycosylation occurs primarily in the Golgi as a **post-translational modification**, N-acetylgalactosamine (GalNAc) groups being attached to Ser and Thr groups by the enzyme **UDP-N-acetyl-D-galactosamine:polypeptide N-acetylgalactosaminyltransferase**. The signals that define the Ser and Thr residues to be glycosylated cannot be described simply by the sequence of the amino acid residues that surround these Ser and Thr residues. It seems probable that the three-dimensional structure of the protein is also important for the definition of the glycosylation sites.

Proteoglycans are composed of protein backbones, to which one or many glycosaminoglycan chains (chondroitin sulphate/dermatan sulphate, heparan sulphate/heparin or keratan sulphate) are covalently attached. Proteoglycans are extensively glycosylated proteins: the large amount of carbohydrate that is attached to the polypeptide chain at many sites usually occupies the bulk of their volume. The glycosaminoglycans are sulphated to various degrees and are usually attached to the protein backbone through a Ser residue. The signal for attachment of chondroitin sulphate chains is the sequence -Ser-Gly-X-Gly- preceded by several acidic residues.

Phosphorylation

A large number of proteins is known to be phosphorylated at specific sites, usually with significant functional consequences. The phosphoryl groups are added by specific **protein kinases** and are removed by specific **phosphatases**, which are under specific control. In signal transduction pathways, many hormones act by increasing the intracellular concentration of second messengers – cyclic AMP, diacylglycerol, or Ca^{2+} – which in turn activate protein kinases that phosphorylate Ser and Thr residues of various proteins. Many growth-factor receptors have protein tyrosine kinase activities that phosphorylate Tyr residues.

It is the primary structure of the substrate protein and the specificity of the kinase that determine which residues are phosphorylated. The cyclic AMP-dependent kinase has a strong preference for Ser residues that occur in the sequence -Arg-Arg-X_{0-2}-Ser-. Other **Ser/Thr protein kinases** recognize Ser residues following one or two basic residues. Tyr phosphorylation by **protein tyrosine kinases** usually involves Tyr residues that occur in the sequence -Lys/Arg-X_3-Asp/Glu-X_3-Tyr.

Methylation

Protein methylation usually occurs on arginine or lysine side chains. Arginine methylation (by **peptidylarginine methyltransferases**) can give rise to monomethylated arginine or asymmetric or symmetric dimethylated arginine. Lysine methylation (by **lysine methyltransferases**) may yield mono-, di- or trimethyllysines. This type of modification of histones plays a major role in regulating the structure of chromatin since methylated chromatin is more tightly packed than unmethylated chromatin. Histone methylation has important roles in regulating gene expression and forms part of the epigenetic memory system that regulates cell fate and identity. Histone methylation is generally associated with transcriptional repression.

For a long time histone methylation was thought to be a permanent modification, however, recently two families of histone demethylating enzymes were discovered. The first of these, **lysine-specific demethylase 1** (LSD1) is a flavin-dependant monoamine oxidase that can demethylate mono- and dimethylated lysines (Holbert & Marmorstein, 2005). A second group of enzymes, the **jumanji domain-containing (JmjC) histone demethylases** are able to demethylate mono-, di-, or trimethylated lysines (Klose, Kallin & Zhang, 2006; Whetstine et al., 2006).

Acetylation

Acetylation (and deacetylation) of lysine side chains of histones plays a major role in gene regulation. The reactions are catalysed by **histone acetyltransferases** or **histone deacetylases**. In histone acetylation the source of the acetyl group is acetyl coenzyme A; conversely, in the deacetylation reaction the acetyl group is transferred to coenzyme A. The biological function of this modification is that acetylation neutralizes the interaction of the N termini of histones with the phosphate groups of DNA therefore the condensed chromatin is transformed into a transiently relaxed structure which allows genes to be transcribed.

Disulphide bond formation

Disulphide bond formation between cysteine residues is common among proteins synthesized in the endoplasmic reticulum. **Protein-disulphide isomerase**, present in the endoplasmic reticulum, catalyses disulphide rearrangements in proteins and thus assists in their folding. Formation of disulphide bonds is intimately linked with three-dimensional folding of the polypeptide chain. It is the conformation of the protein that actually determines which cysteine residues come into appropriate spatial proximity to permit the formation of a disulphide bond.

Hydroxylation of proline and lysine

Proline is the major site of hydroxylation in proteins, primarily in collagens. Hydroxylation of the $C\gamma$ (or the $C\beta$ atom) gives rise to hydroxyproline (Hyp). Lysine may also be hydroxylated on its $C\delta$ atom, forming hydroxylysine (Hyl). These hydroxylation reactions are catalysed by **prolyl 4-hydroxylase**, **prolyl 3-hydroxylase** and **lysyl 5-hydroxylase**, respectively. Since these reactions require iron and ascorbic acid, deprivation of ascorbate (e.g. in scurvy) leads to deficiencies in proline hydroxylation and this in turn leads to less stable collagens (see below).

Tyrosine sulphation

Protein tyrosine sulphation is an important post-translational modification where a sulphate group is added to a tyrosine residue of a protein molecule. The reaction is mediated by a Golgi enzyme, **tyrosylprotein sulphotransferase** that catalyses the transfer of sulphate from 3'-phosphoadenosine 5'-phosphosulphate to tyrosine residues within acidic motifs of polypeptides. Only secreted proteins and extracellular parts of membrane proteins that pass through the Golgi apparatus are sulphated. Although no clear-cut acceptor motif can be defined, there appear to be a few basic rules that define the consensus sites of tyrosine sulphation: the presence of an acidic residue within two residues of the tyrosine (typically at the −1 position), the presence of at least three acidic residues from positions −5 to +5, the presence of turn-inducing amino acid (Pro or Gly) between positions −7 to −2 and positions +1 to +7 etc. The biological importance of tyrosine sulphation is that it may strengthen protein–protein interactions.

Vitamin K-dependent γ-carboxylation of glutamic acid side chains

This type of post-translational modification leads to the carboxylation of certain glutamate residues in proteins to form gamma-carboxyglutamate residues (Gla residues). The biological role of vitamin K in this process is that it is the coenzyme for the **vitamin K-dependent carboxylase** that catalyses this reaction. Gla residues are usually involved in binding calcium, and are essential for the biological activity of all known Gla-containing proteins. The majority of proteins containing Gla residues is involved in blood coagulation (prothrombin, factors VII, IX, X, protein C, protein S) or bone metabolism (e.g. osteocalcin).

Attachment of C-terminal glycosylphosphatidylinositol (GPI) anchors

Chemical linking of the *glycosylphosphatidylinositol* moiety to the C-terminal residue (ω site) of the polypeptide chain of a GPI-anchored protein occurs by a transamidation reaction after proteolytic cleavage of a C-terminal propeptide from the proprotein. To be recognized as substrate of the GPI modification enzyme complex, a specific C-terminal sequence motif in the proprotein sequence appears necessary and sufficient, provided that the protein is exported from the cytoplasm to the endoplasmic reticulum. This sequence motif includes four sequence elements: (i) an unstructured linker region of about 11 residues upstream of the cleavage site ($\omega - 11 \ldots \omega - 1$); (ii) a region of four small residues surrounding the cleavage site (positions $\omega - 1 \ldots \omega + 2$); (iii) a moderately polar spacer region downstream of the cleavage site ($\omega + 3 \ldots \omega + 9$) and (iv) a hydrophobic tail beginning at positions $\omega + 9$ or $\omega + 10$ up to the C-terminal end. Although GPI-anchored proteins are similar to membrane proteins inasmuch as they are also anchored into the plasma membrane, they have some important features that distinguish them from proteins with transmembrane domains. For example, the anchor can be removed by the action of specific phospholipases converting the protein into a water-soluble form.

2.3.2 Nonenzymatic chemical modifications

The covalent enzymatic modifications affect only a few specific residues of a certain protein and frequently fulfil very specific roles. Many chemical modifications occur spontaneously, in the absence of enzymes. A common and biologically significant chemical modification of proteins is the spontaneous deamidation of Asn and Gln residues. Deamidation of Asn residues (changing an uncharged residue to a charged residue) can have severe effects on protein structure and thus may lead to the unfolding and degradation of the protein. The life span of proteins may be controlled by this mechanism.

Another common cause of chemical modification is oxidation by O_2 and by other oxidants such as peroxides, superoxide and hydroxyl radicals that are generated during metabolism. Although many such oxidants are scavenged by enzymes such as superoxide dismutase, peroxidase and catalase, oxidation of proteins does occur. The sulphur atoms of Cys and Met residues are most susceptible to oxidation. The cysteine thiol groups of intracellular proteins are generally protected by the high concentrations of glutathione that are present in cells. Met residues are readily oxidized to the sulphoxide, sometimes with drastic functional consequences. For example, oxidation of a Met residue in the active site of α_1-proteinase inhibitor inactivates the inhibitor. Since one of the functions of this inhibitor is to inhibit serum elastase that could digest and destroy the connective tissue proteins of the lung, inactivation of α_1-proteinase inhibitor may cause pulmonary emphysema. Oxidation of the inhibitor and the severity of emphysema are increased by smoking.

2.4 Interactions that govern protein folding and stability

2.4.1 Noncovalent interactions

Short-range repulsions

Atoms of a protein may be regarded as spheres with definite volumes (van der Waals volume) that are impenetrable to other atoms, therefore the most basic type of interaction between atoms of a protein is the repulsion as they approach each other.

Electrostatic forces

The most important noncovalent interaction in proteins is between electrostatic charges. If the charges are of the same sign, there is repulsion; if the two charges are of opposite sign there is attraction between them. Electrostatic interactions between very close, oppositely charged groups in a protein (e.g. salt bridges involving α-amino and α-carboxyl groups, side chains of Lys, Arg, His, Asp and Glu residues) usually also include hydrogen bonding of the participants.

Electrostatic interactions can also occur between groups that do not have a net charge, only partial charges. For example, the electron density can be localized to the atoms that have greater electronegativities. In such cases the atoms with greater electronegativity will have a partial negative charge, the atoms with lower electronegativities will have partial positive charge (the relative electronegativities of the atoms found in proteins are $O > N > C > S > H$).

The separation of charge in a molecule may be characterized by its **dipole moment**, the product of the magnitude of the separated excess charge and the distance of separation. Dipoles may interact with other dipoles in a manner that is determined by the relative orientations of the interacting groups. For example, two side-by-side dipoles repel each other when parallel, whereas there is attraction between them when they are antiparallel. Maximum interactions occur in a head-to-tail orientation of dipoles.

The importance of interaction between partial charges may be best illustrated by the peptide bond, which has partially double-bonded character due to resonance with a form in which the more electronegative oxygen atom acquires a net partial negative charge, whereas the –NH– group has a partial positive charge.

The importance of electrostatic interactions between partial charges may also be illustrated by the side chains of Phe, Tyr and Trp residues. The π electrons of the aromatic rings of these residues are localized on the face of their aromatic rings, hence the face of the rings has a partial net negative charge, whereas the hydrogen atoms on the edge of the rings have a corresponding partial positive charge. The interactions between aromatic rings are primarily defined by the electrostatic interactions between these partial charges as reflected by the fact that aromatic rings usually interact with the positively charged edge of one aromatic ring pointing at the negatively charged face of another.

Van der Waals interactions

Charged groups can induce a dipole in another spherical molecule by polarizing it. The general importance of polarizability is that there is an attraction between the induced dipole and the field that has induced it. These weak and close-range attractions are known as van der Waals interactions. Van der Waals interactions can arise not only between two permanent dipoles, between a permanent and an induced dipole, but even between two mutually induced dipoles. Interactions between mutually induced dipoles, known as **dispersion forces**, are of great general importance since they may occur among all types of atoms and molecules.

Hydrogen bonds

When two electronegative atoms compete for the same hydrogen atom a hydrogen bond is formed. In some exceptionally strong, short hydrogen bonds, the hydrogen atom is symmetrically placed between the two competing electronegative atoms. More typically the hydrogen atom of the hydrogen bond is covalently bonded to one of the atoms, the donor, but it also interacts favourably with the other electronegative atom, the acceptor.

Hydrogen bonds in proteins most frequently involve the C=O and N–H groups of the polypeptide backbone. In this type of hydrogen bond, N–H...O=C the typical H...O distance is 1.9–2.0 Å, whereas the covalent N–H distance is about 1.0 Å.

2.4.2 The hydrophobic interaction

Nonpolar groups do not interact favourably with water since they cannot participate in hydrogen bonding with this solvent. It is the absence of interactions between nonpolar groups and water that causes interactions among the nonpolar groups themselves to be much more favourable: the nonpolar molecules will greatly prefer nonpolar environments. Their favourable interaction is primarily the result of their exclusion from water. This preferential interaction of nonpolar groups with other nonpolar groups is usually called the hydrophobic interaction. This hydrophobic interaction is a major factor in the folding and stability of proteins that exist in an aqueous environment. In other words, the hydrophobic interaction results from a tendency of nonpolar atoms to interact with each other rather than with water.

The relative hydrophobicities of amino acid side chains (their relative preferences for aqueous vs. nonpolar environments) vary significantly, depending primarily on whether or not polar groups are present in their side chains. In general, ionized and polar side chains interact strongly with water, whereas nonpolar side chains prefer nonpolar environments. It is important to emphasize that the amino acid side chains having both polar and nonpolar segments are **amphiphilic**: they tend to interact in aqueous solution in such a way that their nonpolar segments interact with other nonpolar groups and their polar groups with polar groups, most frequently with water.

2.5 Secondary structural elements

Primary structure usually refers to the chemical structure of a protein, primarily defined as the sequence of amino acids. Secondary structure refers to **local spatial structures** of linear segments of polypeptide chains (such as a helix, an extended strand, a turn), as opposed to tertiary structure, which defines the overall spatial topology of these structures.

A randomly generated polypeptide chain is most likely to adopt a **random coil** structure. A unique conformation may have sufficient stability and it might predominate over all other possible conformations only if a huge number of weak bonds (hydrogen bonds, salt bridges, van der Waals interactions) cooperate to stabilize a single unique conformation. The basic types of regular conformations supported by numerous cooperating weak interactions are known as secondary structures. Surveys of the known structures of natural globular proteins show that roughly 90% of the residues are located in α-helices, β-strands and reverse turns; other types of regular conformations are much less frequent.

2.5.1 The α-helix

A typical α-helix usually contains 10–15 residues, with the backbone carbonyl oxygen of each residue hydrogen bonded to the backbone –NH of the fourth residue along the chain. There are 3.6 residues per turn of the helix, and the side chains of the component residues project outwards into the solution. In the case of the classical α-helix, the hydrogen bonds are nearly parallel to the helix axis. Since all the hydrogen bonds and peptide groups point in the same direction, the dipoles of each peptide bond are cumulative; thus the NH_2 end of the helix has a positive charge, the COOH end has a negative charge.

The various amino acids have markedly different tendencies to form α-helices (Chou & Fasman, 1974). One major factor that affects helical propensities of amino acids is that polar groups on some side chains (Ser, Thr, Asp and Asn) can hydrogen bond to the backbone peptide groups and this interferes with the hydrogen bonding of the α-helix; therefore these amino acids are found less frequently in α-helices. Since the Pro side chain is bonded to the backbone nitrogen atom, there is no –NH for hydrogen bonding; therefore Pro residues are practically incompatible with the α-helix conformation.

It is important to emphasize that although α-helices are structurally ordered (within proteins when they interact with other parts of proteins), they are usually only marginally stable when isolated. Many α-helices are amphipathic in the sense that the side chains projecting from the α-helix are predominantly nonpolar along one side of the helical cylinder and polar along the remainder of the helix surface. Such helices often aggregate with each other via their hydrophobic surfaces or bind to other nonpolar surfaces.

2.5.2 β-sheets

The basic element of β-sheets is the β-strand, usually consisting of 3–10 consecutive residues. The various amino acids have markedly different propensities to form β-sheets (Chou & Fasman, 1974).

In β-strands the polypeptide is almost fully extended, with 2.0 residues per turn. This extended conformation of a single strand is unstable and may be stabilized only when it is part of a β-sheet, where hydrogen bonds are formed between the peptide backbones of adjacent β-strands and the dipole moments of the strands also interact favourably. There is a general tendency that strands in a sheet are also adjacent in the primary structure.

Adjacent β-strands can be either parallel or antiparallel, but antiparallel sheets are usually more stable than parallel sheets; antiparallel sheets often consist of just two strands, whereas parallel sheets always have at least four. Purely parallel or antiparallel sheets of six or eight strands often curve around to close into a continuous β-barrel. The best known example of barrels is the triose phosphate isomerase (TIM) barrel with eight parallel β-strands, with an α-helix on the outside of the barrel connecting each pair of β-strands.

There is one major difference between β-sheets and α-helices. In α-helices interactions take place exclusively between consecutive residues of the amino acid sequence; in β-sheets interactions take place among residues of different strands that may be distant in the sequence or may even come from β-strands of different molecules. Because intermolecular β-sheets can grow indefinitely, sometimes this forms the basis of the formation of large intermolecular aggregates (e.g. amyloid formation).

2.5.3 Reverse turns

In globular proteins about one-third of the residues are involved in tight turns that reverse the direction of polypeptide chains at the surfaces of the molecules and thus make possible the overall globular structure. These reverse turns or loops may be regarded as a third type of ordered secondary structure. Reverse turns are usually classified according to the number of residues they contain and the types of secondary structures that they connect. The best characterized are the **β-hairpins** that link adjacent strands in an antiparallel β-sheet.

If only one residue in a chain is not involved in the hydrogen-bonding pattern of the sheet, it is a very **tight γ-turn**. **β-Turns**, in which two residues are not involved in the hydrogen bonding of the β-sheet, are much more common; the two residues on either side of the nonhydrogen-bonded residues also participate in the definition of the β-turn, therefore it is defined by four residues, i to $i + 3$. The conformations of short loops, such as γ-turns and various β-turns, depend primarily on the positions of certain residues in the loop – usually Gly, Asn or Pro – that allow the chain to take up an unusual conformation. The type I β-turn is compatible with any amino acid at positions i through $i + 3$, except that Pro cannot occur at position $i + 2$. Gly predominates at position $i + 3$, and Pro predominates at position $i + 1$ of both type I and type II turns. Asp, Asn, Ser and Cys residues frequently occur at position i, where their side chains often form hydrogen bonds to the –NH of residue $i + 2$. Gly and Asn occur most frequently at position $i + 2$ of type II turns because they can readily adopt the required peptide backbone angles. In all types of reverse turns, the peptide backbone atoms are not paired by regular hydrogen bonds.

Reverse turns occur on the protein surface and are accessible to the solvent. Larger **loops** connecting α-helices or β-strands usually have less well-defined conformations than reverse turns; frequently they are random coil.

2.6 Supersecondary structures

Certain prototypes of assemblies of a number of secondary-structure elements have been recurrently observed in a number of proteins. These elements of a higher level of protein structural organization are usually called supersecondary structures. Since such prototypical assemblies are found in a number of unrelated proteins, it seems that their popularity reflects some general principles of the structural organization of proteins rather than common ancestry.

It appears that certain types of supersecondary structures result from the general tendency that secondary structure elements that are close in the amino acid sequence are also adjacent in the tertiary structure. In one common supersecondary structure, the **β-α-β structure**, two β-strands are parallel in a β-sheet, and an α-helix, which is antiparallel to the β-strands, occurs in the connecting segment. In the case of antiparallel β-sheets, a common topology is the **β-meander**. The widespread occurrence of this motif results from the tendency that adjacent β-strands in antiparallel β-sheets are usually those that are sequential in the primary structure of the polypeptide chain, and that adjacent β-strands are connected by β-turns. A third frequently observed supersecondary structure consisting of antiparallel β-strands has the '**Greek key**' topology.

2.7 Tertiary structures of proteins

2.7.1 Globular proteins

The three-dimensional structures of several hundred, mostly globular proteins are now known to atomic resolution. The most important general feature of small globular proteins is that they are compact and roughly spherical. The compactness of folded globular proteins may be illustrated by the fact that their total accessible surface areas are only 23–45% of the surface area predicted for the unfolded polypeptide chain. The interiors of globular proteins are densely packed: based on the van der Waals radii of atoms, about 75% of the interior volume is filled with atoms. In the protein interior usually all polar groups are paired in hydrogen bonds; the vast majority of the polar groups of the polypeptide backbone are hydrogen bonded in secondary structure. Water molecules are generally excluded from protein interiors.

All ionized groups in water-soluble proteins are on the surface of the molecule, exposed to the solvent. For example, the charged Asp, Glu, Lys and Arg residues are sevenfold less abundant in the interior than on the protein surface. Nonpolar side chains predominate in the protein interior; Val, Leu, Ile, Phe, Ala and Gly residues constitute about two-thirds of the interior residues. The hydrophobic residues are primarily involved in packing together secondary structure elements.

As discussed above, some amino acid side chains have a dual nature: they have both polar and nonpolar characters. In such cases it is frequently observed that their polar moiety is exposed, whereas their nonpolar region is buried. For example, the ionized terminal groups in the long side chains of Lys and Arg are almost always exposed to the solvent, but their hydrophobic $-CH_2-$ carbons are usually buried in the interior.

The polypeptide backbone of a globular protein usually crosses the domain, then makes a reverse turn on the surface and recrosses the domain to the other side, then makes a reverse turn, etc. Consequently, the overall topology of globular proteins is usually characterized by polypeptide segments (α-helices, β-strands) linked by tight turns or bends that are almost always on the molecule's surface.

Most globular proteins with more than about 200–300 residues appear to consist of two or more **globular units, structural domains**. The individual domains of such protein molecules usually interact less extensively with each other than do structural elements within the domains. Usually a single polypeptide segment links these domains, and each domain consists of a single continuous stretch of polypeptide chain. Some **multidomain proteins** may contain dozens of structural domains (see also Chapter 8).

Basic classes of topologies

Globular protein structures are usually divided into four classes on the basis of their secondary structures:

(α)   proteins containing only α-helices;

(β)   proteins containing primarily β-sheet structure;

(α + β) proteins that contain both helices and sheets but in separate parts of the structure;

(α/β)   proteins in which helices and sheets interact (and frequently alternate) along the polypeptide chain.

In (α) proteins, about 60% of the residues are in α-helices, and the helices are usually in contact with each other. In typical (β) proteins, there are two β-sheets, both usually antiparallel, that pack against each other. In the (α + β) proteins, there may be a single β-sheet; the helices often cluster together at one or both ends of the β-sheet. The (α/β) proteins have one major β-sheet of primarily parallel strands; a helix usually occurs in each of the segments of polypeptide chain connecting the β-strands. In the case of the (α/β) barrel, the α-helices pack around the outside of the barrel (cf. TIM barrel; p. 31).

Detailed classifications of proteins

The past few years have seen a rapid growth in the number of proteins whose structures have been solved by X-ray crystallography or NMR spectroscopy and deposited in the Protein Data Bank (PDB). Despite this rapid growth in the number of entries, it has become increasingly apparent that their protein folds are of a limited variety: the vast majority of new entries are clearly related to structures of earlier entries. Based on these observations, recent estimates indicate that the large majority of proteins come from no more than 1000 protein folds, nearly half of which are already known (Brenner, Chothia & Hubbard, 1997).

Various types of hierarchic structural classification systems have been developed to analyse the relationships among proteins. The SCOP, CATH and Dali databases all distinguish different levels according to different degrees of structural and evolutionary relatedness (Murzin et al., 1995; Holm & Sander, 1996; Orengo et al., 1997). For example, in the CATH (Class/Architecture/

Topology/Homologous superfamily) hierarchic classification system, Class (C-level) categorizes proteins according to the secondary-structure elements and general aspects of the arrangements (mainly α, mainly β, α–β). The Architecture level (A-level) distinguishes proteins within the same class according to the arrangement of their secondary structural elements. Structures grouped at T-level (Topology) have the same overall fold: similar arrangement and connectivity of secondary structure elements. Finally, the Homologous superfamily level (H-level) groups those proteins where there is convincing evidence that the similar topology reflects common ancestry. In some other classifications of proteins (SCOP) the homology level has additional complexity, distinguishing families, superfamilies and hyperfamilies according to the degree of relatedness (Brenner et al., 1997).

Classification of protein structures has highlighted a surprising fact: a small number of folds are far more common than others, a few superfamilies are far more populous than others. One of the best-known, very populous hyperfamilies is that of the α/β (TIM) barrel, with 13 superfamilies.

2.7.2 Fibrous proteins

Fibrous proteins have regular, extended structures that represent a level of complexity that is intermediate between pure secondary structure and the tertiary structures of globular proteins. The basis for their regular conformations is a direct consequence of the regularities in their amino acid sequences. For example, silk **fibroin**, the structural protein of silkworm cocoons, has a repetitive sequence, the protein consists of antiparallel β-sheets, and the segments in the β-sheets consist of about 50 repeated -(Gly-Ala)$_2$-Gly-Ser-Gly-Ala-Ala-Gly-(Ser-Gly-Ala-Gly-Ala-Gly)$_8$-Tyr sequences.

Collagens form a major group of structural proteins that illustrate that regularities in amino acid sequence are reflected in regularities of folded structure. Collagens are the main constituents of the extracellular matrix of animals; there are more than a dozen different types of collagen polypeptide chains in vertebrates. Collagens have the common, distinctive feature of having a repetitive sequence in which every third residue is glycine (-Gly-X-Y-), Pro being found most frequently in X and Y positions. Many of the Pro residues at Y are hydroxylated in a post-translational modification step. In the triple helical structure of collagens three polypeptide chains are coiled together with a twisted, left-handed helical conformation. At every third position the Gly residues come into close proximity with the other two chains in the triple helix. The absence of a side chain in Gly is therefore critical; no other residue can substitute for a glycine in this position since its side chain would be incompatible with this close proximity. In fact, a number of mutant collagens have been identified in which replacement of the essential Gly residues by other amino acids causes disruption of the triple helix and consequent severe abnormalities.

The three polypeptide chains of collagens are linked by hydrogen bonds between the backbone –NH of the Gly residues and the backbone carbonyl group of residue X of another chain. The Pro residues at positions X and Y of collagens lend rigidity and stability to the structure. Furthermore, the hydroxyl groups on hydroxyPro residues (Hyp) are involved in hydrogen bonding between chains, which further stabilize the triple helix.

A third major group of fibrous proteins are the **coiled coil proteins** (e.g. myosin, intermediate filament proteins, etc.) in which two or three α-helices are wound around each other to form a left-handed superhelix. The individual α-helices interact through hydrophobic residues that form an apolar surface along one side of each helix. The interactions are also stabilized by electrostatic interactions between side chains that flank the apolar surface of the helices. The regularities of this structural feature in a coiled coil protein is apparent from its amino acid sequence: the regularity of the α-helix structure is based on repeats of seven residues (a-b-c-d-e-f-g), the so-called **heptad repeat**. The apolar interacting surfaces of helices are defined by hydrophobic side chains of residues a and d of the heptad repeat, and the electrostatic interactions involve primarily residues e and g. Position a is most frequently Leu, Ile or Ala, and d is primarily Leu or Ala. Residues e and g are often Glu or Gln, with Arg and Lys predominant at position g. As mentioned earlier, isolated α-helices are fairly unstable, but they are very stable in coiled coils because of the many interactions between them.

2.7.3 Unusual structures of internally repeated proteins

A large proportion of protein sequences contain repetitive sequence patterns and their three-dimensional structures show a corresponding internal symmetry (Kobe & Kajava, 2000). The leucine-rich repeat (LRR) is a sequence motif that provides a striking example of this phenomenon.

The LRR motif has been found in over 60 proteins with important cellular functions. The repeat of these proteins is defined by a consensus sequence, with a characteristic pattern of XLXXLXLXXNXLXXXXXXXLXXXLXXXXX where L denotes leucine, N denotes asparagine, and X denotes any amino acid (the consensus leucines can be replaced by other aliphatic residues, and the consensus asparagine can sometimes be replaced by a cysteine or a threonine). Most LRRs contain between 20 and 29 residues and usually several LRRs are present in 8–30 tandem repeats, with the highest known number being 30, as in the *Drosophila* protein chaoptin. The crystal structure of ribonuclease inhibitor, containing 15 LRRs, revealed the three-dimensional architecture of such proteins (Kobe & Deisenhofer, 1995). The individual repeats correspond to β–α units, consisting of a short β-strand and an α-helix approximately parallel to each other. A single leucine-rich motif does not appear to fold into a defined structure; several LRRs are needed to form a stable LRR domain. The individual repeats are arranged consecutively and parallel to a common axis, forming a horseshoe-like structure with a protein-binding site

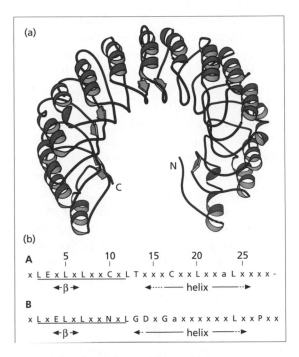

Fig. 2.1 (a) Structure of the ribonuclease inhibitor. (b) In the consensus sequences for the two types (A and B) of leucine-rich repeats 'x' denotes any amino acid, 'a' denotes an aliphatic amino acid. (From Kobe, B. & Deisenhofer, J. (1995) Proteins with leucine-rich repeats. *Current Opinion in Structural Biology* **5**, 409–16.) Reprinted from Current Opinion in Structural Biology, 5, Kobe, B. & Deisenhofer, J., Proteins with leucine-rich repeats, 409–416, Copyright 1995, with permission from Elsevier.

located in the cavity. The α-helices line the outer surface of the horseshoe; the β-strands form a parallel β-sheet along its inner surface (Fig. 2.1).

LRR proteins are widespread, being present in all types of organisms, including bacteria, yeast, plants and animals. Moreover, they are found in the cytoplasm as well as in extracellular proteins, and show functional versatility: they occur in proteins involved in signal transduction, cell adhesion, development, DNA repair, recombination, transcription and RNA processing.

The widespread occurrence of LRR-containing proteins in both pro-karyotes and eukaryotes may argue for a common origin of these proteins. However, considering its simple structure, it is equally possible that not all LRRs share a common ancestor; the same type of structures may have emerged independently.

2.7.4 Secreted proteins and membrane proteins

Secreted proteins and membrane proteins have features that reflect the properties of the membrane through which they pass or in which they reside. In the case of secreted proteins, the presence of a secretory **signal peptide** at the N-terminus of a nascent protein is usually sufficient to cause the protein to be inserted into the endoplasmic reticulum. Although the signal peptides of different proteins have diverse sequences, they have a number of common characteristics, reflecting the fact that there is a single cellular apparatus that recognizes them. There is a charged region at the N-terminus after the initiating Met residue; this region generally has a net positive charge of +2 (the terminal α-amino group included). The positive charge is thought to assist insertion into the membrane by interacting electrostatically with the negatively charged head groups of the membrane lipids. The positively charged region is followed by a stretch of 7–15 primarily hydrophobic residues, predominantly Leu, Ile, Val, Ala and Phe. The hydrophobic segment is of greatest importance for the signal peptide; it functions primarily to interact with the nonpolar interior of the membrane, entering and spanning it. The hydrophobic region is followed by approximately five predominantly polar residues that define the end of the signal peptide. This region serves as a recognition site for the signal peptidase that cleaves off the signal peptide.

Membrane proteins that remain integral parts of the membrane may undergo translocation across the membrane using the same type of signal sequence as for secreted water-soluble proteins. However, in the case of **integral membrane proteins** with signal peptides there is at least one additional transmembrane segment that interrupts the translocation of the polypeptide chain across the membrane. Part of the completed protein is on the extracytoplasmic side, one (or more) segments are within the membrane, and the remainder of the protein is on the cytoplasmic side. Although the hydrophobic nature of the membrane-spanning segment may be sufficient to keep the polypeptide within the membrane, the presence of charged residues flanking the hydrophobic segment is important for determining the orientation of the segment across the membrane.

Some integral membrane proteins are held by a single polypeptide segment that spans the membrane; others have more than one membrane-spanning segment. Proteins with a cleavable N-terminal signal peptide have their N-terminal part on the extracytoplasmic side of the membrane (type I transmembrane proteins). The internal stop-transfer sequence halts translocation, and the C-terminal part of the polypeptide chain remains on the cytoplasmic side of the membrane. Proteins with the opposite orientation (type II transmembrane proteins) are synthesized with N-terminal or internal signal peptides that are not cleaved. The polarity of the transmembrane segment appears to be determined by the flanking polar residues: positively charged residues at one end of a transmembrane segment tend to keep that end on the cytoplasmic side of the membrane. Membrane proteins with multiple transmembrane segments contain a signal sequence before each extracellular loop and a stop-transfer sequence after it.

The polypeptide chains of some integral membrane proteins are almost entirely within the membrane, with only a few residues exposed to the aqueous solvent. In other membrane proteins, the segments accessible to the

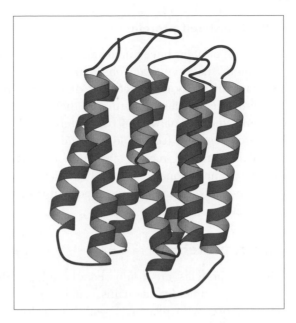

Fig. 2.2 The seven transmembrane α-helices of bacteriorhodopsin. (From Grigorieff et al. (1996) Electron-crystallographic refinement of the structure of bacteriorhodopsin. *Journal of Molecular Biology* **259**, 393–421.) Reprinted from the Journal of Molecular Biology, 259, Grigorieff, N., Ceska, T.A., Downing, K.H., Baldwin, J.M. & Henderson, R., The seven transmembrane α-helices of bacteriorhodopsin, 393–421, Copyright 1996, with permission from Elsevier.

solvent may be extensive and may even correspond to one or more protein domains. It must be emphasized that membrane proteins differ from water-soluble proteins only in their membrane-spanning segments; the loops and domains exposed to the aqueous environment have features characteristic of water-soluble proteins.

The best-studied integral membrane proteins with multiple membrane-spanning segments are bacteriorhodopsin (with seven parallel transmembrane helices; Fig. 2.2) and components of the photosynthetic reaction centre from *Rhodopseudomonas* that function in photosynthesis. The photosynthetic reaction centre consists of four different polypeptide chains, three of which (H, M and L) traverse the membrane. The H chain has a single transmembrane segment, the M and L chains have 5–5 transmembrane segments of 19–30 residues. The membrane-spanning segments are in the α-helical conformation, and they are packed together tightly. The surfaces of the interacting transmembrane regions that are in contact with the lipid bilayer are extremely nonpolar and consist primarily of the side chains of Leu, Ile, Val and Phe residues. The nonpolar transmembrane segments of the reaction centre protein are demarcated by charged residues at both surfaces of the membrane: the residues on the cytoplasmic side of the membrane tend to be positively charged, those on the other side are negatively charged. These residues are thought to be important in orientating the protein in the membrane.

2.7.5 Intrinsically disordered proteins

Although the majority of proteins have unique and stable three-dimensional structures, some proteins have large regions that are natively disordered and there are many **intrinsically unstructured/disordered proteins** that have no single well-defined tertiary structure in their native state (Dyson & Wright, 2005; Fink, 2005). Analyses of completely sequenced genomes suggest that disordered residues account for around one-fifth of a typical eukaryote proteome. Disordered regions are often characterized by low sequence complexity, compositional bias and high flexibility. Intrinsically disordered proteins are involved in numerous important regulatory processes such as transcription etc.

2.8 Multidomain proteins

The majority of larger proteins are composed of multiple **domains**. In multidomain proteins the individual domains usually show a high degree of structural independence, with relatively weak interactions between the domains.

As will be discussed in Chapter 8, one of the most important mechanisms for the creation of multidomain proteins is internal tandem duplication. A large group of multidomain proteins consists of multiple copies of the same type of protein fold (e.g. the cytoskeletal spectrins, immunoglobulins, ovomucoids, etc.).

Fusion of different protein-coding genes (by exon shuffling or by exonic recombination, etc.) is the most important mechanism for the creation of multidomain proteins that consist of different types of protein folds. By these mechanisms chimeras with all sorts of domain combinations may be created:

- proteins with multiple globular domains (e.g. tissue-plasminogen activator);
- proteins with globular and fibrous domains (e.g. modular collagens);
- integral membrane proteins with LRR domains (e.g. some receptor proteins);
- integral membrane proteins with globular domains (e.g. some receptor tyrosine kinases).

2.9 Multisubunit proteins

Many proteins exist as tight noncovalent complexes of two or more poly-peptide chains, either identical or different (e.g. globins, lactate dehydrogenase). In the **quaternary structure** of multisubunit proteins, each polypeptide chain is usually folded into an independent globular conformation, which inter-acts with other subunits. The interfaces between monomers involve primarily hydrophobic interactions between nonpolar side chains and other inter-actions between elements of secondary structure. At the periphery of the interface there are many hydrogen bonds and salt bridges between ionized side chains. An important aspect of all interacting surfaces is that they are complementary in shape and in interacting groups.

References

Brenner, S.E., Chothia, C. & Hubbard, T.J.P. (1997) Population statistics of protein structures: lessons from structural classifications. *Current Opinion in Structural Biology* **7**, 369–76.

Chou, P.Y. & Fasman, G.D. (1974) Prediction of protein conformation. *Biochemistry* **13**, 222–45.

Cosgrove, M.S. (2006) PHinDing a new histone 'effector' domain. *Structure* **14**, 1096–8.

Dhalluin, C., Carlson, J.E., Zeng, L., He, C., Aggarwal, A.K. & Zhou, M.M. (1999) Structure and ligand of a histone acetyltransferase bromodomain. *Nature* **399**, 491–6.

Dyson, H.J. & Wright, P.E. (2005) Intrinsically unstructured proteins and their functions. *Nauret Reviews in Molecular and Cellular Biology* **6**, 197–208.

Fink, A.L. (2005) Natively unfolded proteins. *Current Opinion in Structural Biology* **15**, 35–41.

Grigorieff, N., Ceska, T.A., Downing, K.H., Baldwin, J.M. & Henderson, R. (1996) Electron-crystallographic refinement of the structure of bacteriorhodopsin. *Journal of Molecular Biology* **259**, 393–421.

Holbert, M.A. & Marmorstein, R. (2005) Structure and activity of enzymes that remove histone modifications. *Current Opinion in Structural Biology* **15**, 673–80.

Holm, L. & Sander, C. (1996) Mapping the protein universe. *Science* **273**, 595–603.

Kobe, B. & Deisenhofer, J. (1995) Proteins with leucine-rich repeats. *Current Opinion in Structural Biology* **5**, 409–16.

Kobe, B. & Kajava, A.V. (2000) When protein folding is simplified to protein coiling: the continuum of solenoid protein structures. *Trends in Biochemical Science* **25**, 509–15.

Klose, R.J., Kallin, E.M. & Zhang, Y. (2006) JmjC-domain-containing proteins and histone demethylation. *Nature Reviews in Genetics* **7**, 715–27.

Murzin, A.G., Brenner, S.E., Hubbard, T. & Chothia, C. (1995) SCOP: a structural classification of proteins database for the investigation of sequences and structures. *Journal of Molecular Biology* **247**, 536–40.

Orengo, C.A., Michie, A.D., Jones, S., Jones, D.T., Swindells, M.B. & Thornton, J.M. (1997) CATH – a hierarchical classification of protein domain structures. *Structure* **5**, 1093–108.

Seet, B.T., Dikic, I., Zhou, M.M. & Pawson, T. (2006) Reading protein modifications with interaction domains. *Nature Reviews in Molecular and Cellular Biology* **7**, 473–83.

Whetstine, J.R., Nottke, A., Lan, F. et al. (2006) Reversal of histone lysine trimethylation by the JMJD2 family of histone demethylases. *Cell* **125**, 467–81.

Useful internet resources

Post-translational modifications

The **NetNglyc** server predicts N-glycosylation sites in human proteins using artificial neural networks that examine the sequence context of Asn-Xaa-Ser/Thr sequons (http://www.cbs.dtu.dk/services/NetNGlyc/).

The **NetOglyc** server produces neural network predictions of mucin-type GalNAc O-glycosylation sites in mammalian proteins (http://www.cbs.dtu.dk/services/NetOGlyc/).

O-GLYCBASE is a database of O- and C-glycosylated proteins (http://www.cbs.dtu.dk/databases/OGLYCBASE/index.php).

The **NetPhosK** server produces neural network predictions of kinase-specific eukaryotic protein phosphorylation sites (http://www.cbs.dtu.dk/services/NetPhosK/).

The **NetPhos 2.0** server produces neural network predictions for serine, threonine and tyrosine phosphorylation sites in eukaryotic proteins (http://www.cbs.dtu.dk/services/NetPhos/).

The **Phospho.ELM** database contains a collection of experimentally verified serine, threonine and tyrosine sites in eukaryotic proteins (http://phospho.elm.eu.org/).

The **Sulfinator** predicts tyrosine sulphation sites in protein sequences. Tyrosine sulphation is an important post-translational modification of proteins that go through the secretory pathway (http://ca.expasy.org/tools/sulfinator/).

PROSITE, a database of biologically significant sites is useful for the detection of motifs, such as the glycosaminoglycan attachment site motif (http://us.expasy.org/cgi-bin/nicedoc.pl?PDOC00002).
Tyrosine sulphation site motif (http://us.expasy.org/cgi-bin/prosite-search-ac?PDOC00003).
Tyrosine kinase phosphorylation site motif (http://us.expasy.org/cgi-bin/prosite-search-ac?PDOC00007).

DiANNA (DiAminoacid Neural Network Application) is a web server that provides cysteine classification prediction (whether a given cysteine is free, half-cystine or ligand-bound) and predicts disulphide connectivity (http://clavius.bc.edu/~clotelab/DiANNA/).

DIpro 2.0 is a web server for protein disulphide bond prediction (http://contact.ics.uci.edu/bridge.html).

CYSREDOX predicts the redox state of cysteines in proteins from multiple sequence alignments (http://manaslu.aecom.yu.edu/cysredox.html).

big-PI Predictor predicts potential GPI modification sites in proprotein sequences (http://mendel.imp.ac.at/sat/gpi/gpi_server.html).

GPI-SOM identifies GPI-anchor signals (http://gpi.unibe.ch/).

Protein structure databases

RCSB PDB is an archive of experimentally determined biological macromolecule 3D structures (http://www.rcsb.org/pdb/home/home.do).

PDBsum is a pictorial database providing an at-a-glance overview of every macromolecular structure deposited in the Protein Data Bank (PDB) (http://www.ebi.ac.uk/thornton-srv/databases/pdbsum/).

Protein Data Bank (http://www.pdb.org).

MSD is a macromolecular structure database (http://www.ebi.ac.uk/msd).

Protein structure classification

CATH provides protein structure classification (http://www.cathdb.info/latest/index.html).

SCOP – Structural Classification of Proteins (http://scop.mrc-lmb.cam.ac.uk/scop/).

The **Dali** server is a network service for comparing protein structures in 3D (http://www.ebi.ac.uk/dali/).

The **Dali Database** is based on exhaustive, all-against-all 3D structure comparison of protein structures in the Protein Data Bank (http://ekhidna.biocenter.helsinki.fi/dali/start).

HOMSTRAD (HOMologous STRucture Alignment Database) provides aligned 3D structures of homologous proteins (http://www-cryst.bioc.cam.ac.uk/~homstrad/).

The **Superfamily** server provides structural assignments to protein sequences at the superfamily level (http://supfam.mrc-lmb.cam.ac.uk/SUPERFAMILY/index.html).

Databases and Tools for 3-D Protein Structure Comparison and Alignment
(http://cl.sdsc.edu/ce.html).

VAST (Vector Alignment Search Tool) is a program used to identify
structurally similar proteins based on statistical criteria
(http://www.ncbi.nlm.nih.gov/Structure/VAST/vast.shtml).

VAST Search is NCBI's structure–structure similarity search service. It compares 3D
coordinates of a newly determined protein structure to those in the MMDB/PDB
database (http://www.ncbi.nlm.nih.gov/Structure/VAST/vastsearch.html).

Prediction of coiled coils

COILS predicts coiled-coil regions in proteins
(http://www.ch.embnet.org/software/COILS_form.html).

Coiled-coil predictions (http://www.russell.embl.de/cgi-bin/coils-svr.pl).

Prediction of signal peptides

SignalP 3.0 server predicts the presence and location of signal peptide
cleavage sites in amino acid sequences from different organisms
(http://www.cbs.dtu.dk/services/SignalP/).

SIG-Pred – Signal Peptide Prediction
(http://bioinformatics.leeds.ac.uk/prot_analysis/Signal.html).

Phobius is a combined transmembrane topology and signal peptide predictor.
It is especially useful in distinguishing N-terminal, cleavable signal peptides
from the transmembrane segments of type II transmembrane proteins
(http://phobius.cgb.ki.se/).

Predotar is a prediction service for identifying putative N-terminal targeting
sequences (http://urgi.infobiogen.fr/predotar/predotar.html).

TargetP predicts the subcellular location of eukaryotic proteins. The location assignment
is based on the predicted presence of any of the N-terminal presequences: chloroplast
transit peptide (cTP), mitochondrial targeting peptide (mTP) or secretory pathway
signal peptide (SP) (http://www.cbs.dtu.dk/services/TargetP/).

PSORT is a computer program for the prediction of protein localization sites
in cells. It analyses the input sequence by applying the stored rules for various
sequence features of known protein sorting signals (http://psort.nibb.ac.jp/).

Prediction of transmembrane segments

TMHMM Server v. 2.0 predicts transmembrane helices in proteins
(http://www.cbs.dtu.dk/services/TMHMM/).

TopPred 2 predicts the topology of membrane proteins, i.e. both the
orientation and location of transmembrane helixes is predicted
(http://www.sbc.su.se/~erikw/toppred2/).

'DAS' – Transmembrane prediction server
(http://www.sbc.su.se/~miklos/DAS/maindas.html).

HMMTOP is an automatic server for predicting transmembrane helices and
topology of proteins (http://www.enzim.hu/hmmtop/).

Prediction of intrinsically disordered proteins

DisEMBL™ Intrinsic Protein Disorder Prediction 1.5 is a computational tool
for prediction of disordered/unstructured regions within a protein sequence
(http://dis.embl.de/).

The DISOPRED2 Disorder Prediction Server (http://bioinf.cs.ucl.ac.uk/disopred).

GlobPlot 2 Intrinsic Protein Disorder, Domain & Globularity Prediction
(http://globplot.embl.de/).

DISpro: Protein Disorder Region Prediction
(http://www.ics.uci.edu/~baldig/dispro.html).

Detection of internal repeats in proteins

RADAR stands for Rapid Automatic Detection and Alignment of Repeats in protein sequences (http://www.ebi.ac.uk/Radar/index.html).

TRUST is a method for ab-initio determination of internal repeats in proteins (http://ibivu.cs.vu.nl/programs/trustwww/).

The program **REPRO** is able to recognize distant repeats in a single query sequence (http://ibivu.cs.vu.nl/programs/reprowww/).

Hhrep – De novo repeat identification in protein sequences by HMM-HMM comparison (http://toolkit.tuebingen.mpg.de/hhrep).

REPPER is a server that detects and analyses regions with short gapless REPeats in protein sequences. It finds PERiodicities by Fourier Transform (FTwin) and internal homology analysis (REPwin) (http://toolkit.tuebingen.mpg.de/repper).

Protein domain databases

InterPro is a database of protein families, domains and functional sites in which identifiable features found in known proteins can be applied to unknown protein sequences (http://www.ebi.ac.uk/interpro/).

Pfam is a large collection of multiple sequence alignments covering many common protein domains and families (http://www.sanger.ac.uk/Software/Pfam/).

SMART (Simple Modular Architecture Research Tool) allows the identification and annotation of genetically mobile domains and the analysis of domain architectures (http://smart.embl-heidelberg.de/).

PROSITE is a database of protein families and domains (http://us.expasy.org/prosite/).

CDD is a database of conserved domains (http://www.ncbi.nlm.nih.gov/entrez/query.fcgi?db=cdd).

Prediction of α-helices and β-strands

The **Chou–Fasman secondary structure prediction method** was one of the first automated structure prediction methods (http://fasta.bioch.virginia.edu/fasta_www2/fasta_www.cgi?rm=misc1).

The **GOR** (Garnier–Osguthorpe–Robson) secondary-structure prediction method is used for protein secondary-structure prediction from single sequence information (http://fasta.bioch.virginia.edu/fasta_www2/fasta_www.cgi?rm=misc1).

Prediction of β-turns

BTPERD is a method for predicting the location and type of β-turns in protein sequences. Predictions are made using a combination of artificial neural networks and simple filtering rules (http://www.biochem.ucl.ac.uk/bsm/btpred/index.html).

Prediction of quaternary structure

Quaternary Structure Predictor predicts the 'mericity', i.e. the number of subunits in a multisubunit protein (http://www.mericity.com/).

Chapter 3: Mutations

During chromosome replication DNA sequences are usually copied with high fidelity; however, sometimes errors occur in the replication that may give rise to changes in the DNA sequences. The DNA sequence may also be altered in processes that are independent of DNA replication. Accordingly, it is useful to distinguish **replication-dependent** and **replication-independent mutations**.

Mutations may occur in somatic or germline cells. Since **somatic mutations** are not inherited, they are of no major consequence for evolution, therefore in this book we will discuss primarily **germline mutations**. Nevertheless, in special cases (generation of antibody diversity, malignant transformation) somatic mutations may have significant functional importance.

3.1 Types of mutations

If a nucleotide of DNA is replaced by a different nucleotide it is called a **substitution**. Changes that affect only a single nucleotide are called **point mutations**. Mutations may involve the **deletion** of one or more base pairs or the **insertion** of one or more base pairs. **Inversions** may result from the reversal of the polarity of a sequence involving several nucleotides. Changes that alter the wild-type structure of DNA are usually called **forward mutations**, whereas **back mutations** that result in restoration of the original state are called **true reversions**.

3.1.1 Substitutions

The most common type of mutation is the substitution of one base pair for another. The more common type of single-base substitution is **transition**, which is the replacement of one purine by another purine (A, G) or of one pyrimidine by another pyrimidine (C, T). Significantly less frequent are the **transversions**, the replacement of a purine by a pyrimidine or of a pyrimidine by a purine.

A point mutation that changes only a single base may result from incorrect replication of DNA if a wrong base is inserted into DNA during its synthesis (replication-dependent mutation). Alternatively, it may result from direct chemical modification of the bases in DNA (replication-independent mutation). Such modification may occur spontaneously (due to errors during replication or as a result of the action of some physiological agent), or may be induced by various physicochemical and chemical agents of the natural environment.

45

Nucleotide substitutions occurring in protein-coding regions may also be categorized according to their effect on the protein. In translated regions a substitution is **synonymous** (or **silent**) if it causes no amino acid change, since the altered codon codes for the same amino acid. Those nucleotide substitutions that alter the amino acid are called amino acid changing or **nonsynonymous** substitutions.

Nonsynonymous mutations are called **missense mutations** if the mutation changes the codon of an amino acid into a codon that specifies a different amino acid. A **nonsense mutation** changes an amino acid-coding codon into one of the termination codons, leading to premature termination of the protein. Conversely, the mutation of the original stop codon of the protein into a sense codon may lead to the addition of a C-terminal extension of a protein.

Because of the structure of the standard genetic code (see Table 1.1), synonymous substitutions occur mainly at the third (wobble) position of codons. In contrast, all the substitutions at the second position and the vast majority of nucleotide changes at the first position of codons are nonsynonymous.

Spontaneous substitution mutations

One major reason for the spontaneous occurrence of replication-dependent transition mutations is that the amino and keto groups of bases can **tauto-merize** (to the imino and enol forms) and these transient tautomers can form nonstandard base pairs. For example, the imino tautomer of adenine can pair with cytosine instead of thymine; this abnormal A–C pairing would allow C to become incorporated into a growing DNA strand in place of T, eventually leading to a transition mutation, replacing the normal AT base pair by a mutant GC base pair.

Another major source of spontaneous mutations is that cytosine spontan-eously **deaminates** at a perceptible rate to form uracil. Since uracil pairs with adenine, this causes a transition mutation, replacing the original GC base pair by an AU (eventually AT) base pair.

Most incorrect base pairs formed in the polymerization step, however, do not become permanently incorporated into DNA. The DNA polymerases themselves proofread the outcome of a polymerization step before proceed-ing to the next one, incorrectly inserted bases being removed by the 3′→5′ exonuclease component of the DNA polymerase. This 3′→5′ exonuclease activity markedly enhances the accuracy of DNA replication.

The significance of this **proofreading** activity is illustrated by the fact that some *Escherichia coli* mutants with abnormally high mutation rates have an altered DNA polymerase III with lowered 3′→5′ exonuclease activity: the inefficiency of the exonuclease activity allows a high proportion of mispaired nucleotides to escape removal. On the other hand, a very efficient 3′→5′ nuclease activity leads to a very low mutation rate: mutant DNA polymerases with a higher than normal ratio of exonuclease to polymerase activity have a lower than normal spontaneous mutation rate.

These observations emphasize that – in an evolutionary sense – there may be an optimal mutation rate. Too high a rate could lead to an excessive proportion of nonviable progeny, whereas mutation rates that are too low may diminish genetic diversity, decreasing the chance of survival in a changing environment. In fact, bacteria are known to respond to environmental challenges with increased mutation rates (see section 3.2). The significance of proofreading activity is also illustrated by the fact that in the case of the mitochondrial genomes, where proofreading activity is missing, high mutation rates are observed.

Incorrectly paired bases that escape detection at the proofreading stage may be corrected at a second line of defence: the **mismatch repair** enzymes scan newly replicated DNA for incorrectly inserted bases and remove a single-stranded segment containing the wrong nucleotide, allowing a DNA polymerase to insert the correct base when it fills the resulting gap. The biological significance of mismatch correction may be illustrated by the fact that mutations in mismatch repair genes lead to genetic predisposition to cancer (Edelmann et al., 1997).

Induced mutations

The frequency of the occurrence of mutations may be increased by certain natural mutagenic agents; the changes they cause are referred to as induced mutations. Natural mutagens either act on the DNA directly to change its template properties or in some way interfere with correct replication so that a wrong base is inserted.

DNA-reactive chemicals and ultraviolet radiation act directly by chemically modifying the bases of DNA. For example, nitrous acid can convert cytosine into uracil, which then pairs with an A instead of G (with which the original C would have paired). Similarly, nitrous acid can also deaminate adenine to hypoxanthine, which pairs with cytosine rather than with thymine: the original AT is replaced by a GC pair.

A major natural source of mutations is ultraviolet radiation. Ultraviolet light is absorbed strongly by the bases of DNA leading to photochemical fusion of two adjacent pyrimidines. The human hereditary skin disease xeroderma pigmentosum is caused by genetic defects in enzymes that remove pyrimidine dimers and other ultraviolet-induced lesions (Taylor et al., 1997). As a result of this deficiency the skin of affected individuals is extremely sensitive to ultraviolet light, and skin cancer usually develops in these patients.

Repair of damaged DNA

As a result of ultraviolet radiation DNA bases become covalently cross-linked through the formation of pyrimidine dimers. Nearly all organisms contain a photoreactivating enzyme called **DNA photolyase** that reverses the photochemical fusion of adjacent pyrimidine bases, restoring the original structure of DNA.

A more universal process is **excision repair**, in which the damaged segment is removed and replaced by new DNA synthesis, using the undamaged strand as template. This repair system is not specific for ultraviolet damage, but senses any kind of serious distortion of the DNA helix, since it recognizes the absence of a normal DNA shape. In *E. coli* several enzymatic activities are essential for this repair process. First, the uvrABC enzyme complex detects the distortion produced by the pyrimidine dimer then cuts the damaged DNA strand at two sites to remove a segment that includes the lesion. The gap is filled by DNA polymerase I, then the newly synthesized DNA and the original DNA chain are joined by DNA ligase. The significance of this repair system is underlined by the fact that cells mutant in the DNA polymerase I gene are very deficient in excision repair.

Removal of uracil from DNA

As mentioned above, cytosine spontaneously deaminates at a perceptible rate to form uracil, and some mutagens can facilitate this conversion. Since uracil pairs with adenine, this chemical change would lead to a transition mutation if it is left uncorrected.

Such mutations are prevented by a repair system that recognizes uracil as an abnormal base in DNA and removes these bases. In the first step an enzyme, **uracil-DNA glycosylase**, hydrolyses the bond between the uracil base and the deoxyribose moiety (leaving an unpaired G residue on the complementary strand). The 'hole' that results in the mutant strand is called an **AP site** (apyrimidinic, apurinic site), since it lacks a base. (Such AP sites may also result from natural loss of bases through breakage of the glycosidic bond.) Irrespective of the cause that created the 'hole', the AP site is recognized by an **AP endonuclease**, which nicks the backbone adjacent to the missing base. DNA polymerase I excises the residual deoxyribose phosphate unit and inserts cytosine (if there is a G on the other strand) and the repaired strand is sealed by DNA ligase, restoring the original sequence of the DNA.

It should be mentioned here that spontaneous deamination of **5-methylcytosine** residues of DNA leaves thymine, not uracil. Since thymine is a normal constituent of DNA (and is not removed by uracil-DNA glycosylase), this repair system cannot operate in these circumstances and a mutation will result. Therefore, 5-methylcytosines are hotspots for spontaneous mutations.

3.1.2 Deletion, duplication, insertion and fusion

In the translated region of a protein-coding gene, deletions, duplications or insertions involving a number of nucleotides that is not three or a multiple of three will cause a shift in the reading frame so that the coding sequence downstream of the deletion will be read in the wrong phase. Such mutations are known as **frameshift mutations**. A frameshift mutation introduces numerous amino acid changes and is likely to bring into phase a new stop codon, thus resulting in a protein of abnormal (most frequently shorter) length. Deletions, duplications and insertions are collectively referred to as **gap events**, because when the mutant sequence carrying a deletion, duplication or insertion is compared with the original wild-type sequence a 'gap' will appear in one of the two sequences. The number of nucleotides involved in a gap event ranges from one or a few nucleotides to contiguous stretches involving thousands of nucleotides.

Chimeric genes, in which different parts originate from different genes, may be formed by fusion of (parts of) different genes or insertion of (parts of) other genes. The classical examples of chimeric genes created by fusion are haemoglobins Lepore and anti-Lepore, which arose by unequal crossing over between genes encoding the haemoglobin β and δ chains. In the chimeric Lepore chain the amino-terminal part came from the δ chain, the carboxyl-terminal part from the β chain. In the reciprocal hybrid, the amino-terminal portion corresponds to the β chain, the carboxyl-terminal segment to the δ chain.

As will be discussed in Chapter 8, formation of chimeric genes encoding multidomain proteins has played a major role in the evolution of novel proteins.

Mechanism of deletion, duplication, insertion and fusion

Endogenous or exogenous polycyclic molecules (present in many foodstuffs) that bind to and intercalate between adjacent bases of DNA can induce the looping out of either the template DNA strand or the growing strand during DNA synthesis, thereby greatly increasing the chance that one or more base pairs will be inserted or deleted. Since these short deletions/insertions are most likely to cause shifts in the reading frame, such mutagens are called **frameshift mutagens**.

Replication errors by DNA polymerase are not limited to single-base substitutions. **Replication slippage** or **slipped-strand mispairing** can occur because of mispairing between neighbouring repeats and can result in either deletion or duplication of a DNA segment, depending on whether the slippage occurs in the $5' \rightarrow 3'$ direction or in the $3' \rightarrow 5'$ direction. DNA regions containing short repeats are most susceptible to this type of replication error since they are most prone to slipped-strand mispairing. In eukaryotic genomes, short tandem repeats and runs of identical bases in the DNA are hotspots for deletions and insertions by this mechanism. Such errors in the process of DNA replication usually create only short gaps (up to 20–30 nucleotides).

A distinct mechanism exists that leads to frequent expansion of certain types of triplets of DNA (Mitas, 1997). Since replication of duplex DNA requires separation of the two parental strands at the replication fork, during this time single-stranded DNA has the opportunity to form stable self-complementary **hairpin structures**. These hairpins interfere with the progression of enzymes involved in DNA replication and may thus cause repeat expansion or deletion, depending on whether they are formed on the nascent DNA strand or on the template strand. Formation of stable hairpin structures is especially likely if the DNA has inverted or triplet repeat sequences. **Triplet repeat expansion diseases** (TREDs) are characterized by the coincidence of disease manifestation with amplification of d(CAG.CTG), d(CGG.CCG) or d(GAA.TTC) repeats found within specific genes (e.g. genes affected in Huntington's disease, fragile X syndrome, myotonic dystrophy). Amplification of triplet repeats continues in offspring of affected individuals, which generally results in progressive severity of the disease, a phenomenon referred to as **anticipation**. Stepwise expansion by this mechanism may create relatively long repeated regions (Wells, 1996). Hairpin formation of single-stranded DNA has also been responsible for repeat expansions during the evolution of the mammalian involucrin genes (Tseng, 1997).

Longer insertions, duplications, deletions or fusions of genomic DNA occur mainly by recombination via **unequal crossing over**, **exon shuffling** or **transposition**. Deletions and insertions of introns may occur via processes that also involve reverse transcription (for details see sections 6.2.1 and 8.1.1). **Chromosome translocation** and gene fusion are frequent events in the human genome and are often the cause of many types of tumour. Chimeric genes may also be formed through fusion of (parts of) neighbouring genes as a result of unequal crossing over. The features of genomic DNA that predispose to such mutations will be discussed in Chapters 6, 8 and 9.

The observation that inappropriately elevated levels of homologous recombination activity may contribute to genomic instability and cancer predisposition in Fanconi anaemia (Thyagarajan & Campbell, 1997) illustrates that, in an evolutionary sense, there may be an optimal recombination rate. Very low recombination rates may diminish genetic diversity, whereas very high rates could diminish viability.

The biological significance of recombination in generating gene fusions is best illustrated by the **somatic recombinations** of the various segments of **immunoglobulin** genes that contribute to the diversity of antibodies. During the development of bone marrow-derived lymphocytes, complete immunoglobulin genes are formed by joining different members of gene segment repertoires (Tonegawa, 1983; Litman et al., 1993).

3.2 Factors affecting rates of mutation

Different sites within the sequence of the DNA of a given organism are not equally susceptible to mutations, therefore mutations do not occur randomly throughout the genome. Sites that gain far more mutations than expected on the basis of statistical probability are called **hotspots**. Different types of mutations have different hotspots, reflecting differences in the underlying mechanisms. For example, a major site of spontaneous substitution mutations is the modified base 5-methylcytosine. 5-Methylcytosine suffers spontaneous deamination at an appreciable frequency, converting it to thymine, thereby converting a wild-type GC pair into a mutant AT pair. Since cytosines in the dinucleotide 5'-CpG-3' are frequently methylated in vertebrate genomes, these are the primary hotspots of spontaneous mutation. With the evolution of the heavily methylated vertebrate genome there was strong selective pressure to suppress CpG dinucleotides, explaining their relative paucity in such genomes (Krawczak & Cooper, 1996). For deletions and insertions by slipped-strand mispairing, DNA regions containing short tandem repeats are the major hotspots, as illustrated by microsatellite expansions (Chakraborty et al., 1997). Triplet repeat sequences that can readily form hairpin structures are hotspots for mutations causing a variety of triplet repeat expansion diseases (Perutz, 1996; Wells, 1996; Mitas, 1997).

One major reason for variation of mutation rates among the genetic material of different organisms, and among their different cellular compartments, may be that their molecular devices for DNA replication, proofreading, repair of DNA damage, etc. may show striking differences with respect to **fidelity** or efficiency (e.g. mitochondrial genomes vs. nuclear genomes).

However, even within a given species, there may be significant differences in mutation rates. High mutability in the male is a general property of human and other vertebrate genes – the male/female ratio of nucleotide substitution rate is estimated as ≈6 in humans. Since this ratio is close to the ratio of the number of male/female **germ cell divisions per generation**, this observation suggests that nucleotide substitutions in the germline are largely replication dependent.

The per year substitution rate is faster for those organisms with a short **generation time** than for those with a long generation time (the generation time effect). In mice and rats, the number of germ cell divisions per year is ≈100, in humans it is ≈10, which is in harmony with the observed faster rate of silent nucleotide substitution in rodents than in humans (see below).

Mutation rates may also vary in response to environmental changes – and not only because they may alter the mutagenicity of the **environment**. There is evidence in bacteria and yeast that a particular environmental stress that induces the increased transcription of a particular gene also leads to a higher mutation rate of that gene. The mutations would be thus 'directed' by the environment in the sense that a specific gene or class of genes that are relevant to the stress are subject to higher rates of mutation (Wright, 1997). The basis of an increased mutation rate for such genes is that the process of transcription increases the concentration of single-stranded DNA, which is especially vulnerable to mutagenesis.

3.3 The fate of mutations

The fate of a new mutation, the outcome of its competition with the original wild-type allele, depends primarily on whether it is **neutral**, **deleterious** or **advantageous** relative to the wild-type form. Although **natural selection** is the major driving force of evolution, chance effects (**random genetic drift**) also play an important role especially in the case of small populations where random fluctuations in allele frequencies are very significant.

Natural selection

Natural selection favours genotypes that have higher success in reproduction than other competing genotypes (because of differences in their viability, mortality, fertility, number of offspring, etc.): the outcome of the competition depends on the relative **fitness** of the competing genotypes. When the competing genotypes differ significantly from each other in fitness, there is strong natural selection and there will be marked changes in allele frequencies in favour of the genotype that has higher fitness value. Deleterious mutations that reduce the fitness of their carriers will be eventually eliminated from the population; this type of selection is usually called **purifying** or **negative selection**. Advantageous mutations (ones that have a higher fitness than other alleles and thus confer a selective advantage on their carriers) will be subjected to **positive selection**. Significantly, even a minor difference in fitness value ($s = 1\%$) may eventually lead to elimination of the allele with lower fitness and fixation of the allele with higher fitness.

A mutation that has the same fitness value as the 'original' allele is **selectively neutral**; in this case the fate of the genotype is not determined by selection, but by chance factors.

Random genetic drift

Changes in allele frequency may occur by chance. The process of change in allele frequency due to chance effects is called random genetic drift. In such cases, although the changes are random from generation to generation, the frequency of an allele will tend to deviate more and more from its initial frequency. Random drift is most pronounced in small populations.

Probability of the fixation of a mutation

The probability that a new mutant allele will become fixed in a population (i.e. the mutant gene completely substitutes the original wild-type allele) depends on its selective advantage, disadvantage or neutrality, as well as the population size.

According to the calculations of Kimura (1962) for a neutral allele the fixation probability (P) equals its frequency in the population. This plausible conclusion reflects the fact that in the case of neutral alleles, fixation occurs by random genetic drift, where neutral alleles have an equal probability of fixation, the outcome of the competition depends only on their frequency. It could be also shown that if an advantageous mutation arises in a large population and its selective advantage (s) over the rest of the alleles is small then the probability of its fixation $P \approx 2s$, i.e. it is approximately twice its selective advantage. In other words, if a mutant has 1% selective advantage, it has about 2% chance of fixation. An important consequence of this conclusion is that an advantageous mutation does not always become fixed in the population but may be lost by chance. The results of Kimura's work are of great theoretical importance, since they show that the earlier views that saw evolution as a process in which advantageous mutations are always fixed and only advantageous mutations are fixed are oversimplified. In fact, the calculations show that neutral and even slightly deleterious mutations may have a definite probability of becoming fixed in a population.

The neo-Darwinian theory vs. the neutral mutation hypothesis

According to classical **neo-Darwinism** natural selection plays the dominant role in the process of evolution, whereas chance factors, including random drift, are of minor importance. The most extreme form of neo-Darwinism – **selectionism** – considers selection as the only force that drives the evolutionary process. According to this view evolution is the result of a positive adaptive process whereby a new allele is fixed only if it improves the fitness of the organism. Moreover, **polymorphisms** in a population are maintained only when the coexistence of two or more alleles is advantageous.

In contrast with the selectionist hypothesis, Kimura has suggested that the majority of molecular changes in evolution are due to the random fixation of neutral or nearly neutral mutations. According to the neutral theory of molecular evolution, the majority of evolutionary changes as well as the polymorphisms within species are caused by random genetic drift of alleles that are selectively neutral or nearly neutral (Kimura, 1968, 1983). In the neutral theory of molecular evolution the emphasis is on the statement that the fate of alleles is determined primarily by random genetic drift. Although it acknowledges that selection does operate, it claims that chance effects are of major importance.

As may be clear from this brief summary, the dispute between neutralists and selectionists is essentially centred around the frequency distribution of fitness values of mutant alleles. Neutralists and selectionists agree that the majority of new mutations are deleterious and that these mutations are quickly removed from the population by purifying selection, consequently they make a negligible contribution to polymorphisms within populations. The key difference between neutralists and selectionists is in their assessment of the relative proportion of neutral vs. advantageous mutations. Selectionists claim that very few mutations are selectively neutral, neutralists maintain that most nondeleterious mutations are neutral and very few are advantageous. Considering the fact that even a minor selective advantage may ensure fixation of a mutant, it is apparent that the boundaries between the selectionist and neutralist camps are sometimes unclear. Nevertheless, the formulation of the neutral mutation hypothesis had a major impact on ideas of evolution (and protein evolution) since it has led to the general recognition that the effect of random drift cannot be neglected.

Natural selection and patterns of amino acid replacements

Since each codon can undergo nine types of single-base substitutions, point mutations in the 61 sense codons can lead to 549 types of single-base substitutions. Of these, 392 result in the replacement of one amino acid by another (nonsynonymous substitutions), whereas 134 result in 'silent' mutations (synonymous substitutions). Here we will be concerned primarily with the probabilities and patterns displayed by nonsynonymous substitutions.

An accepted amino acid replacement is the result of two distinct processes: the first is the occurrence of a mutation in the protein-coding gene; the second is the acceptance (fixation) of the mutation by the population (species) as the new predominant form. Accordingly, there could be two main reasons why the various nonsynonymous substitutions would not occur with equal probability.

In principle, one major source of bias in nonsynonymous mutations could be the structure of the genetic code itself: those interchanges that require two or three single-base substitutions have a much lower chance of occurring than those that require single-base substitutions. Of the 190 possible interchanges of the 20 amino acids, only 75 can be achieved by single-base substitutions, 101 amino acid interchanges can occur by two-base substitutions, whereas there are 14 interchanges that can occur only if all three bases of the codon are changed. Surveys of the 75 single-base interchanges have shown that there is only a slight preference for interchanges between similar amino acids. Even more striking is that some interchanges between chemically similar amino acids – e.g. tyrosine↔tryptophan (UAU/C↔UGG) and phenylalanine↔tryptophan (UUU/C↔UGG) interchanges – require two base changes. In other words, the pattern that might be due to the structure of the genetic code is distinct from the pattern of the chemical similarities of the amino acids.

Another cause of preferences might be that the new amino acid must function in a way similar to the old one (or better than the old one) otherwise the mutation is rejected by natural selection. Obviously, the synonymous mutations are likely to be selectively neutral at the protein level since they do not change the amino acid sequence. According to the neutral theory such neutral synonymous mutations have a definite probability of being accepted. Similarly, of the replacement mutations, **conservative changes** to chemically and physically similar amino acids are likely to be nearly neutral and therefore are likely to be accepted. It may be expected that there will be a strong bias against **radical changes** to chemically dissimilar amino acids since such mutations are most likely to be deleterious.

To characterize the actual mutational preferences in proteins Dayhoff has tabulated nonsynonymous mutations observed in several different groups of alignments of related protein sequences (ones that are at least 85% identical) from which mutation data matrices could be derived (Dayhoff, Schwartz & Orcutt, 1978; George, Barker & Hunt, 1990). The observed mutational patterns have two distinct aspects: the resistance of an amino acid to change and the pattern observed when it is changed (Table 3.1(a)). The data collected on a large number of protein families have revealed striking differences between the **relative mutabilities** of the different amino acids: on average, Asn, Ser, Ala, etc. are most mutable (the lowest figures on the diagonal), whereas Trp, Cys, Tyr and Phe are the least mutable (the highest figures on the diagonal).

The relative immutability of cysteine can be interpreted as a reflection of the fact that it has several unique, indispensable functions that no other amino acid side chain can mimic (e.g. it is the only amino acid that can form disulphide bonds). Clearly, the low mutability of Trp is not due to the structure of the genetic code, since all single-base mutations affecting its single codon would change it to another amino acid (or to a stop codon). Its high degree of conservation reflects the unique role of this bulky aromatic residue in protein folding. The low mutabilities of Tyr and Phe may also be explained by the importance of these hydrophobic residues in protein folding (see Chapter 2).

When the distribution of accepted amino acid replacement mutations observed between closely related sequences was analysed it was clear that the majority of replacements were the result of single-base substitutions. However, about 20% of the interchanges – far more than one would expect on the basis of chance – involved amino acids whose codons differ by more than one nucleotide. Conversely, many of the changes expected from the mutations of one nucleotide in a codon are seldom observed (e.g. exchanges between Trp and most other amino acids). All these findings indicated that some of the changes are rejected by selection, whereas multiple changes at some of the mutable sites may be favoured by selection.

It is obvious from an analysis of the data shown in Table 3.1(a) that favoured interchanges of amino acids have something to do with their physicochemical similarities. In general, the groups of chemically similar amino acids tend to replace one another: the aliphatic group (M, I, L, V); the aromatic group (F, Y, W); the basic group (R, K); the acid-amide group (N, D, E, Q);

Table 3.1 Mutation data matrices. The amino acids are arranged in clusters based on their physicochemical properties. The neutral score is zero, positive values represent conservative replacements (shown in **bold**). (a) General data set for 250 PAMs (Per cent Accepted point Mutations). (From George et al., 1988. Reprinted by permission of John Wiley & Sons, Inc.) (b) Mutation data matrix for 250 PAMs for transmembrane proteins. (Reprinted from *FEBS Letters* **339**, 269–275. Jones et al. A mutation data matrix for transmembrane proteins. Copyright 1994, with permission from Elsevier Science.) (*Continued*)

(a)

	C	S	T	P	A	G	N	D	E	Q	H	R	K	M	I	L	V	F	Y	W
C	12																			
S	0	2																		
T	−2	1	3																	
P	−3	1	0	6																
A	−2	1	1	1	2															
G	−3	1	0	−1	1	5														
N	−4	1	0	−1	0	0	2													
D	−5	0	0	−1	0	1	2	4												
E	−5	0	0	−1	0	0	1	3	4											
Q	−5	−1	−1	0	0	−1	1	2	2	4										
H	−3	−1	−1	0	−1	−2	2	1	1	3	6									
R	−4	0	−1	0	−2	−3	0	−1	−1	1	2	6								
K	−5	0	0	−1	−1	−2	1	0	0	1	0	3	5							
M	−5	−2	−1	−2	−1	−3	−2	−3	−2	−1	−2	0	0	6						
I	−2	−1	0	−2	−1	−3	−2	−2	−2	−2	−2	−2	−2	2	5					
L	−6	−3	−2	−3	−2	−4	−3	−4	−3	−2	−2	−3	−3	4	2	6				
V	−2	−1	0	−1	0	−1	−2	−2	−2	−2	−2	−2	−2	2	4	2	4			
F	−4	−3	−3	−5	−4	−5	−4	−6	−5	−5	−2	−4	−5	0	1	2	−1	9		
Y	0	−3	−3	−5	−3	−5	−2	−4	−4	−4	0	−4	−4	−2	−1	−1	−2	7	10	
W	−8	−2	−5	−6	−6	−7	−4	−7	−7	−6	−3	2	−3	−4	−5	−2	−6	0	0	17

(b)

	C	S	T	P	A	G	N	D	E	Q	H	R	K	M	I	L	V	F	Y	W
C	6																			
S	1	3																		
T	0	2	3																	
P	−4	−1	−1	11																
A	0	2	1	0	2															
G	−1	1	0	−2	1	6														
N	−1	2	1	−2	−1	−2	11													
D	−3	0	0	−2	0	3	6	12												
E	−3	0	−1	−3	0	3	1	8	13											
Q	−3	−1	−2	0	−2	−1	3	2	7	11										
H	−1	−2	−2	−4	−3	−3	3	3	2	7	11									
R	−1	−1	−1	−3	−1	0	2	1	2	6	5	7								
K	−3	−1	−2	−4	−2	−1	5	3	1	6	4	9	12							
M	−1	−2	0	−3	−1	−3	−2	−3	−3	−2	−3	0	−1	3						
I	−1	−1	0	−3	0	−2	−3	−3	−4	−2	−4	−3	−4	1	2					
L	−1	−2	−1	−1	−2	−4	−4	−5	−5	−2	−4	−3	−4	1	1	3				
V	0	−1	0	−3	0	−1	−3	−3	−2	−4	−4	−2	−4	1	2	1	2			
F	1	−1	−2	−4	−2	−4	−4	−6	−6	−4	−3	−4	−5	0	1	1	−1	5		
Y	3	0	−3	−5	−3	−5	−1	−2	−5	0	6	−1	1	−2	−4	−3	−4	2	10	
W	1	−3	−4	−6	−4	−2	−3	−4	−3	0	−1	5	3	−2	−3	−2	−2	−3	−2	12

Continued on p. 57

Table 3.1 (*Continued*) (c) BLOSUM62 substitution matrix from conserved protein blocks. (From Henikoff & Henikoff, 1992.)

(c)	C	S	T	P	A	G	N	D	E	Q	H	R	K	M	I	L	V	F	Y	W
C	9																			
S	−1	4																		
T	−1	1	5																	
P	−3	−1	−1	7																
A	0	1	0	−1	4															
G	−3	0	−2	−2	0	6														
N	−3	1	0	−2	−2	0	6													
D	−3	0	−1	−1	−2	−1	1	6												
E	−4	0	−1	−1	−1	−2	0	2	5											
Q	−3	0	−1	−1	−1	−2	0	0	2	5										
H	−3	−1	−2	−2	−2	−2	1	−1	0	0	8									
R	−3	−1	−1	−2	−1	−2	0	−2	0	1	0	5								
K	−3	0	−1	−1	−1	−2	0	−1	1	1	−1	2	5							
M	−1	−1	−1	−2	−1	−3	−2	−3	−2	0	−2	−1	−1	5						
I	−1	−2	−1	−3	−1	−4	−3	−3	−3	−3	−3	−3	−3	1	4					
L	−1	−2	−1	−3	−1	−4	−3	−4	−3	−2	−3	−2	−2	2	2	4				
V	−1	−2	0	−2	0	−3	−3	−3	−2	−2	−3	−3	−2	1	3	1	4			
F	−2	−2	−2	−4	−2	−3	−3	−3	−3	−3	−1	−3	−3	0	0	0	−1	6		
Y	−2	−2	−2	−3	−2	−3	−2	−3	−2	−1	2	−2	−2	−1	−1	−1	−1	3	7	
W	−2	−3	−2	−4	−3	−2	−4	−4	−3	−2	−2	−3	−3	−1	−3	−2	−3	1	2	11
	C	S	T	P	A	G	N	D	E	Q	H	R	K	M	I	L	V	F	Y	W

the hydroxylic amino acids (S, T), etc. Cysteine practically stands alone, primarily reflecting the fact that it has a unique feature that no other amino acid has, namely the ability to form disulphide bonds. Glycine–alanine interchanges seem to be driven by selection for small side chains, proline–alanine interchanges by selection for small aliphatic side chains, etc. Some of these groups overlap: the basic, acid and amide groups tend to replace one another to some extent (histidine is as likely to be replaced by asparagine as by arginine), and phenylalanine often interchanges with the aliphatic group.

These patterns are imposed principally by natural selection against drastic changes: they reflect the similarity of the functions of the amino acid residues in the three-dimensional conformation of proteins. Some of the key properties of an amino acid residue that determine its role and replaceability are size, shape, polarity, electric charge and its ability to form salt bridges, hydrophobic bonds, hydrogen bonds and disulphide bonds.

It should be emphasized that the database used for the generation of the matrix of Table 3.1(a) was biased inasmuch as globular, water-soluble and extracellular proteins were overrepresented. The extent of such a bias is best appreciated if we compare this matrix with a mutation data matrix defined for transmembrane segments of membrane proteins (Jones, Taylor & Thornton, 1994). As a reflection of the environment of transmembrane segments of integral membrane proteins, the most commonly occurring residues in transmembrane helices are leucine, valine and isoleucine, whereas polar residues are not frequent in these segments. The transmembrane protein mutation data matrix (Table 3.1(b)) is quite different from the matrix calculated from a general sequence set. The most obvious feature of the matrix is the high relative mutability of the hydrophobic aliphatic residues: isoleucine, methionine, valine and leucine. The explanation for the high mutability of residues that are most important in defining trans-membrane helices is that it is not the actual amino acid side chain but the helical structure and overall hydrophobicity that is conserved. Although polar residues in general are less mutable in transmembrane protein segments than their counterparts in globular proteins, serine and threonine are as mutable as aliphatic residues. Serine and threonine are unusual in that they are cap-able of satisfying the hydrogen-bonding capacity of their single hydroxyl groups by interacting with the main chain carbonyl group of residue $i - 3$ or $i - 4$ in the previous turn of the helix, and are thus compatible with the lipid environment. Polar residues (N, D, E, Q, H, R, K) are fairly highly conserved, reflecting the fact that in proteins with multiple transmembrane segments these residues are generally associated with specific functions (e.g. they form ion channels, stabilize the helical bundles by forming ion pairs, etc.). As dis-cussed in Chapter 2, arginine and lysine play an important role as topogenic signals in transmembrane proteins and this explains their conservation. R and K tend to exchange between themselves, indicating that they are equally satisfactory in directing membrane insertion. Proline residues appear to be highly conserved in transmembrane segments, presumably due to the special role of proline residues in 'kinking' transmembrane helices. A striking observation is that the hydrophobic W and Y are frequently replaced by the basic R and K. The possible explanation is that the side chains of R and K, and of W and Y have both polar and nonpolar characters. Perhaps the most notable change in mutability is that observed for cysteine, which changes from being the second least mutable residue in the general sequence set (Table 3.1(a)) to being one of the least conserved in the transmembrane protein set (Table 3.1(b)). It is most frequently replaced by hydrophobic residues (Y, F, W, etc.) or the quasi-hydrophobic residue S. The plausible explana-tion is that in the general set it fulfils the unique role of forming disulphide bonds, whereas in the transmembrane segments it plays the role of being a nonpolar residue.

The most important conclusion is that matrices calculated for general sequence sets do not adequately describe the conservation patterns for special types of proteins (in this case transmembrane segments). Although in both data sets the chemically most similar amino acids are clustered in similar subgroups (e.g. I, L, V, M), the relative importance of these properties is very different for transmembrane segments. In the general set, alanine, serine, threonine and proline cluster with the polar residues, while in trans-membrane segments they are more closely related to the hydrophobic group.

Hydrophobicity is of course by far the most significant factor for trans-membrane segments, but the next most important aspect is whether the side chain is charged, and whether it is negatively charged or positively charged.

Whereas the mutation data matrices of Dayhoff were based on substitution rates derived from alignments of protein sequences that are at least 85% identical, Henikoff and Henikoff (1992) have derived substitution matrices from conserved blocks of aligned amino acid sequence segments characterizing more distantly related proteins. The primary difference is that in the case of distantly related proteins the Dayhoff matrices make predictions based on observations of closely related sequences, whereas the 'blocks approach' makes direct observations on blocks of distantly related proteins. The matrices derived from a database of blocks in which sequence segments are identical at 45% and 80% of aligned residues are referred to as BLOSUM 45, BLOSUM 80, etc. The BLOSUM matrices (Table 3.1(c)) show some consistent differences when compared with Dayhoff matrices. According to the Dayhoff matrices (Table 3.1(a)), hydrophilic, polar amino acids (S, T, N, D, E, Q, H, R, K) are significantly more tolerant of substitutions (average score 4) than the hydrophobic, apolar amino acids (M, I, L, V, F, Y, W; average score 8.1); the ratio of average scores for polar/apolar conservation is 0.49. This low ratio probably reflects the fact that the hydrophobic interior plays a critical role in the folding and stability of water-soluble globular proteins.

In the case of BLOSUM matrices there is a significant shift in this ratio of average scores for polar/apolar conservation: it is 0.93, since there is less tolerance to substitutions involving hydrophilic amino acids. Since the blocks were derived from the most highly conserved regions of proteins, the differences between BLOSUM and Dayhoff matrices arise from the different constraints on conserved regions. A telling example is the case of asparagine. In the Dayhoff matrices asparagine is the most mutable residue, whereas in BLOSUM asparagine is involved in substitutions at an average frequency. The explanation for this is that highly mutable (surface) regions of proteins are usually not represented in blocks; asparagines located in conserved regions show an average tendency to be involved in substitutions. The differences between the BLOSUM and Dayhoff matrices are primarily due to the fact that the most variable, surface-exposed regions of proteins (loops, β turns) are underrepresented, and the highly conserved regions (secondary structure elements) that form the conserved core of protein folds are overrepresented. The relatively weak conservation of polar residues in the Dayhoff matrices is due to the fact that in the case of residues of surface loops it is the hydrophilicity rather than the actual residue that is conserved.

3.4 The molecular clock

The idea of a 'molecular clock' was based on the initial observation that the number of amino acid or nucleotide substitutions separating **orthologous**

genes (i.e. the 'same' genes) of different species is roughly proportional to the time that passed since these species diverged from a common ancestor. This is what we would expect if we assumed that mutations occur and substitutions are fixed at a constant rate in the case of a given type of gene. If the molecular clock hypothesis were generally valid then it could serve as a chronometer to estimate the times of divergence of species.

Another important observation was that different types of genes change at vastly different rates, the rates being inversely proportional to the extent of **structural and functional constraints** imposed on them by their biological importance for the organism. This is understandable if we assume that in proteins which are under more stringent constraint (e.g. histones) a much smaller (but more or less constant) proportion of mutations can be accepted and fixed, and a larger proportion of substitutions are disruptive and are rejected by natural selection. In the case of such highly constrained proteins the 'clock' is expected to run at a slower rate then in the case of less constrained proteins. Clearly, if we can delineate the factors that affect the speed and constancy of the clock then we can get a fuller understanding of the mechanisms of protein evolution. In the following we will briefly summarize why the actual molecular clock is not a smooth-running clock.

First of all, in the actual molecular clock the 'ticks' (mutations, substitutions) do not occur at regular intervals but rather at random time points (Gillespie, 1991). This fact may be properly taken into account by statistical models in which events occur at random times; the first such model (Zuckerkandl & Pauling, 1962) assumed a Poisson distribution of mutations. If the *average* rate of substitution were constant per unit time then the molecular clock could still provide a reliable time scale for evolution. Statistical analyses using such models, however, have clearly shown that the actual variation in rates is significantly greater than expected under the Poisson clock, indicating that the variations in evolutionary rates are larger than expected by chance. There are several reasons why the molecular clock does not follow a simple Poisson process. First, there is strong evidence that the **mutation rates** (expressed in **substitutions per unit time**) may vary among different evolutionary lineages (see section 3.2). One major reason for this is that since inherited changes are associated with gamete replication, species with significantly longer generation times would have fewer chances for change during a unit period of time. Consequently, the number of generations may be a more pertinent parameter than time, and in fact DNA divergence can be shown to correlate better with **generation time** than with **historical time**. Moreover, in higher animals a 'generation' often involves 30–50 rounds of **gamete replication** during gametogenesis; the true 'zygote-to-zygote' generation must take such differences into account (see section 3.2).

Several examples illustrate that the **rate of substitutions** may be subject to drastic alterations as a result of **changes in functional constraints** and **changes in selection**. One of the most striking examples is the speed up in the rate of replacement substitution in the insulin gene of hystricomorph rodents, where the rates varied by as much as 30-fold. This acceleration could be due to adaptive changes as part of a general evolution of the gastroenteropancreatic hormonal system (for details see section 5.3). Another example is the acceleration in the rate of replacement substitution in the lysozyme of langurs, which was associated with the recruitment of lysozyme to digest bacteria in the stomach (for details see section 5.2). Similarly, in the case of the visual pigment genes, preceding the emergence of the three colour pigments, there was an acceleration in the replacement rate associated with a shift in the absorption spectrum. These examples show clearly that the accelerations were due to positive selection for adaptive changes.

Another major source of fluctuations in rates of substitution is that the substitutions at different sites within a protein-coding gene are not independent. There is convincing evidence to show that the occurrence of a substitution at one site affects the likelihood of substitutions at other sites. For example, the three residues constituting the catalytic triad of serine proteinases are not independent; substitution of any one of them increases the chances of accepted substitutions of the others (see section 7.2.2). In other words, substitutions themselves alter the rate of evolution causing significant fluctuations. The nonindependence of sites in proteins necessarily leads to the view that their evolution may be episodic, with bursts of substitutions separated by periods of relative quiescence.

The environment may also cause changes in the rate of substitutions in two different ways: it may directly alter the mutation rate, or it may alter substitution rate by changing functional constraints for the given protein (cf. haemoglobins of animals living at high altitudes, lysozymes in foregut fermenters; see section 5.2). This has great general importance in the case of duplicated genes that are frequently expressed in different cellular and tissue environments (for examples see section 7.1.2) or in different developmental stages (e.g. foetal haemoglobins) where they are subject to different functional constraints. As a reflection of this fact, the rate of evolution often accelerates following gene duplication and protein sequences usually evolve much more rapidly at times of adaptive radiation (for examples see Chapters 5 and 7).

References

Chakraborty, R., Kimml, M., Stitvers, D.N. et al. (1997) Relative mutation rates at di-, tri-, and tetranucleotide microsatellite loci. *Proceedings of the National Academy of Sciences of the USA* **94**, 1041–6.

Dayhoff, M.O., Schwartz, R.M. & Orcutt, B.C. (1978) A model of evolutionary change in proteins. In M.O. Dayhoff (ed.), *Atlas of Protein Sequence and Structure*, Vol. 5 (Suppl. 3), pp. 345–52. Washington, DC: National Biomedical Research Foundation.

Edelmann, W., Yang, K., Umar, A. et al. (1997) Mutation in the mismatch repair gene Msh6 causes cancer susceptibility. *Cell* **91**, 467–77.

George, D.G., Hunt, L.T. & Barker, W.C. (1988) In D.H. Schlesinger (ed.), *Macromolecular Sequencing and Synthesis*. New York: A.R. Liss.

George, D.G., Barker, W.C. & Hunt, L.T. (1990) Mutation Data Matrix and its uses. *Methods in Enzymology* **183**, 333–51.

Gillespie, J.H. (1991) *The Causes of Molecular Evolution*. Oxford: Oxford University Press.

Henikoff, S. & Henikoff, T.G. (1992) Amino acid substitution matrix from protein blocks. *Proceedings of the National Academy of Sciences of the USA* **89**, 10915–19.

Jones, D.T., Taylor, W.R. & Thornton, J.M. (1994) A mutation data matrix for transmembrane proteins. *FEBS Letters* **339**, 269–75.

Kimura, M. (1962) On the probability of fixation of mutant genes in populations. *Genetics* **47**, 713–19.

Kimura, M. (1968) Genetic variability maintained in a finite population due to mutational production of neutral and nearly neutral isoalleles. *Genetics Research* **11**, 247–69.

Kimura, M. (1983) *The Neutral Theory of Molecular Evolution*. Cambridge: Cambridge University Press.

Krawczak, M. & Cooper, D.N. (1996) Mutational processes in pathology and evolution. In M. Jackson, T. Strachan & G. Dover (eds), *Human Genome Evolution*, pp. 1–33. Oxford: Bios Scientific.

Litman, G.W., Rast, J.P., Shambott, M.J. et al. (1993) Phylogenetic diversification of immunoglobulin genes and the antibody repertoire. *Molecular Biology and Evolution* **10**, 60–72.

Mitas, M. (1997) Trinucleotide repeats associated with human disease. *Nucleic Acids Research* **25**, 2245–53.

Perutz, M. (1996) Glutamine repeats and inherited neurodegenerative diseases: molecular aspects. *Current Opinion in Structural Biology* **6**, 848–58.

Taylor, E.M., Broughton, B.C., Botta, E. et al. (1997) Xeroderma pigmentosum and trichithiodystrophy are associated with different mutations in the XPD (ERCC2) repair/transcription gene. *Proceedings of the National Academy of Sciences of the USA* **94**, 8658–63.

Thyagarajan, B. & Campbell, C. (1997) Elevated homologous recombination activity in Fanconi anemia fibroblasts. *Journal of Biological Chemistry* **272**, 23328–33.

Tonegawa, S. (1983) Somatic generation of antibody diversity. *Nature* **302**, 575–81.

Tseng, H. (1997) Complementary oligonucleotides and the origin of the mammalian involucrin gene. *Gene* **194**, 87–95.

Wells, R.D. (1996) Molecular basis of genetic instability of triplet repeats. *Journal of Biological Chemistry* **271**, 2875–8.

Wright, B. (1997) Does selective gene activation direct evolution? *FEBS Letters* **402**, 4–8.

Zuckerkandl, E. & Pauling, L. (1962) Molecular disease, evolution and genetic heterogeneity. In M. Kasha & B. Pullman (eds), *Horizons in Biochemistry*, pp. 189–225. New York: Academic Press.

Useful internet resources

Chromosome translocation and gene fusion

ChimerDB is a database for fusion sequences (http://genome.ewha.ac.kr/ChimerDB/).

Chapter 4: Evolution of protein-coding genes

In the previous chapter we discussed the factors that affect the rate of mutations and the probability of their fixation. In this chapter we will survey how the nucleotide changes accumulate at different sites of the DNA sequence of a protein-coding gene as a function of time. We will treat protein-coding (translated) regions, transcribed untranslated regions, introns and 5′ and 3′ flanking sequences separately, since they usually evolve at markedly different rates. Within the translated region it is important to distinguish synonymous and nonsynonymous substitutions since they are known to be accepted at different rates.

In this chapter the emphasis will be on comparison of **orthologous** pairs of genes, i.e. genes that diverged through speciation and fulfil essentially identical functions in the different species. The special features of those orthologous protein-coding genes that fulfil a significantly modified function in some species (e.g. lysozymes of foregut fermenters, haemoglobins of species living in low-oxygen environments) or paralogous proteins that fulfil drastically different functions in the same species (e.g. various members of the trypsin family) will be discussed principally in Chapters 5 and 7.

4.1 Alignment of nucleotide and amino acid sequences

Comparison of the sequences of homologous protein-coding genes requires that the nucleotides (and amino acids) from the 'same' positions must be brought into register; the procedure that aligns sites of common ancestry is referred to as **sequence alignment**. In the simplest case this involves **pairwise alignment** of two homologous nucleic acid (or amino acid) sequences. In some cases this requires the identification of the locations of deletions and insertions that might have occurred in either of the two lineages since their divergence from a common ancestor. In the alignments, insertions and deletions (**indels**) show up as **gaps**, therefore both types of event are collectively referred to as **gap events**. In the case of two aligned homologous nucleotide sequences a base from one sequence may be aligned with an identical base of the other sequence: there is a **match**. If a site has undergone a point mutation in one or both homologous genes there will be a **mismatch**; it may be of the **transition type** (e.g. A↔G) or **transversion type** (e.g. A↔C, A↔T). If a base deletion has affected a site, then there will be a gap (a null base) in the altered sequence. Conversely, if a mutation has inserted a base in one of the genes, alignment of the sequences will introduce a gap in the other sequence. In the case of closely related

63

sequences a matched pair usually implies a site that has not changed since the divergence of the two sequences (unless there were parallel substitutions in both lineages, etc.), a mismatched pair denotes a single substitution, and a null pair indicates that a deletion (or an insertion) has occurred at this position in one (or the other) of the two sequences. In the case of more distantly related sequences there may have been multiple hits at the same site.

In the case of alignment of amino acid sequences, the same principles hold: their correct alignment is the one in which amino acid positions of common ancestry are aligned. In this case the alignment also means matched amino acids as well as mismatched amino acids and gaps (null positions). A major difference between nucleotide and amino acid sequences is that in the case of nucleotide sequences matches may be of four types and mismatches may be of six types (two sets of transitions, four sets of transversions), whereas in the case of amino acid sequences matches may be of 20 types and mismatches may be of 190 types (190 possible interchanges of the 20 amino acids). In the case of nucleotide sequences, transitions and transversions are known to have different probabilities, while in the case of amino acid sequences, most of the 190 possible amino acid interchanges have different probabilities (see section 3.3). The significance of these differences between nucleotide and amino acid sequences is that the degree of similarity may be measured in a more refined way for aligned amino acid sequences than for nucleotide sequences. As a consequence, amino acid sequences of protein-coding genes may be aligned with greater reliability than their nucleotide sequences, especially in the case of distantly related sequences.

Optimizing alignments

It is important to emphasize that only correct alignments can be used for the correct reconstruction of evolutionary events. The correct or **true alignment** of two sequences is the one in which only sites of common ancestry are aligned, and all sites of common ancestry are aligned. As we have seen above this may mean matched bases (or amino acids) and mismatched bases (or amino acids), as well as gaps. In most cases the true alignment coincides with the one in which matches are maximized, and mismatches and gaps are minimized: the **alignment score** is maximal. This may be achieved by using some type of **scoring system** that rewards similarity. However, the proper criteria for selecting scores for matches and mismatches are not always known, hence finding the true alignment is not a trivial task. This problem is especially serious in the case of distantly related sequences when gaps are common. For example, the number of mismatches may be minimized and the number of matches may be maximized at the expense of introducing a large number of gaps. To prevent incorrect alignments of sequences with an unrealistic abundance of gaps, a **gap penalty** (a negative score) must be introduced. The numerical value of gap penalties, however, is somewhat arbitrary, even though they may be based on certain *a priori* assumptions of how frequently mutations causing deletions and insertions of various sizes occur and are fixed relative to point mutations.

The use of **linear gap penalty functions** (penalty rises monotonously with length of gap) would be adequate only if the probability of a three-base deletion/insertion were the same as that of the consecutive deletion of three adjacent bases. Most current alignment methods employ **affine gap penalty functions** that assign lower penalty values for opening than for extending a gap. The biological justification for this approach is that different mutational mechanisms may be behind single-base deletions/insertions and deletions/insertions of multiple contiguous bases (see section 3.1.2) as well as the fact that the probability of the acceptance of gaps may depend more on the position than on the size of the gap (see chapter 7). **Terminal gaps** at the ends of protein-coding sequence alignments are treated differently to internal gaps (i.e. they are scored with lower or no gap-penalty values) since N-terminal and C-terminal regions of proteins are usually quite tolerant to extensions/deletions.

It is also important to remember that sequences may not be homologous over their entire length: some multidomain mosaic proteins may share just a single domain that is homologous (cf. Chapter 8). It is therefore useful to distinguish **global alignment** procedures, which align two or more sequences over their full lengths, from **local alignment** tools that align only the most similar consecutive segments of two or more sequences.

When counting mismatches in nucleic sequence alignments, it may be useful to assign different scores to transition mismatches and transversion mismatches since the former are known to have greater probability. In the case of amino acid sequences, the best alignment is identified in a similar way: matches are maximized, and mismatches and gaps are minimized. In this case, however, it may be very useful to assign different scores to each of the 20 different matches and the 190 types of amino acid interchanges since they are known to have different probabilities. This can be achieved by using scoring systems based on appropriate mutation data matrices (see section 3.3).

Most of the commonly used sequence-alignment programs are based on the Needleman and Wunsch algorithm that identifies the **'best' alignment** as the one with maximum similarity among all possible alignments (Needleman & Wunsch, 1970). However, it must be remembered that the resulting alignment depends on the choice of the scoring system (especially gap penalties), therefore it must not be taken for granted that the mathematically 'best' alignment is also a historically true alignment.

4.2 Estimating the number of nucleotide substitutions

When two orthologous protein-coding genes start to diverge from each other (as a result of speciation), each of them will start accumulating nucleotide substitutions independently. Since the number of nucleotide substitutions that have occurred is a function of the time since divergence that was spent in separation, this parameter is used most frequently in studies on the evolution of protein-coding genes. If the two sequences have diverged relatively recently (the difference in their nucleotide sequences is small), the chance that more than one substitution occurred at any site is negligible. In such cases the **number of differences** observed between the two sequences is close to the actual **number of substitutions**. At greater evolutionary distances, when there is high degree of divergence, the observed number of differences is likely to be smaller than the actual number of substitutions since multiple substitutions or multiple 'hits' at the same site may be counted as single differences. (In such cases appropriate statistical corrections can be made to take into account the effect of multiple hits at the same site.)

The simplest mathematical scheme describing the **dynamics of nucleotide substitution** is the Jukes–Cantor model, which is based on the following assumptions: substitutions at different sites occur randomly and independently; all the nucleotides are equivalent; each nucleotide will change to the three others with equal rate; and there is no bias in the direction of change (Jukes & Cantor, 1969). Since this simplest model assumes only one parameter for the various interchanges it is also called the **one-parameter model**. If this assumption was valid then the probability that the initial nucleotide and the nucleotide at time t differ by one of the two types of transversion would be twice the probability that it differs by a transition since each nucleotide is subject to two types of transversion (e.g. A→C and A→T), but only one type of transition (e.g. A→G). Obviously, the above assumptions of the Jukes–Cantor model are not always valid. As discussed in the previous chapter, there may be hotspots of mutations, where the probability of mutation is much higher than at other sites and there is a strong bias in the direction of the change. For example, 5-methylcytosines in the dinucleotide 5′-CpG-3′ suffer spontaneous deamination at a high frequency, converting them to thymine by a C→T transition.

In general, transitions are known to be more frequent than transversions, therefore Kimura (1980) has proposed a **two-parameter model** that takes this fact into account. In this more realistic model the probability of each transitional substitution is higher than the probability of each type of transversional substitution. It should be emphasized that neither the one-parameter nor the two-parameter model can take into account the fact that the rate of nucleotide substitutions at different sites of a protein-coding gene may be drastically different, either because the rate of mutation is different (hotspots), or because different sites are under different functional constraints. Clearly, the translated regions, promoter regions, untranslated regions and introns of protein-coding genes are subject to markedly different functional constraints, therefore it is important to treat them separately.

4.2.1 Substitutions in translated regions

In translated regions, mutations that cause the replacement of one amino acid by another (nonsynonymous substitutions) are usually accepted at a significantly lower rate than mutations that do not change the amino acid (synonymous substitutions), therefore it is useful to distinguish synonymous and nonsynonymous nucleotide substitutions (Li, Wu & Luo, 1985). In the first step of this method we first classify all the nucleotide positions according to whether substitutions in that position would change the amino acid (**nonsynonymous position**) or would remain silent (**synonymous position**). For example, in the case of the four codons of Gly (GGT, GGC, GGA, GGG) the first two positions are counted as nonsynonymous sites, since all possible substitutions here would change Gly to another amino acid (see Table 1.1). The third position is counted as a fully synonymous site because all possible changes at this position would leave Gly unaltered. In the two codons of Phe (TTT, TTC), the first two positions are nonsynonymous since here all possible substitutions would change Phe to another amino acid. However, in this case the third position is counted as one-third synonymous and two-thirds nonsynonymous since only one of the three possible changes at this position is synonymous. Obviously, in the case of the codons of Met and Trp all three positions are counted as nonsynonymous.

When comparing two sequences, first the **number of synonymous sites** (N_S) and the **number of nonsynonymous sites** (N_A) are counted in each of the compared sequences, then the actual nucleotide differences between the compared sequences are classified into **synonymous** (M_S) and **nonsynonymous mutations** (M_A). In the case of codons that differ by only one nucleotide, the mutation is easily identified. For example, a T→C transition may be inferred from the silent substitution GGT→GGC that converted one glycine codon to another. In the case of codons that differ by more than one nucleotide there may be several alternative evolutionary pathways of multiple single-base substitutions. In such cases the most probable route must be selected: i.e. the one with the lowest number of substitutions, synonymous substitutions being preferred over nonsynonymous substitutions, and transitions being preferred over transversions. For example, the conservative interchange between the codons AGT (Ser) and ACG (Thr) requires a minimum of two substitutions. One hypothetical pathway would be through an ACT (Thr) intermediate [AGT (Ser) ↔ ACT (Thr) ↔ ACG (Thr)]. Alternatively, the intermediate could be AGG [AGT (Ser) ↔ AGG (Arg) ↔ ACG (Thr)]. Both routes require two transversions; the first route requires one synonymous and one nonsynonymous change, whereas the second requires two nonsynonymous changes (with a nonconservative Arg as an intermediate). Since synonymous substitutions are more frequent, we may assume that the first pathway is the more likely pathway. Once the number of all synonymous and nonsynonymous mutations has been calculated, the number of synonymous differences per synonymous site (M_S/N_S) and the number of **nonsynonymous differences per nonsynonymous site** (M_A/N_A) may be calculated.

Alternatively, nucleotide sites of translated regions may be classified into nondegenerate, twofold degenerate and fourfold degenerate sites. A site is nondegenerate if all possible changes at this site are nonsynonymous (like the first two positions in the above examples of Gly and Phe, and all three positions of Met and Trp); it is twofold degenerate if one of the three possible changes is synonymous (like the third position in the above example of Phe); and it is fourfold degenerate if all possible changes at the site are synonymous (the third position in the above example of Gly). Ile is the only amino acid where a position (its third position) is threefold degenerate. By definition, all substitutions at nondegenerate sites are nonsynonymous, and all substitutions at fourfold degenerate sites are synonymous. Based on the classification of nucleotide sites into degeneracy classes the number of substitutions between two coding sequences for the different types of sites may be calculated separately. As a consequence of the structure of the standard genetic code, at twofold degenerate sites (with a few exceptions) transitional changes (C↔T and A↔G) are usually synonymous, whereas transversions are usually nonsynonymous.

4.2.2 Substitutions in untranslated regions, introns and 5′ and 3′ flanking regions of protein-coding genes

In untranslated regions, introns and 5′ and 3′ flanking regions of protein-coding genes mutations are usually accepted at a significantly higher rate than in translated regions. These regions are sometimes collectively called 'noncoding' regions to distinguish them from the translated 'coding' regions of protein-coding genes. When comparing the nucleotide sequences of protein-coding genes these sites are usually referred to as **noncoding sites** (N_N) to distinguish them from synonymous sites (N_S) and nonsynonymous sites (N_A) of the translated region. Accordingly, the actual nucleotide differences may be classified into **noncoding** (M_N), synonymous (M_S) and nonsynonymous **mutations** (M_A).

4.3 Rates and patterns of nucleotide substitution

It is important to define the rates of nucleotide substitutions in different regions of protein-coding genes since it can provide insights into the factors controlling nucleotide substitution and protein evolution. Furthermore, if we understand the factors that influence the rate of nucleotide substitutions it may also improve the reliability of dating evolutionary events, such as the divergence between species, gene duplications, etc.

4.3.1 Rates of nucleotide substitution

The rate of nucleotide substitution (r) is usually defined as the number of substitutions per site per year. The rate of nucleotide substitution may be calculated by dividing the **number of substitutions per site** (K) between two homologous sequences, by $2T$, where T is the time of divergence between the two sequences: $r = K/(2T)$. For example, if we compare orthologous genes from two different species, the divergence time (T) corresponds to the time of divergence of these species (T may be obtained from palaeontological data). The number of substitutions per nonsynonymous site (K_A), per synonymous site (K_S) and per noncoding site (K_N), is usually counted separately.

In this section the emphasis will be on the causes of rate variations among different regions of a protein-coding gene and among different types of protein-coding genes. A useful principle in understanding these rate variations is the general principle of functional constraint: the rate of accepted mutations reflects the importance of the affected sites for the expression of a functional protein. Intergenic regions, 5′ and 3′ flanking regions of genes, untranslated regions and introns, which are under weaker functional constraints, are usually characterized by similar, high rates of accepted mutations. In translated regions the rates of accepted mutations correlate with the degree of degeneracy of a site, silent mutations being accepted at the highest rate. Nonfunctional pseudogenes accept mutations at the highest rates, similar to those of introns, untranslated regions and fourfold degenerate sites of functional genes. Figure 4.1 summarizes the relative susceptibility of different regions of protein-coding genes to nucleotide substitutions.

Translated region of different protein-coding genes

The rates of synonymous and nonsynonymous substitutions for selected orthologous human and rodent protein-coding genes are summarized in Table 4.1. It should be noted that the rate of nonsynonymous substitution is extremely variable among genes. It ranges from zero in histones 3 and 4 (functionally highly constrained proteins) to about 2×10^{-9} substitutions per nonsynonymous site per year in immunoglobulins, such as Ig k (highly variable proteins). The moderately constrained α- and β-globins evolve at intermediate rates ($\approx 6 \times 10^{-10}$ substitutions per nonsynonymous site per year).

The rate of synonymous substitution for these proteins varies much less than their nonsynonymous rates, the rates are within a twofold range (between 3×10^{-9} and 6×10^{-9} substitutions per synonymous site per year; see Table 4.1). Note that the rate of synonymous substitution is always higher than the nonsynonymous rate of the same protein. This is most obvious from the last column of Table 4.1 where the K_A/K_S ratios are shown. In general, for functionally constrained proteins K_A/K_S ratios are significantly lower

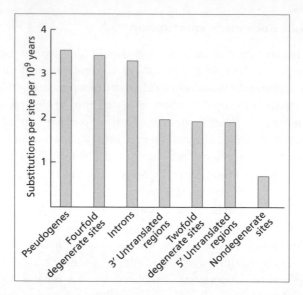

Fig. 4.1 Average rates of substitutions in different parts of protein-coding genes and pseudogenes. (Modified from Li & Graur, 1991.) Reprinted by permission of Sinauer Associates, Inc. Publishers.

Table 4.1 Rates of synonymous and nonsynonymous substitutions in various mammalian protein-coding genes. Rates are in units of substitutions per site per 10^9 years. (Modified from Li & Graur, 1991.)

Gene	Nonsynonymous rate	Synonymous rate	K_A/K_S
Histone 3	0.00	6.38	0
Histone 4	0.00	6.12	0
Actin α	0.01	3.68	0.002
Actin β	0.03	3.13	0.010
Aldolase A	0.07	3.59	0.020
Insulin	0.13	4.02	0.030
GAPDH	0.20	2.84	0.070
LDH A	0.20	5.03	0.040
α-Globin	0.55	5.14	0.107
Myoglobin	0.56	4.44	0.126
β-Globin	0.80	3.05	0.262
Ig V_H	1.07	5.66	0.189
Urokinase	1.28	3.92	0.326
Ig k	1.87	5.90	0.316

than 1.0, reflecting strong purifying selection. In the most extreme cases, such as histones, the synonymous ('silent') rate is very high, although their amino acid sequences are practically unchanged: their K_A/K_S ratios are zero.

The 'avoidance' of amino acid-changing mutations may also be illustrated if the rates of substitutions at fourfold degenerate sites are compared with those at the twofold degenerate sites of the same proteins. Based on studies of a large number of proteins, Li and Graur (1991) have shown that the average rate at fourfold degenerate sites (3.7×10^{-9} substitutions per synonymous site per year) is significantly higher than at twofold degenerate sites (2.2×10^{-9} substitutions per site per year). The fact that the rate at the 'silent' fourfold degenerate sites is similar to that observed in the case of pseudogenes (see below) suggests that fourfold degenerate sites are more or less free of functional constraints (see Fig. 4.1).

Noncoding regions

Pseudogenes, DNA sequences that were derived from functional protein-coding genes but have become nonfunctional, are the examples *par excellence* of noncoding regions where all substitutions are neutral. Pseudogenes show very high rates of substitution (about 4×10^{-9} substitutions per site per year), consistent with the fact that they are subject to no functional constraints. Since they are not encoding functional proteins, the distinction between synonymous and nonsynonymous sites becomes meaningless. As a reflection of this fact, their virtual K_A/K_S ratios are usually close to 1.0. Since all mutations in pseudogenes are selectively neutral and become fixed in the population with equal probability they are particularly useful for studying the pattern of spontaneous mutations in the absence of selection. Analysis of the pattern of substitution inferred from mammalian pseudogene sequences indicates that the direction of mutations is nonrandom: transitions occur more often than transversions, and their observed frequency is about twice the value expected for random mutations (see section 3.1.1).

The substitution rates in 5' and 3' untranslated regions are relatively high: the average rates for these regions are $\approx 2 \times 10^{-9}$ substitutions per year. The fact that these rates are lower than the average rate for their fourfold degenerate sites indicates that these regions contain sites that fulfil some essential functions (e.g. sites affecting translation, polyadenylation, interactions with noncoding RNAs etc.).

In general, the rates of substitutions in vertebrate introns is very high, similar to the values observed in the case of fourfold degenerate sites, suggesting that most of the mutations affecting introns are selectively neutral (Fig. 4.1). The 3' flanking regions of protein-coding genes are nearly as tolerant of substitutions as introns, whereas the 5' flanking regions of genes accept mutations at a lower rate than introns. This reflects the fact that these regions contain important signals regulating gene expression.

4.4 Variation in substitution rates

When looking for the causes of variation in substitution rates among DNA regions, it should be remembered that the rate of substitution is determined by the rate of mutation and the probability of fixation of a mutation. The latter clearly depends on whether the mutation is advantageous, neutral or deleterious. As discussed above, substitution rates are highest in nonfunctional pseudogenes where all substitutions are, by definition, neutral.

In the case of functional protein-coding genes, substitution rate is highest at fourfold degenerate sites, nearly as high as in pseudogenes. This observation suggests that fourfold degenerate sites for most protein-coding genes are nearly neutral, despite the existence of codon bias (see section 1.3). Substitution rate is only slightly lower in vertebrate introns, reflecting the fact that the major part of such introns (except for the short signals that are crucial for splicing) is 'junk' DNA. Substitution rates are intermediate for the 3' untranslated region, the 5' flanking and untranslated regions, and for twofold degenerate sites, and lowest at nondegenerate sites, the lower rates reflecting higher degrees of functional constraints (Fig. 4.1).

4.4.1 Variation among different sites of the translated region

The largest differences in substitution rate are those observed in translated regions. Since within a single protein-coding gene the chances of mutation at synonymous and nonsynonymous sites are likely to be the same, the large differences in substitution rates can be attributed only to differences in the intensity of selection between the two types of sites.

Obviously, silent mutations that do not lead to amino acid substitution have much lower chances of causing deleterious effects on the protein-coding gene than do amino acid replacements. Consequently, synonymous changes have the highest chance of being neutral, and many of these will be fixed in a population, whereas the majority of nonsynonymous mutations will be deleterious and will be eliminated from the population by selection. The net result is a reduction in the rate of substitution at nonsynonymous sites (Table 4.1). The contrast between synonymous and nonsynonymous rates in a gene demonstrates the general principle of protein evolution: the stronger the **functional constraints** on a gene (or a position of the gene), the stronger the purifying (negative) selection will be and this will slow down the rate of its evolution. Obviously, on average, functional constraints are much higher for nonsynonymous sites than for synonymous sites, therefore the former are expected to evolve more slowly than the latter. The fact that in general nonsynonymous substitutions are 'suppressed' relative to synonymous substitutions illustrates that – for proteins with well-established functions – amino acid-changing mutations are much more likely to be deleterious than advantageous.

The principle of functional constraints also predicts that the highest rate is expected to occur in a sequence that does not have any function, so that all mutations are neutral. Indeed, pseudogenes have the highest rate of nucleotide substitution. The fact that 5′ and 3′ untranslated regions have lower substitution rates than the rate of silent substitutions in coding regions is also consistent with this principle: these regions contain important signals for initiation and termination of transcription and translation, regulation of gene expression, etc. (see Chapter 1). The principle of functional constraints also explains the observed patterns of nonsynonymous substitutions (see section 3.3): in general preferred substitutions are those that are most likely to be selectively neutral or nearly neutral (interchanges among chemically similar amino acids). As will be discussed in Chapter 5, within a single protein the different amino acid positions (having different structural and/or functional significance) are subject to differential functional constraints and thus accept nonsynonymous mutations at vastly different rates.

In summary, the differences in rates of nucleotide substitutions in protein-coding genes can usually be explained as an outcome of: (i) **purifying (negative) selection** against deleterious mutations; and (ii) random genetic drift of neutral or nearly neutral mutations. Since synonymous mutations are most likely to be neutral, and nonsynonymous mutations are most likely to be deleterious, K_A/K_S ratios are usually significantly lower than 1.0.

Nevertheless, it must not be forgotten that only nonsynonymous substitutions have a chance of improving the function of a protein. If **advantageous selection** played a major role in the evolution of a protein, the rate of its nonsynonymous substitutions should exceed that of its synonymous substitution. Indeed, there are cases of **adaptive evolution** where the frequency of nonsynonymous substitutions (K_A) is significantly higher than that of synonymous substitutions (K_S) or of substitutions in noncoding regions (K_N). For example, in the case of some immunoglobulin genes, the nonsynonymous rate in those regions that define antibody specificity is higher than the synonymous rate, reflecting positive selection for advantageous substitutions that increase antibody diversity (Tanaka & Nei, 1989).

Positive selection for advantageous mutations has played a major role in the evolution of novel functions or novel functional features in proteins (for examples see Chapters 5 and 7). It must be remembered, however, that positive selection for advantageous mutations is expected to favour non-synonymous substitutions over synonymous substitutions only at sites that are critical for the novel, advantageous function. For these sites (and these sites only) the K_A/K_S ratio is expected to be significantly greater than 1.0, reflecting positive selection. If there are many such sites in a protein, the overall K_A/K_S ratio for the entire protein may also be significantly higher than 1.0, otherwise the effect of positive selection on the K_A/K_S ratio may be over-ridden by purifying selection at other sites. Since identification of cases of positive selection sometimes requires a deeper understanding of structure–function aspects of proteins, discussion of cases of positive selection is postponed until Chapters 5 and 7.

4.4.2 Variation among genes

In principle, the large variation in the rates of nonsynonymous substitution among genes (Table 4.1) may be due to differences in their exposure to mutations and/or the intensity of selection. The first factor should not be neglected, since more actively transcribed genes may be more prone to mutate (see section 3.2), and different regions of a genome may have different propensities to mutate (see section 9.5). Indeed, there is evidence that different regions of the mammalian nuclear genome may differ from each other by a factor of two in their rates of mutation (Wolfe, Sharp & Li, 1989). Nevertheless, it is clear that this moderate difference cannot account for the enormous differences in nonsynonymous substitution rates of histones vs. immunoglobulins. It is quite obvious that the most important factor that determines the rate of nonsynonymous substitutions is intensity of selection, which is determined by functional constraints.

The classic examples for the role of functional constraints are the histones, the principal packaging proteins of DNA. DNA in chromatin is organized in arrays of nucleosomes containing two copies of each histone protein – H2A, H2B, H3 and H4 – and 145–147 base pairs of DNA. Recently the crystal structure of the nucleosome core particle has been determined; this shows in atomic detail how the histone protein octamer is assembled and illustrates why these proteins are so conservative (Luger et al., 1997). Each histone has the characteristic histone-fold, which packs to form the heterodimers H2A-H2B and H3-H4, which in turn oligomerize to form the histone octamer. Both histone/histone and histone/DNA interactions critically depend on the conserved histone-fold domains and additional ordered structure elements extending from this fold. Since nucleosomes are the primary determinants of DNA accessibility, the structural organization of nucleosomes affects the vital processes of recombination, replication, mitotic condensation and transcription. During the cell cycle the amino termini of histones are modified by acetyltransferases and deacetylases and these specific modifications control histone assembly and thereby help regulate the unfolding and transcription of genes (Grunstein, 1997). Most residues of histones are thus critical either for the formation of the histone-fold, or for the interaction of histones with DNA or with other core histones to form the nucleosomes, other regions must undergo cycles of highly specific post-translational modifications that control nucleosome formation. Very few nonsynonymous substitutions can occur at any site within the histone molecule without impeding some of these vital functions. Consequently, all parts of histones are intolerant of mutations that would result in the change of amino acids. Histones are among the slowest evolving proteins known; they are practically unchanged since they were first established in an ancestral eukaryote (see also section 9.6.2). It is thus safe to conclude that proteins with the most extensive functional interactions are the ones that are expected to show the lowest rate of nonsynonymous mutations.

It is less obvious why the rate of synonymous substitution also varies (although moderately) from gene to gene (Table 4.1). As already mentioned, one source of rate variation is that the rate of mutation may differ among different regions of the genome, thus the variation in rates of synonymous substitution may reflect the chromosomal position of the genes (Wolfe, Sharp & Li, 1989). The genome of eukaryotes is made up of segments of distinct GC content (called isochores), which may be replicated independently and consequently may exhibit different rates of mutation (see section 9.5). An alternative explanation is that synonymous positions are not really neutral, and that there is selection even at these sites. As already discussed in Chapter 1, not all synonymous codons are equivalent in fitness; in the case of highly expressed genes some synonymous codons may be strongly preferred. As a reflection of the relative abundance of isoaccepting tRNA species some synonymous substitutions may be selected against (or selected for). Accordingly, genes with little or no bias in codon usage exhibit higher K_S values than genes with high codon usage bias (Powell & Moriyama, 1997). There are additional reasons why substitutions at synonymous sites may not be neutral: for example in regions corresponding to exon–exon boundaries, exonic splicing enhancers or exonic splicing silencers (recognized by spliceosomes), the chance of the acceptance of mutations is influenced by the importance of the conservation of these nucleic acid sequence motifs.

4.4.3 Constancy and variation in substitution rates of orthologous genes

In comparative studies of haemoglobin and cytochrome c sequences from different species, Zuckerkandl and Pauling (1962, 1965) noted that the rates of amino acid substitution in these proteins were approximately the same among various mammalian lineages. They have suggested that for any given type of protein the rate of molecular evolution is approximately constant over time in all lineages (the 'molecular clock'; see section 3.4). If we use the principle of functional constraints, this would imply that the mutation rate is practically constant and the functional constraints for the above proteins are practically unchanged in all lineages.

This is not always the case. As for functional constraints, there is evidence that protein sequences evolve much more rapidly at times of adaptive radiation when new functions are acquired (e.g. lysozyme; see section 5.2). As for constancy of mutation rates, there is evidence that this may be very different in different lineages (see section 3.2). For example, there is evidence that the substitution rates are significantly higher in rodents than in primates (Li, 1997).

The validity of the rate-constancy assumption is a crucial issue especially if it is used in the reconstruction of phylogenetic trees, or in the estimation of species divergence or gene duplication times. To decide whether or not two lineages evolved at similar or different rates, relative-rate tests may be used. The essence of this approach is that to compare the rates in lineages A and B, a third species (C) is used as reference, if it is known that this reference species branched off before A and B diverged from an ancestral species, O.

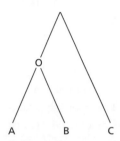

Given this evolutionary relationship (i.e. A and B are more closely related to each other than they are to C) it is expected that the number of substitutions between species A and B (K_{AB}) would be lower than the number of substitutions between species A and C (K_{AC}) or species B and C (K_{BC}). If the two lineages evolved at an equal rate, they must be at an equal 'distance' from the more distantly related C, hence $K_{AC} = K_{BC}$. Conversely, if lineage B evolved at a higher rate than lineage A then: $K_{AC} > K_{BC}$, etc.

Using such **relative-rate tests**, with human or rabbit (C species) as references, Li (1997) has shown that in mice and rats (A and B species) the rates of synonymous substitutions are nearly equal and that the same is true for the rates of nonsynonymous substitutions. Conversely, using the relative-rate test (with New World monkeys as reference) it was shown that the rate of nucleotide substitution in the Old World monkey lineage is about 1.4–2 times higher than in the human lineage (Li, 1997).

The higher substitution rates in rodents than in primates, and in monkeys than in humans is usually explained by differences in their generation times. The generation time in rodents (and monkeys) is much shorter than in man, so the number of germline DNA replications per year is significantly higher in rodents (and monkeys) than in humans (see section 3.2). However, when related organisms with similar generation times are compared, the rate-constancy assumption seems to be valid (e.g. for mice and rats).

4.4.4 Nonrandom substitutions at synonymous positions

Since synonymous mutations do not cause any change in amino acid sequence and since natural selection operates predominantly at the protein level, synonymous mutations are usually considered silent, selectively neutral mutations. If all synonymous mutations were indeed selectively neutral then synonymous codons for an amino acid should be used with equal frequency. In contrast with this expectation, in both prokaryotic and eukaryotic genes, there is a striking bias in synonymous codon usage.

In *Escherichia coli*, ribosomal-protein genes were found to use preferentially those synonymous codons that are recognized by the most abundant tRNA species (see Chapter 1). The plausible explanation for this preference is natural selection: translational efficiency and accuracy is better guaranteed if codons are translated by an abundant tRNA species. Conversely, translational efficiency and accuracy would be endangered by codons that are translated by rare tRNA species.

The generality of this correlation is supported by the fact that in both *E. coli* and yeast there is a positive correlation between the relative frequencies of the synonymous codons in a gene and the relative abundances of their cognate tRNA species. Even more striking is this correlation if we consider the fact that different species have different codon biases. In *E. coli*, where tRNA recognizing CUG is the most abundant leucine tRNA species, CUG is much more frequently used than the other five codons. Conversely, in yeast, where the most abundant leucine tRNA species recognizes UUG, this UUG codon is the predominant leucine codon.

The correlations are strongest for highly expressed genes, i.e. those that are translated in large quantities. In contrast with this, for genes with low levels of expression, the correlation between tRNA abundance and codon bias is much weaker in both *E. coli* and yeast. This difference is readily explained in terms of natural selection. In highly expressed genes, where the proteins are produced in large quantities, translation may become limited by the availability of the appropriate tRNA species, if that tRNA species is not sufficiently abundant. In this case selection for translation efficiency and accuracy is strong, the codon usage bias is therefore pronounced. Conversely, in the case of genes where the protein is produced in small amounts, translational efficiency and accuracy are less likely to be limited by the availability of the appropriate tRNA species, so there is weaker selection for favoured codons. It seems that in *E. coli*, yeast and *Drosophila* (Powell & Moriyama, 1997) codon usage is under the influence of tRNA availability, resulting in selection against codons recognizing rare tRNA species, and selection for codons recognizing the most abundant tRNA species.

Irrespective of this correlation between tRNA availability and codon usage for highly expressed genes, there is evidence that, in general, codon usage is a reflection of the overall GC content of different genomes. Recent studies on 30 nematode species have revealed highly derived patterns of codon usage and it was shown that the major factor affecting differences is the genomic GC content, which is likely the product of directional mutation pressure (Mitreva et al., 2006). Similarly, the extreme A+T-richness of the genome of *Dictyostelium discoideum* was found to determine codon usage, inasmuch as codons of the form NNT or NNA are favoured over their NNG or NNC synonyms (see Chapter 9).

4.5 Molecular phylogeny

The purpose of this section is to explain how phylogenetic trees may be reconstructed from analysis of nucleotide and protein sequences of protein-coding genes, from analyses of protein structures and from analyses of exon–intron structures of genes. Such analyses enable the evolutionary relationships among organisms and their genes to be deduced. Although the major goal of evolutionary studies is to reconstruct the correct genealogical relationships between organisms or genes, and to estimate the time of their divergence from a common ancestor (either by speciation, gene duplication or lateral gene transfer), these studies can also provide important insights into changes in functional constraints and structure–function aspects of proteins.

4.5.1 Phylogenetic trees

A **phylogenetic tree**, describing the evolutionary relationships among organisms, species or genes (in general, **taxonomic units**), consists of **branches** that connect adjacent **nodes** (the taxonomic units). In the case of rooted phylogenetic trees the **root** is the common ancestor of all the nodes, the direction leading from this root to the nodes corresponds to the direction of evolutionary time. The branching pattern, the **topology** of the tree, defines the phylogenetic relationships among the nodes. The unique phylogenetic tree (with a unique topology) that properly represents the true evolutionary history of the taxonomic units is called the **true tree**.

For example, the evolutionary relationship among human (H), mouse (M) and chicken (C) myoglobins may be represented by a tree such as:

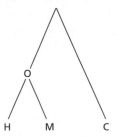

where H, M and C (external nodes, or **operational taxonomic units, OTUs**) represent the myoglobin sequences from extant species, and the internal node (O) represents ancestral units (e.g. O, the myoglobin of the last common mammalian ancestor of mouse and human).

In most tree-building programs the branch lengths are proportional to the number of changes (nucleotide substitutions, amino acid substitutions, etc.) that have occurred in that branch. Since the number of changes along a branch depends on whether different lineages evolve at similar or different rates, branch lengths contain very important information on the rates of evolution in different lineages.

4.5.2 Tree reconstruction

Tree-building methods attempt to reconstruct ancestral events (ancestral sequences, ancestral gene structures or protein structures) by analysis of extant sequences, and extant gene and protein structures.

Phylogeny based on nucleotide and amino acid sequences

It is important to emphasize that reconstruction of a true tree is possible only if **historically correct multiple alignments** are used. A historically correct multiple alignment is the one in which the aligned positions are of common ancestry. In the case of closely related sequences (which are less likely to have a gap problem), such a unique multiple alignment may be usually found as the one in which the sums of matches are maximized and the sums of mismatches and gaps are minimized for the entire multiple alignment. However, in the case of alignments of distantly related sequences where mismatches predominate and gaps are common, multiple alignments may be ambiguous, and will critically depend on the choice of scoring matrices used to score matches, mismatches and, especially, gaps. As a consequence of these problems, a mathematically optimal multiple alignment is not necessarily identical with the historically correct multiple alignment.

The best **multiple alignments** may be obtained by various progressive alignment procedures, in which the sequences are first clustered according to their distances. The common rationale for these approaches is that the reliability of a pairwise alignment decreases with the evolutionary distance of the aligned sequences. For example, if we align two closely related members of a gene family, one of which has a gap relative to the other, the position of this gap may be determined unambiguously (since the matches around the gap restrict the position of the gap). If, however, we compare two distantly related sequences their alignment will have a greater number of mismatches and the position of gaps becomes significantly more ambiguous: the best alignment may not be the true alignment.

It is a well-known fact that different segments of a protein fold do not accept gaps (or amino acid substitutions) with equal probability. Surface loops, reverse turns connecting secondary structural elements (α-helices, β-strands), and peptide regions connecting domains of multidomain proteins (linker regions) are significantly more tolerant to gap events than segments involved in α-helices and β-sheets (see Chapters 5 and 7). A consequence of this is that for each protein family (whose members have the same protein fold) there are preferred locations of gaps (or amino acid variations). In other words, in multiple alignments of homologous sequences gaps are likely to occur in similar positions. Since the most closely related sequences can be aligned with the least ambiguity, the gaps introduced at this stage are correctly located in the sense that they are most likely to identify regions tolerant to gaps. Consequently, in the case of more distantly related sequences where the placement of gaps is more ambiguous, those alignments should be preferred which place the gap in the same position as in the case of the more closely related sequence. Accordingly, progressive alignments start with the most closely related sequences, sequentially proceeding in the direction of more and more distantly related sequences (Feng & Doolittle, 1987, 1990; Patthy, 1987; Higgins & Sharp, 1989; Thompson, Higgins & Gibson, 1994).

Multiple alignments of distantly related sequences must also exploit the fact that residues (positions) critical for the maintenance of a protein fold are most highly conserved (for details see Chapters 5 and 7) and thus most characteristic of a protein family. Such critical positions must be scored differently than the positions that are known to be tolerant to variation. Various consensus sequence procedures employ progressive iterative alignment procedures to force the alignment of such conserved residues (Patthy, 1987, 1996). One peculiarity of such iterative progressive alignment procedures is that they contain multiple cycles of alignment steps (Patthy, 1987, 1996). First they create distance matrices based on pairwise alignments of the sequences, using standard scoring systems. In this first step a 'preliminary tree' is created (usually a UPGMA (unweighted pair group method with arithmetic mean) tree) then the sequences are aligned in a progressive fashion starting with the most closely related sequences (with the most reliable alignments). At the same time the characteristic features (conserved positions, variable positions, gaps) of the aligned sequences are also identified in a progressive fashion, permitting the modification of the scoring system in a way that reflects the individual features of the given family. The resulting multiple alignment may differ significantly from the individual pairwise alignments, therefore the multiple alignment may define a significantly modified distance matrix, a modified tree and a modified scoring system permitting further improvements in the quality (historical correctness) of the multiple alignment.

Phylogeny based on protein structures

Multiple structural alignments of homologous proteins are usually carried out progressively, the most similar structures being aligned first; the gaps introduced do not change in later stages (Johnson, Sutcliffe & Blundell, 1990; Johnson, Sali & Blundell, 1990). The best tree is usually defined as the one that minimizes the distance, i.e. the dissimilarity among structures. The value of structure-based phylogenies is that evolutionary trees for divergent proteins, where the sequence relationships are not statistically significant, can be derived from structures because tertiary structure is more conserved in evolution than sequence.

Phylogeny based on exon–intron structures

The rationale for phylogenies based on exon–intron structures is that intron insertions and intron removal are relatively rare events, therefore similarity of gene structure may be preserved over great evolutionary distances. Alignment of exon–intron structures is carried out by aligning homologous introns (Patthy, 1988; Brown et al., 1995). Introns are considered to be homologous if they are located in homologous positions of aligned nucleic acid and amino acid sequences (in translated regions they split the reading frame in the same phase). 'Noncoding' introns are those present in the 5' and 3' noncoding regions, while phase 1, 2 or 0 introns are those found in translated regions. The best tree may be defined as the most parsimonious, i.e. the one that minimizes the number of intron events (insertion or removal of introns). The value of intron-based phylogenies is that evolutionary trees can be derived for very distantly related protein-coding genes.

4.5.3 Tree-making methods

A rooted phylogenetic tree that describes the historically true pathways and the actual events that led from a common ancestor to various descendants is called a **cladogram**. Conversely, a tree that simply describes the observed similarity relationships among descendants is called a **phenogram**. Accordingly, tree-making methods that attempt the reconstruction of the ancestral events are cladistic methods (e.g. the maximum parsimony method), whereas tree-making methods that cluster descendants according to similarities are called phenetic methods (e.g. the UPGMA method).

There are several types of tree-making methods that are based on comparison of nucleic acid or amino acid sequences, or of gene structures or protein structures. In the case of the various distance matrix methods the evolutionary distances are calculated for all pairs of taxonomic units and a tree is constructed by using an algorithm based on the distance values. In the maximum parsimony method the shortest pathway connecting different states is chosen as the best tree, while the maximum likelihood method chooses the tree with the largest maximum likelihood value as the best tree.

The distance matrix methods

In distance matrix analyses evolutionary distances are computed for all pairs of sequences and the resulting matrix is used to construct a phylogenetic tree. In general, the best tree is the tree whose branching order and branch lengths are most consistent with the measured distances. Different types of distance matrix methods (UPGMA method, neighbour-joining (NJ) method, etc.) use different criteria to find the best tree.

The **UPGMA method** is the simplest method for tree reconstruction. It can be used to construct phylogenetic trees only if the rates of evolution are constant among the different lineages, i.e. there is a perfect linear relationship between evolutionary distance and time of divergence. The UPGMA method employs sequential clustering, local topological relationships being identified in the order of decreasing similarity (the tree is built from twigs to branches). (As mentioned above, progressive multiple alignment programs usually employ UPGMA methods.) For example, for the following tree:

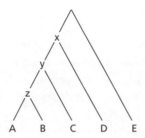

the first clustering identifies A and B sequences as being most similar to each other. In the next step a new distance matrix is computed in which A and B are treated as a new single taxonomic unit (a composite OTU, [AB]). For example, the [AB]C distance will be computed as the average of the AC and BC distances: $d[AB]C = (dAC + dBC)/2$. The new matrix will identify [AB] and C as being most closely related, etc. and the process is continued until only two OTUs are left: [ABCD] and E.

The justification for averaging distances for closely related OTUs is the implicit assumption that different lineages evolved at the same constant rate (e.g. A,B are at the same distance from C; A,B,C are at the same distance from D, etc.). If the assumption of rate constancy among lineages does not hold, UPGMA may give a tree with an erroneous topology (obviously the branch lengths are also incorrect). In other words, if the topology of a UPGMA tree conflicts with the topologies obtained by cladistic approaches, it may call our attention to the fact that the rate constancy assumption is not valid.

The **NJ method** is based on the assumption that the 'shortest' tree is the most probable tree: its goal is to find neighbours sequentially in order to minimize the total length of the tree. Among all possible pairs of sequences, the pair is chosen that gives the smallest sum of branch lengths (A and B). This pair is then regarded as a single OTU, and the average distances between OTUs are computed to form a new distance matrix. The next pair of OTUs that gives the smallest sum of branch lengths is then chosen, etc.

In the **minimum-evolution method**, for each possible alternative tree the length of each branch is calculated from the pairwise distances between sequences and then the sum (S) of all branch lengths is computed. From all possible trees, the minimum-evolution method chooses the tree with the smallest S value as the best tree. The confidence in the 'best' tree may be assessed by testing the statistical significance of the differences in S values among candidate trees. Note that the minimum-evolution method and NJ method are based on the same notion: the shortest tree is most likely to be the true tree.

The maximum parsimony method

Maximum parsimony analysis attempts to reconstruct the course of events that converted an assumed ancestral sequence into its various descendants. The basic assumption of the maximum parsimony method is that evolution proceeds by the parsimony principle, i.e. it makes the fewest changes possible to go from an ancestral sequence to any descendant sequence. In this method the shortest pathway leading from ancestral sequences to descendant sequences is chosen as the best tree. Accordingly, methods based on the principle of maximum parsimony identify trees that require the smallest number of evolutionary changes to explain the differences among taxonomic units.

The maximum likelihood method

The maximum likelihood method describes the evolutionary process by a probabilistic model, i.e. it evaluates the transition probability from one nucleotide state to another in each branch. The model assumes that the nucleotide sites evolve independently, therefore the probability for all sites of a DNA sequence is the product of the likelihoods for individual sites. This method identifies the tree with the highest maximum likelihood value as the best tree.

Comparison of tree-making methods

A general source of error that equally affects all methods of tree construction based on sequences is that if some sequences are only distantly related to the others this may introduce serious ambiguity in the multiple alignment on which the tree construction is based. Thus all tree-building methods using sequences as taxonomic units critically depend on a historically correct multiple alignment of the sequences.

Another frequent source of problems is that if the taxonomic units compared are too closely related, the information that can be used for tree construction is limited, therefore the trees are subject to large statistical errors. Conversely, if the number of differences between sequences is high there is a high probability that positions in a sequence alignment have undergone more than one change. (In a minor fraction of such cases, the so called homoplastic events – parallel, convergent and back substitutions – these multiple substitutions will appear as zero substitutions.) As a result of multiple hits, the observed difference between two distantly related sequences always underestimates the actual number of changes that have taken place. A consequence of this is that – in both parsimony and distance matrix analyses – the most distantly related sequences will appear to be more closely related than they actually are. In the case of distance matrix analysis it is possible to correct statistically for this underestimation. This is, however, not possible for parsimony analysis since it identifies actual changes. The maximum parsimony method is therefore much more sensitive to the distortions for distantly related lineages than distance matrix methods.

If the assumption of rate constancy does not hold then rapidly evolving lineages may appear to have branched off earlier than they actually did. The UPGMA trees are most sensitive to 'violation' of rate constancy.

In summary, since each of the above methods has its limitations, none of them works well under all conditions. It is important to remember the basic assumptions of these methods since it helps us realize when and why a certain method is likely to give an erroneous tree.

A tree is usually considered to be correct if its topology (branching pattern) is identical with that of the true tree, and much less attention is paid to the correctness of branch lengths. Nevertheless branch lengths contain important information, for example about rates of evolution of a protein in different lineages. The UPGMA method has the implicit assumption of a constant rate for all branches, i.e. it disregards this question. Since violation of rate constancy seems to be the rule rather than the exception, both the topology and the branch lengths of a UPGMA tree may be different from those of the true tree. The other distance matrix methods, which do not assume rate constancy (e.g. NJ method, minimum evolution method), are not subject to this type of error. The explicit assumption of the maximum parsimony method is that a tree that requires the fewest substitutions is the best tree. An inevitable consequence of this assumption is that it minimizes the number of multiple, parallel, convergent and back mutations and thus underestimates the branch lengths.

Finally, in the case of very distantly related protein-coding genes, where all the noncoding and synonymous sites and the majority of nonsynonymous sites have been saturated with substitutions, homology may be apparent only at the amino acid sequence level, using special techniques (see Chapters 5 and 7). In such cases it may be virtually impossible to reconstruct the actual nucleotide substitution pathways, and distance matrix methods based on amino acid sequence similarities (using position-dependent scoring systems) may give more reliable trees than maximum parsimony or maximum likelihood methods.

Evaluation of trees

Various types of statistical methods have been developed to evaluate the statistical confidence of a tree. For example, statistical tests are used to decide whether the best maximum parsimony tree is significantly more parsimonious than all of the other possible trees, whether the best minimum-evolution tree has a total tree length that is significantly shorter than the total length of any other tree, etc. The **bootstrap technique** is a special statistical technique that is frequently used to estimate the confidence level of trees. In the bootstrap technique a large number of pseudosamples are created from the original data set and the pseudosamples are then used to generate a large number of trees to evaluate the reproducibility of the tree topology.

4.5.4 Estimation of species-divergence times

As discussed in section 4.3.1, the following equation holds for the rate of nucleotide substitution (r, the number of substitutions per site per year), the number of substitutions per site (K) between two homologous sequences, and the time of divergence between the two sequences (T): $r = K/(2T)$. Using this equation, the dates of divergence of various species can be estimated from comparison of the sequences of orthologous protein-coding genes of the given species. For example, if the rate of evolution for a given type of DNA sequence is known to be r substitutions per site per year and we have determined the number of substitutions per site (K) between species A and B, then the divergence time can be calculated as $T = K/(2r)$.

Since the rate of nucleotide substitution determined in one group of organisms may not be applicable to other lineages there is some uncertainty as to the value of r. To avoid this problem, we may estimate the substitution rate by using as reference a third species, C, whose divergence time (T_1) is known and which diverged before the divergence of species A and B (T_2), i.e. T_2 is unknown but $T_1 > T_2$. The unknown divergence time between species A and B (T_2) may be estimated from the ratio:

$$T_2/T_1 = 2K_{AB}/(K_{AC} + K_{BC})$$

This equation assumes rate constancy. As discussed earlier, this assumption often does not hold, and so the estimated divergence time usually must be treated with caution.

References

Brown, N.P., Whittaker, A.J., Newell, W.R. et al. (1995) Identification and analysis of multigene families by comparison of exon fingerprints. *Journal of Molecular Biology* **249**, 342–59.

Feng, D.F. & Doolittle, R.F. (1987) Progressive sequence alignment as a prerequisite to correct phylogenetic trees. *Journal of Molecular Evolution* **25**, 351–60.

Feng, D.F. & Doolittle, R.F. (1990) Progressive sequence alignment and phylogenetic tree construction of protein sequences. *Methods in Enzymology* **183**, 375–87.

Grunstein, M. (1997) Histone acetylation in chromatin structure and transcription. *Nature* **389**, 349–52.

Higgins, D.G. & Sharp, P.M. (1989) Fast and sensitive multiple sequence alignments on a microcomputer. *Computer Applications in the Biosciences* **5**, 151–3.

Johnson, M.S., Sutcliffe, M.J. & Blundell, T.L. (1990) Molecular anatomy: phyletic relationships derived from three-dimensional structures of proteins. *Journal of Molecular Evolution* **30**, 43–59.

Johnson, M.S., Sali, A. & Blundell, T.L. (1990) Phylogenetic relationships from three-dimensional protein structures. *Methods in Enzymology* **183**, 670–90.

Jukes, T.H. & Cantor, C.R. (1969) Evolution of protein molecules. In H.N. Munro (ed.), *Mammalian Protein Metabolism*, pp. 21–132. New York: Academic Press.

Kimura, M. (1980) A simple method for estimating evolutionary rate of base substitution through comparative studies of nucleotide sequences. *Journal of Molecular Evolution* **16**, 111–20.

Li, W.H. (1997) *Molecular Evolution*. Sunderland, MA: Sinauer Associates.

Li, W.H. & Graur, D. (1991) *Fundamentals of Molecular Evolution*. Sunderland, MA: Sinauer Associates.

Li, W.H., Wu, C.I. & Luo, C.C. (1985) A new method for estimating synonymous and nonsynonymous rates of nucleotide substitution considering the relative likelihood of nucleotide and codon changes. *Molecular Biology and Evolution* **2**, 150–74.

Luger, K., Mader, A.W., Richmond, R.K., Sargent, D.F. & Richmond, T.J. (1997) Crystal structure of the nucleosome core particle at 2.8Å resolution. *Nature* **389**, 251–60.

Mitreva, M., Wendl, M.C., Martin, J. et al. (2006) Codon usage patterns in Nematoda: analysis based on over 25 million codons in 32 species. *Genome Biology* 7, R75.

Needleman, S.B. & Wunsch, C.D. (1970) A general method applicable to the search of similarities in the amino acid sequence of two proteins. *Journal of Molecular Biology* **48**, 443–53.

Patthy, L. (1987) Detecting homology of distantly related proteins with consensus sequences. *Journal of Molecular Biology* **198**, 567–77.

Patthy, L. (1988) Detecting distant homology of mosaic proteins. Analysis of the sequences of thrombomodulin, thrombospondin, complement components C9, C8a, C8b, vitronectin and plasma cell membrane protein PC-1. *Journal of Molecular Biology* **202**, 689–96.

Patthy, L. (1996) Consensus approaches in detection of distant homologies. *Methods in Enzymology* **266**, 184–98.

Powell, J.R. & Moriyama, E.N. (1997) Evolution of codon usage bias in *Drosophila*. *Proceedings of the National Academy of Sciences of the USA* **94**, 7784–90.

Tanaka, T. & Nei, M. (1989) Positive Darwinian selection observed at the variable region genes of immunoglobulins. *Molecular Biology and Evolution* **6**, 447–59.

Thompson, J.D., Higgins, D.G. & Gibson, T.J. (1994) Clustal W: improving the sensitivity of progressive multiple sequence alignment through sequence weighting, position-specific gap penalties and weight matrix choice. *Nucleic Acids Research* **22**, 4673–80.

Wolfe, K.H., Sharp, P.M. & Li, W.H. (1989) Mutation rates differ among regions of the mammalian genome. *Nature* **337**, 283–5.

Zuckerkandl, E. & Pauling, L. (1962) Molecular disease, evolution and genic heterogeneity. In M. Kash & B. Pullman (eds), *Horizons in Biochemistry*, pp. 189–225. New York: Academic Press.

Zuckerkandl, E. & Pauling, L. (1965) Evolutionary divergence and convergence in proteins. In V. Bryson & H.J. Vogel (eds), *Evolving Genes and Proteins*, pp. 97–166. New York: Academic Press.

Useful internet resources

Pairwise sequence alignments

SIM Local similarity program is an alignment tool for nucleic acid sequences (http://www.expasy.org/tools/sim-nucl.html).

SIM Local similarity program is an alignment tool for protein sequences (http://au.expasy.org/tools/sim-prot.html).

Align is a pairwise global and local alignment tool for protein and nucleic acid sequences (EMBOSS) (http://www.ebi.ac.uk/emboss/align/).

Pairwise Alignment tools for global and local alignment of protein and nucleic acid sequences (http://gcnomc.cs.mtu.edu/align/align.html).

Multiple sequence alignments

ClustalW is a general-purpose multiple sequence alignment program for DNA or proteins (http://www.ebi.ac.uk/clustalw/).

MultAlin is a multiple sequence alignment tool (http://bioinfo.genopole-toulouse. prd.fr/multalin/multalin.html).

T-COFFEE is a multiple alignment tool (http://www.ch.embnet.org/software/ TCoffee.html).

Visualization of multiple sequence alignments

CINEMA is a Colour INteractive Editor for Multiple Alignments (http://www.bioinf.man.ac.uk/dbbrowser/CINEMA2.1/).

Jalview is a Java multiple alignment editor (http://www.ebi.ac.uk/~michele/jalview/contents.html).

BOXSHADE provides Pretty Printing and Shading of Multiple-Alignment files (http://www.ch.embnet.org/software/BOX_form.html)

Species divergence times

TIMETREE is a public knowledge-base for information on the timescale of life (http://www.timetree.net/).

R8S (analysis of rates ('r8s') of evolution) is a program for estimating divergence times and rates of evolution based on a phylogenetic tree with branch lengths using maximum-likelihood and semi- and nonparametric methods (http://ginger.ucdavis.edu/r8s/).

Detection of positive selection

Selecton is a server for the identification of site-specific positive selection and purifying selection (http://selecton.bioinfo.tau.ac.il/index.html).

SWAKK (Sliding Window Analysis of KA and KS) is a bioinformatic web server for detecting amino acid sites under positive selection using a sliding window substitution rate analysis (http://oxytricha.princeton.edu/SWAKK/).

Datamonkey Adaptive Evolution Server for fast detection of positive or negative selection (http://www.datamonkey.org/).

Phylogenetic trees

PHYLIP package of phylogeny programs:

   dnadist computes distance matrices from nucleotide sequences
      (http://bioweb.pasteur.fr/seqanal/interfaces/dnadist-simple.html)

   dnapars is a DNA parsimony program
      (http://bioweb.pasteur.fr/seqanal/interfaces/dnapars-simple.html).

protdist computes distance matrices from protein sequences
(http://bioweb.pasteur.fr/seqanal/interfaces/protdist-simple.html).

protpars is a protein sequence parsimony method
(http://bioweb.pasteur.fr/seqanal/interfaces/protpars-simple.html)

neighbor provides neighbour-joining and UPGMA methods
(http://bioweb.pasteur.fr/seqanal/interfaces/neighbor-simple.html).

fitch provides Fitch–Margoliash and least-squares distance methods
(http://bioweb.pasteur.fr/seqanal/interfaces/fitch-simple.html).

kitsch provides Fitch–Margoliash and least-squares methods with evolutionary
clock (http://bioweb.pasteur.fr/seqanal/interfaces/kitsch-simple.html).

drawtree plots an unrooted tree diagram
(http://bioweb.pasteur.fr/seqanal/interfaces/drawtree-simple.html).

drawgram plots a cladogram- or phenogram-like rooted tree
(http://bioweb.pasteur.fr/seqanal/interfaces/drawgram-simple.html).

consense is a consensus tree program
(http://bioweb.pasteur.fr/seqanal/interfaces/consense-simple.html).

PALI provides Phylogeny and ALIgnment of homologous protein structures
(http://pauling.mbu.iisc.ernet.in/~pali/).

fastDNAml allows construction of phylogenetic trees of DNA sequences using
maximum likelihood (http://bioweb.pasteur.fr/seqanal/interfaces/fastdnaml-
simple.html).

MEGA (Molecular Evolutionary Genetics Analysis) is an integrated tool for
automatic and manual sequence alignment, inferring phylogenetic trees, mining
web-based databases, estimating rates of molecular evolution and testing
evolutionary hypotheses (http://www.megasoftware.net/).

Molphy is a computer program package for molecular phylogenetics used for
inferring evolutionary trees from protein sequences via maximum likelihood
(http://bioweb.pasteur.fr/seqanal/interfaces/prot_nucml.html).

PAML provides Phylogenetic Analysis by Maximum Likelihood
(http://abacus.gene.ucl.ac.uk/software/paml.html).

PAUP (Phylogenetic Analysis Using Parsimony) is a program for phylogenetic
analysis using parsimony, maximum-likelihood and distance methods
(http://paup.csit.fsu.edu/).

PHYLIP (PHYLogeny Inference Package) is a package of programs for various computer
platforms to infer phylogenies or evolutionary trees, freely available from the web
(http://evolution.genetics.washington.edu/phylip/phylipweb.html).

Taxonomic classification of organisms

The NCBI Taxonomy homepage (http://www.ncbi.nlm.nih.gov/Taxonomy/tax.html/).

The **UCMP Taxon lift** (http://www.ucmp.berkeley.edu/help/taxaform.html).

The structure of the **Tree of Life Web Projec**t illustrates the genetic connections
between all living things (http://tolweb.org/tree/phylogeny.html).

Chapter 5: Evolution of orthologous proteins

The previous chapter discussed in general terms how functional constraints influence the rate of the evolution of protein-coding genes, why coding and noncoding regions evolve at different rates, why synonymous and non-synonymous substitutions have different probabilities, and what dictates the observed preferences among nonsynonymous substitutions. In this chapter we will have a closer look at individual families of protein-coding genes and will examine the influence of protein structure and function on their evolutionary behaviour. Here we will concentrate on **orthologous genes** of different species (paralogous genes will be discussed in Chapter 7). Genes of different species are said to be orthologous if they arose through speciation, i.e. they derive from a single ancestral gene in the last common ancestor of the respective species.

There are three major types of methods for determining **orthology**. The **reciprocal best-hit** approach is based on the principle that in any two pairs of genomes orthologous pairs of genes (the 'same genes' in different species) should give best reciprocal hits, provided that gene birth or death (or drastic change of function of the proteins) has not occurred since the speciation event linking the two genomes. In other words, this approach is reliable only in the case of the so-called 1:1 orthologs, but the exact relationship among homologous genes of different species may be quite confused if, following speciation, the orthologous genes were duplicated in one or both species. An additional practical limitation of this approach is that it can identify **orthologs** only if the genomes analysed are completely sequenced and all their proteins are full length and correctly predicted.

The 'conservation of gene order' or **'conserved synteny'** approach for determining orthology is based on the principle that orthologous genes arise through speciation, therefore they are located in the same chromosomal environments in different species. Accordingly, if a series of related genes is found in the same order in chromosomal segments of different genomes, it is likely that these segments are actually orthologous. It should be noted, however, that the physical linkage (**synteny**) of genes may not persist over longer evolutionary times (cf. Chapter 9), therefore this conserved synteny approach may be used only for relatively recently diverged species to infer orthology.

The **phylogenomics methods** decide the orthology vs. paralogy issue by reconstruction of phylogeny from a multiple alignment of a gene/protein family. Comparison of the molecular phylogeny with the taxonomic tree permits the establishment of orthologous and paralogous relationships.

Obviously, the reliability of this approach is a function of the quality of the multiple alignment and the reliability of the phylogenetic tree(s) obtained by various tree-builiding approaches.

Following speciation, two (1:1) orthologous protein-coding genes fulfilling the same biological function in the two diverging species start to accumulate mutations independently. Despite their independence, acceptance of mutations usually follows the same pattern in both nascent species: the pattern dictated by the same structural and functional constraints of the proteins. For example, myoglobins of different species must fold into similar three-dimensional structures and must fulfil a similar function. The relative probability of mutations that are deleterious, neutral or advantageous is thus expected to be identical for the two orthologous genes. In the case of orthologous genes with well-established functions, nonsynonymous substitutions are predominantly deleterious, whereas synonymous substitutions are predominantly neutral, therefore usually $K_A/K_S \ll 1$ (see Table 4.1). In most cases the functional constraints of orthologous proteins remain essentially the same even in different species, thus the ratio of nonsynonymous to synonymous substitutions is usually a constant, low value.

Major deviation from this K_A/K_S constancy rule may be observed if the diverging species adapt to different environments and the functional or structural requirements for the given proteins (biological activity, specificity, stability, etc.) are altered. During such adaptations, there is positive selection for advantageous mutations that make the protein better adjusted to the new needs: there will be an increase in substitutions at nonsynonymous relative to synonymous sites. A diagnostic sign of positive selection for advantageous point mutations is that the K_A/K_S ratio is much higher than unity at the sites involved in the adaptive changes. It must be emphasized that adaptive changes may occur by substitutions at surprisingly few sites and the K_A/K_S ratio will increase only at these sites. For example, the biological specificity or stability of a protein may be 'improved' by substitutions at a few critical sites, while the majority of sites (those involved in maintaining the 'original' three-dimensional structure of the protein fold) are subject to the same structural constraints: purifying selection will thus keep the K_A/K_S ratio at low values at the majority of sites. Consequently, unless there is a major functional and structural adaptation, the overall K_A/K_S ratio for the entire protein may remain dominated by purifying selection and may still be significantly lower than unity. Once the protein has adapted to the new function, even this local and temporary rate acceleration comes to an end, and the new functional constraints will lower the rates of nonsynonymous relative to synonymous substitutions at the adapted sites.

A dramatic increase in the overall K_A/K_S ratio may also occur in cases when the protein loses its biological importance, i.e. if the protein is liberated from all its original functional constraints. If all mutations are selectively neutral the ratio K_A/K_S will be close to unity for the entire protein, the gene will accept nonsynonymous mutations at an accelerated rate, and eventually the gene may become a pseudogene.

In the following section the influence of functional and structural constraints is illustrated with a few examples.

5.1 Orthologous proteins with the same function in different species

Evolutionary divergence of species results in many variants of the same protein, all with essentially the same structure and biological function but with very different amino acid sequences. It is now clear that the three-dimensional structures of very distant members of a family (with very dissimilar sequences) are practically superimposable. In other words, despite significant differences in amino acid sequences they can adopt essentially the same three-dimensional structure and perform the same biological function. The identities (similarities) of the amino acid sequences of many variants thus identify the regions that are absolutely critical for the structure and function of the given group of orthologous proteins.

For example, it has become clear from early studies on globins from various species that the constraints on their amino acid sequences result from the necessity that they must function as globins. The most highly conserved residues interact directly with the haem iron and with oxygen (and are thus critical for globin function); other conserved residues are critical for maintaining the folded conformation of globins (Fig. 5.1).

Similarly, in the case of the proteinase chymotrypsin, the most highly conserved residues are the His-57, Asp 102 and Ser-195 residues of the catalytic triad that is indispensable for the enzyme's activity; other conserved residues are critical for maintaining the folded conformation of the protein and the proper positioning of the catalytic residues (Fig. 5.2). Even in regions that show sequence variations, the choice of nonsynonymous mutations is not random. As discussed in Chapter 3, a survey of the evolution of many different protein families (primarily globular proteins) has identified some common features of nonsynonymous mutations. For example, the most prevalent replacements occur between amino acids with similar side chains, such as Ile/Val/Leu/Met, Asp/Glu, Lys/Arg, Tyr/Phe, and so on. Another common feature is that the large hydrophobic Trp residue and the disulphide bond-forming Cys residue are replaced least often. It is clear from these studies that the biases in conservation or in amino acid replacements reflect the role of purifying selection. A Cys residue involved in a disulphide bond that is essential for the stability of a protein cannot be substituted by any other residue since it would disrupt that disulphide bond and would destabilize the protein. On the other hand, substitutions of chemically very similar residues (Phe↔Tyr, Arg↔Lys, Asp↔Glu, etc.) in most cases are without major effect on the structure and function of the protein. Despite these general tendencies,

```
                   HHHHHHHHHHH                HHHHHHHHHHHHHHHHH HHHH               HHHHHHHHHHHHHHHHHHHH HHHHH
hbb_human   VHLTPEEKSAVTALWGKV..NVDEVGGEALGRLLVVYPWTQRFFESFGDLSTPDAVMGNPKVKAHGKKVLGAFSDGLAHLDNLKGTFATLSELHCDKLHV
hbb_pig     VHLSAEEKEAVLGLWGKV..NVDEVGGEALGRLLVVYPWTQRFFESFGDLSNADAVMGNPKVKAHGKKVLQSFSDGLKHLDNLKGTFAKLSELHCDQLHV
hbb_horse   VQLSGEEKAAVLALWDKV..NEEEVGGEALGRLLVVYPWTQRFFDSFGDLSNPGAVMGNPKVKAHGKKVLHSFGEGVHHLDNLKGTFAALSELHCDKLHV
hbb_bovin   ~MLTAEEKAAVTAFWGKV..KVDEVGGEALGRLLVVYPWTQRFFESFGDLSTADAVMNNPKVKAHGKKVLDSFSNGMKHLDDLKGTFAALSELHCDKLHV
hbb_chick   VHWTAEEKQLITGLWGKV..NVAECGAEALARLLIVYPWTQRFFASFGNLSSPTAILGNPMVRAHGKKVLTSFGDAVKNLDNIKNTFSQLSELHCDKLHV
hba_horse   ~VLSAADKTNVKAAWSKVGGHAGEYGAEALERMFLGFPTTKTYFPHF.DLSH.....GSAQVKAHGKKVGDALTLAVGHLDDLPGALSNLSDLHAHKLRV
hba_human   ~VLSPADKTNVKAAWGKVGAHAGEYGAEALERMFLSFPTTKTYFPHF.DLSH.....GSAQVKGHGKKVADALTNAVAHVDDMPNALSALSDLHAHKLRV
hba_bovin   ~VLSAADKGNVKAAWGKVGGHAAEYGAEALERMFLSFPTTKTYFPHF.DLSH.....GSAQVKGHGAKVAAALTKAVEHLDDLPGALSELSDLHAHKLRV
hba_pig     ~VLSAADKANVKAAWGKVGGQAGHGAEALERMFLGFPTTKTYFPHF.NLSH.....GSDQVKAHGQKVADALTKAVGHLDDLPGALSALSDLHAHKLRV
hba_chick   ~VLSAADKNNVKGIFTKIAGHAEEYGAETLERMFTTYPPTKTYFPHF.DLSH.....GSAQIKGHGKKVAALIEAANHIDDIAGTLSKLSDLHAHKLRV
Consensus   ------K-------K------G-E-L-R-----P-T---F--F--LS------------HG-KV------D------LS-LH----L--L-V
                                                                            *                                  *

                   HHHHHHHHHHH HHHHHHHHHHHHHHHHHHHH
hbb_human   DPENFRLLGNVLVCVLAHHFGKEFTPPVQAAYQKVVAGVANALAHKYH
hbb_pig     DPENFRLLGNVIVVVLARRLGHDFNPDVQAAFQKVVAGVANALAHKYH
hbb_horse   DPENFRLLGNVLVVVLARHFGKDFTPELQASYQKVVAGVANALAHKYH
hbb_bovin   DPENFRLLGNVLVVVLARNFGKEFTPVLQADFQKVVAGVANALAHRYH
hbb_chick   DPENFRLLGDILIIVLAAHFSKDFTPECQAAWQKLVRVVAHALARKYH
hba_horse   DPVNFKLLSHCLLSTLAVHLPNDFTPAVHASLDKFLSSVSTVLTSKYR
hba_human   DPVNFKLLSHCLLVTLAAHLPAEFTPAVHASLDKFLASVSTVLTSKYR
hba_bovin   DPVNFKLLSHLLLVTLASHLPSDFTPAVHASLDKFLANVSTVLTSKYR
hba_pig     DPVNFKLLSHCLLVTLAAHHPDDFNPSVHASLDKFLANVSTVLTSKYR
hba_chick   DPVNFKLLGQCFLVVVAIHHPAALTPEVHASLDKFLCAVGTVLTAKYR
Consensus   DP-NF-LL-------A-----P--A---K---V---L--Y-
```

Fig. 5.1 Multiple alignment of the amino acid sequences of α and β chains of human, pig, horse, bovine and chick haemoglobins. The consensus sequence shows residues identical in all sequences. The asterisks indicate the histidine residues involved in binding oxygen and haem iron. The residues involved in α-helices of horse α-globin are indicated by H. Note that gaps in the alignment are between α-helices and the majority of conserved residues are within α-helices. The ribbon diagram of horse deoxyhaemoglobin is also shown.

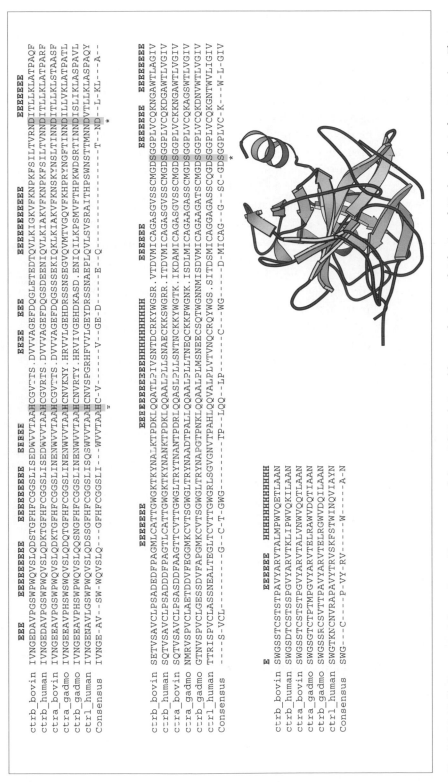

```
                EEE      EEEEEEEE    EEEEEEEE   EEEE       EEEE      EEE     EEEEEEEEE   EEE        EEEEEEE
ctrb_bovin IVNGEDAVPGSWPWQVSLQDSTGFHFCGGSLISEDKVVTAAHCGVTS.DVVVAGEFDQGLTEDTQVLKIKGKVFKNPKFSILTVRNDITLLKLATPAQF
ctrb_human IVNGEDAVPGSWPWQVSLQDKTGFHFCGGSLISEDWVTAAHCGVRTS.DVVVAGEFDQGSDEENIQVLKIAKVFKNPKFSILTVNNDITLLKLATPARF
ctra_bovin IVNGEEAVPGSWPWQVSLQDKTGFHFCGGSLINENWVVTAAHCGVTS.DVVVAGEFDQGSSEKIQKLIAKVFKNSLTINNDITLLKLSTAASF
ctra_gadmo IVNGEEAVPHSWSWQVSLQDQTGFHFCGGSLINENWVVTAAHCNVKZY.HRVLGEHDRSSNSEGVQVMTVGQVFKHPRYNGFTINNDILLVKLATPATL
ctrb_gadmo IVNGEEAVPHSWPWQVSLQQSNGFHFCGGSLINENWVVTAAHCNVRTY.HRVIVGEHDKASD.ENIQILKPSMVFTHPKWDSRTINNDISLIKLASPAVL
ctrl_human IVNGENAVLGSWPWQVSLQDSSGFHFCGGSLISQSWVVTAAHCNVSPGRHFVVLGEYDRSSNAEPLQVLSVSRAITHPSWNSTTMNNDVTLLKLASPAQY
Consensus  IVNGE-AV--SW-WQVSLQ--GFHFCGGSLI--WVVTAAHC-V-----V-GE-D----E--Q------T--ND--L-KL--A--
                                                                                                      *

               EEEEE    EEEEEEEEEHHHHHHHHH               EEEEE     EEEEEE    EEEEEEE
ctrb_bovin SETVSAVCLPSADEDFPAGMLCATTGWGKTKYNALKTPDKLQQATLPIVSNTDCRKYWGSR.VTDVMICAGASGVSSCMGDSGGPLVCQKNGAWTLAGIV
ctrb_human SQTVSAVCLPSADDFPAGTLCATTGWGKTKYNANKTPDKLQQAALPLLSNAECKKSWGRR.ITDVMICAGASGVSSCMGDGGPLVCQKDGAWTLVGIV
ctra_bovin SQTVSAVCLPSASDDFAAGTTCVTTGWGLTRYTNANTPDRLQQASLPLLSNTNCKKYWGTK.IKDAMICAGASGVSSCMGDSGGPLVCKKNGAWTLVGIV
ctra_gadmo NMRVSPVCLAETDDVFEGMKCVTSGWLTRYNAADTPALLQQAALPLLTNEOCKKFWGNK.ISDLMICAGAGASSCMGDSGGPLVCQKAGSWTLVGIV
ctrb_gadmo GTNVSPVCLGESSDVFAPGMKCVTSGWGLTRYNAPGTPNKLQQAALPLMSNEECSQTWGNNMISDVMICAGAGATSCMGDSGGPLVCQKDNVWTLVGIV
ctrl_human TTRISPVCLASSNEALTEGLTCVTTGWGRLSGVGNVTPAHLQQVALPLVTVNQCRQYWGS.SITDSMICAGGAGASSCQGDSGGPLVCQKGNTWVLIGIV
Consensus  ----S-VCL-------G--C-T-GWG-----TP--LQQ--LP------C----WG----D-MICAG--G--SC-GDSGGPLVC-K---W-L-GIV
                                                                  *

               E    EEEEEE  HHHHHHHHHHHH
ctrb_bovin SWGSSTCSTSTPAVYARVTALMPWVQETLAAN
ctrb_human SWGSDTCSTSSPGVYARVTKLIPWVQKILAAN
ctra_bovin SWGSSTCSTSTPGVYARVTALVNWVQQTLAAN
ctra_gadmo SWGSGTCTPTMPGVYARVTELRAWVDQTIAAN
ctrb_gadmo SWGSSRCSVTTPAVYARVTELRGWVDQLAAN
ctrl_human SWGTKNCNVRAPAVYTRVSKFSTWINQVIAYN
Consensus  SWG---C----P-VY-RV--W----W-----A-N
```

Fig. 5.2 Multiple alignment of the amino acid sequences of the catalytic chains of human, bovine and Atlantic cod (*Gadus morhua*) chymotrypsins. The consensus sequence shows residues identical in all sequences. The asterisks indicate the residues of the catalytic triad. The residues involved in β-strands or α-helices of chymotrypsin are indicated by E and H, respectively. Note that the majority of conserved residues are in segments comprising the catalytic residues, in segments involved in secondary structure elements or in disulphide bonds. Also note that the gaps in the alignment are between secondary structural elements. The ribbon diagram of bovine α-chymotrypsin is also shown.

the substitution patterns may be drastically different in different groups of orthologous proteins as well as in different segments of orthologous proteins.

Since the α-helices, β-strands and reverse turns accept substitutions that are compatible with the integrity of these secondary structural elements their substitution patterns will be different (see section 3.3). As a reflection of the fact that surface loops are significantly less constrained than α-helices and β-strands (which constitute the core of protein folds), the 'conserved blocks' will show a substitution pattern that deviates from the pattern averaged for the entire protein (compare substitution patterns in conserved blocks with the substitution pattern for entire proteins; Table 3.1(a) vs. Table 3.1(c)). The difference between these substitution patterns primarily reflects the fact that residues in surface loops are less conserved than those in α-helices and β-sheets. Note that (as a result of the lower conservation of surface areas) it is the peripheral part of the protein folds of distantly related homologs that may adopt different conformations (Chothia & Lesk, 1986).

The importance of the position factor may be illustrated by looking at the catalytic residues of enzymes. For example, whereas Ser→Thr or Asp→Glu substitutions are quite common in all proteins (including serine proteinases) these substitutions are not tolerated in the case of the active site Ser-195 and Asp-102 residues of serine proteinases since either type of substitution would disrupt the catalytic triad and would eliminate catalytic activity (Fig. 5.2).

In summary, each family of orthologous proteins fulfilling identical roles in different species may be characterized by a unique pattern of conserved residues, variable positions and gap positions. Residues critical for **biological activity** and for the maintenance of the **three-dimensional structure** are highly conserved, whereas nonconservative amino acid substitutions and gap events identify regions that are less critical for structure and function.

5.2 Orthologous proteins with modified function in different species

Orthologous proteins, although they fulfil similar roles, may undergo major functional adaptations in some species.

Adaptation of haemoglobin to low oxygen pressure

The major role of haemoglobin is to bind oxygen from the lungs and to transport and release it to tissues and cells. In some species that live in environments where oxygen pressure is low, haemoglobins with increased oxygen affinity have evolved, permitting the evolutionary adaptation of these species. For example, birds that are able to fly at a very high altitude, and mammals living in mountainous regions, at 4000–6000 m above sea level, have haemoglobins with unusually high oxygen affinities, significantly higher than that of their lowland relatives. A somewhat surprising conclusion from comparison of haemoglobins of closely related species living at different altitudes is that increased oxygen affinities are usually due to just a few (functionally critical) amino acid substitutions.

Llama, alpaca, guanaco and vicuna live at high altitudes (≈4000 m) in the Andes. These animals have adapted to life under hypoxic conditions thanks to minor changes in their haemoglobin sequences. The haemoglobin of these mammals has a high affinity for oxygen, whereas the haemoglobins of their closest lowland relatives, camels, have 'normal' oxygen affinity. Comparison of the (closely related) amino acid sequences of the α- and β-globins of these species identified only nine sites where globins of llama, alpaca, guanaco or vicuna differ from those of camels (Kleinschmidt et al., 1986; Piccinini et al., 1990). Of these the β2His→Asn substitution appears to be primarily responsible for the higher oxygen affinity of llama, alpaca, guanaco and vicuna haemoglobins. In the case of vicuna an additional substitution (α130Ala→Thr) leads to the highest oxygen affinity of mammalian haemoglobins, permitting these animals to live permanently at 4000–5000 m.

Bird haemoglobins provide many similar examples of adaptation to low oxygen pressures. Again, the general conclusion from studies on avian haemoglobins is that increased oxygen affinity could be achieved by one or just a few amino acid substitutions. One of the most clear-cut cases is the case of the barheaded goose. These birds live and nest in Tibet at altitudes of 4000–6000 m but they migrate across the Himalayas at an altitude of 9000 m to winter in the plains of India. Comparison of the haemoglobin sequences of the barheaded goose with those of its closest lowland relative, the greylag goose, revealed that their haemoglobin differs from the greylag goose haemoglobin by only four amino acids. Of these, the high oxygen affinity of the barheaded goose haemoglobin could be assigned primarily to a single α119Pro→Ala substitution (Perutz, 1983).

Adaptation of lysozyme to foregut fermentation

Lysozyme is a bacteriolytic enzyme that can cleave the glycosidic bonds in the cell-wall peptidoglycans of bacteria. Thanks to its lytic activity, the enzyme is part of the antibacterial defence mechanisms of many animals; it is found primarily in the tears and saliva of mammals and in the eggs of birds. In foregut-fermenting animals, where ingested plant material is subject to bacterial fermentation, lysozyme is also secreted in the digestive system, permitting the retrieval of the nutrients from lysed bacterial cells. The ruminant artiodactyls (e.g. cow, deer) and the leaf-eating colobine monkeys (e.g. langur) have independently recruited lysozyme as a means of digesting bacteria (Stewart & Wilson, 1987; Jolles et al., 1989).

Independent exposure of lysozyme to the 'novel' environment of the stomach (lower pH, proteinase-rich environment, reactive chemicals produced by fermentation) has favoured similar amino acid substitutions in the two groups. These mutations have made ruminant and langur lysozymes more resistant to proteinases, have readjusted the pH optimum of the enzyme to increase its activity in an acidic environment and made it more resistant to some chemical modifications. The change in pH optimum is also illustrated by the fact that at higher pH, langur and cow lysozymes are less active than human lysozyme. These observations clearly show that a similar functional challenge to lysozyme has favoured parallel occurrence of advantageous substitutions in different evolutionary lines, resulting in parallel adaptations.

Analysis of the evolution of ruminant lysozymes has shown that there was an apparent acceleration in the rate of evolution of lysozyme in the first half of the ruminant lineage followed by a deceleration in the second half (Jolles et al., 1989). In the early part of ruminant evolution the substitution rate was about three times the typical mammalian rate of substitutions, whereas the average rate in the later part of ruminant evolution is about half the typical substitution rate. Significantly, this rate variation was specific for lysozyme, since no similar rate variation was observed for several other proteins of ruminants. Analysis of these substitutions suggested that many of these may be adaptive for an enzyme that functions at low pH. For example, the number of aspartyl and amide bonds (known to be sensitive to acid) is reduced in the ruminants relative to the pig, endowing it with increased stability at a lower pH. The number of arginine residues is reduced in the ruminants making the enzyme more resistant to attack by pancreatic trypsin, as well as to destruction by diacetyl, a product of fermentation.

The evolution of langur lysozyme shows striking similarities to that of ruminants (Stewart & Wilson, 1987). First, there appears to be a similar acceleration in the rate of evolution of lysozyme in langurs: in the langur line the rate is 2.5 times higher than in the related baboon line, the acceleration being associated with the recruitment of lysozyme for use in the stomach. Second, five of the 10 substitutions on the langur lineage are to amino acids that are found in comparable positions of the cow lysozyme. The unusual lysozymes of ruminants and colobine monkeys have evolved at higher than the normal rate during adaptation, suggesting that most (if not all) of the amino acid changes are adaptive and were selected for. This conclusion is further supported by the fact that parallel substitutions occurred independently in both lineages. These studies on the convergent evolution of lysozymes of foregut fermenters were recently extended to the hoatzin, the only known foregut-fermenting bird (Kornegay, Schilling & Wilson, 1994). Sequence comparisons have shown that similar adaptive structural changes occurred at five positions in the cow, langur and hoatzin lysozymes, permitting a better performance of lysozyme at low pH values and increased resistance to inactivation by diacetyl or digestion by trypsin.

Messier and Stewart (1997) have examined the evolution of primate lysozymes in greater detail and have pinned down the proposed adaptive changes to a very short period during the early emergence of colobine monkeys. Their analysis has shown that this brief episode of positive selection (in the lineage leading to the common ancestor of the foregut-fermenting colobine monkeys) was followed by an episode of negative selection. Since the group of colobine monkeys, as a whole, are leaf-eaters, it makes good biological sense that most of the adaptive substitutions in lysozyme occur during the early evolution of the group. The leaf-eating colobine monkeys are unique among Old World primates in having a complex, fermenting stomach, whereas other Old World primates, including hominoids and even the most closely related cercopithecines, have simple stomachs. Comparison of colobine lysozyme DNA sequences with those from other cercopithecines and hominoids has shown that there are more nonsynonymous nucleotide differences than synonymous nucleotide differences, suggesting that the colobine lysozymes evolved under positive selection. Pairwise comparisons of the lysozyme DNA sequences of colobines with other cercopithecines showed K_A/K_S ratios significantly greater than 1.0, reflecting positive selection. The between-group average K_A/K_S ratio for all colobine lysozymes vs. all cercopithecine lysozymes, $K_A/K_S - 3.0$, is one of the highest known for any whole protein. Similarly, all comparisons between the colobine and the hominoid lysozymes gave a high average K_A/K_S ratio ($K_A/K_S - 3.0$). The fact that the K_A/K_S ratio for the whole lysozyme sequence is significantly higher than unity probably indicates that adaptation to an acidic environment required rather extensive readjustments of the lysozyme sequence.

In contrast to these between-group comparisons, all within-group averages of catarrhine lysozymes gave K_A/K_S ratios less than 1.0, suggesting that purifying selection has been active during the recent evolution of primate lysozymes. The striking contrast of the purifying selection within groups and the positive selection between groups suggests that the adaptive evolution of lysozyme occurred early in catarrhine evolution.

To test this hypothesis, Messier and Stewart (1997) have reconstructed ancestral DNA sequences representing all internal nodes on the tree of primate lysozymes using maximum-likelihood and maximum-parsimony methods; the reconstructed DNA sequences of primate ancestors were then used in pairwise K_A/K_S calculations. This ancestral analysis has pinpointed an episode of adaptive lysozyme sequence evolution ($K_A/K_S = 4.7$) on the ancestral colobine lineage. The maximum-likelihood analysis also suggests that there were nine amino acid replacements during the brief adaptive episode on the ancestral colobine lineage, five of which are among those originally identified to have occurred in parallel or convergently along the lineage leading to cow and other ruminant stomach lysozymes.

5.3 Orthologous proteins with major modification of function

Insulin of hystricomorphs

In most mammals insulin associates into hexamers, each of which is stabilized by two zinc atoms bound to His-10 of the insulin B chains. Guinea pig and other hystricomorph rodents are exceptions: they store insulin as monomers rather than as hexamers. Although the biological significance of this is unclear at present (see below), the evolutionary readjustments that have accompanied this change have resulted in a very rapid evolution of insulin in the hystricomorphs as compared with its slow and conservative evolution in other mammals (Blundell & Wood, 1975). The rate of amino acid substitution in insulins is more than 10 times higher in the hystrico- morph lineage than in other mammals.

Comparison of insulins from hystricomorphs with those of other mam- mals has revealed one critical substitution. His-10 of the B chain which binds to zinc and thus is essential for the stability of the hexamer has been replaced by Asn (in guinea pig and cuis) or by Gln (in coypu and caisiragua). There are several other changes that also contributed to the disruption of the hexamer. Most of the substitutions occurred on that surface of the insulin molecule where the monomers of nonhystricomorph insulins are held together by hydrophobic bonds. With the changeover to the monomeric storage form, this patch has become more hydrophilic in hystricomorphs so that the monomers remain apart.

What the hystricomorph insulins clearly illustrate is the nonindependence of sites: all those sites that are subject to the constraint of maintaining the hexameric structure were relaxed during transition to a monomeric form. In this sense, what we see in the case of the hystricomorph insulins is the relaxation of a structural constraint that exists in other mammals. Why hystricomorphs could relax this constraint is not fully understood at present, but it may be related to the fact that there is a much higher level of insulin in the blood of guinea pigs, and that they have a higher density of insulin receptors. Thus, it may well be that the shift from hexamer storage to monomer storage was accompanied by a change in insulin function or that the unusu- ally low potency of guinea pig insulin was compensated by readjustments in insulin expression and insulin-receptor density.

5.4 Orthologous proteins that have lost their function

α-Crystallin of the blind mole rat

There are not many cases when a protein important for the majority of species loses its function in some species. α-Crystallin is a major structural component of the eye lens, but it is deprived of its functional importance in the blind mole rat, which has lost its sight during evolution. Comparison of the blind mole rat α-crystallin with orthologous sequences from other rodents revealed an acceleration of substitution rate in nonsynonymous positions in this lineage. It thus seems that the relaxation of functional constraints on this protein has permitted an increased rate of accepted nonsynonymous mutations (Hendriks et al., 1987).

5.5 Orthologous proteins that have gained additional functions

Gene sharing

A special situation arises when a gene product is used to serve an additional function without losing its original function. This phenomenon, termed **gene sharing**, was discovered in the case of lens crystallins, which are used in the eye lens to maintain transparency (Tomarev & Piatigorsky, 1996). Many types of crystallins were found to be identical with catalytically active enzymes, such as α-enolase (τ-crystallin), glyceraldehyde-3-phosphate dehydrogenase (π-crystallin), lactate dehydrogenase (ε-crystallin), aldehyde dehydrogenase (η-crystallin), argininosuccinate lyase (δ-crystallin), and NADPH:quinone oxidoreductase (ζ-crystallin). Gene sharing is expected to decrease the rate of evolution of 'shared' genes since it places them under more than one selective pressure, i.e. maintenance of activity and suitability as a lens protein. Such gene sharing also requires a change in the tissue specificity of the expression of the shared gene.

5.6 Prediction of the function of orthologous proteins

The term **protein function** is very broad and can be characterized by the molecular function, subcellular localization of the protein, by the partners with which it interacts, the cells and tissue in which it is expressed, by the changes in its expression during development and in response to various external or physiological stimuli, by its involvement in pathways and biological processes etc. Gene ontology provides a hierarchical set of keywords called GO terms which describe different aspects of protein function, ranging from **molecular function** to the **cellular component** in which it is present and the **biological process** in which it is involved (Ashburner et al., 2000). For example, in this system the protein function of chymotrypsinogen is characterized by the molecular function term 'chymotrypsin activity', the biological process term 'proteolysis', and the cellular component term 'extracellular region'. GO terms are widely used for the automatic prediction of the function of proteins and proteome annotation.

Prediction of the function of orthologous proteins is usually based on the assumption that they have similar functions at all levels, therefore the **function-prediction protocols** usually transfer the annotation from a gene/protein where these functions are known to its orthologs. In the absence of such information nonhomology-based methods can be used to predict the function of proteins; these will be discussed in Chapter 7.

5.7 The three-dimensional structure of orthologous proteins

Orthologous proteins have invariably been found to have very similar folded conformations. For example, the three-dimensional structures of horse and human haemoglobins are virtually identical, even though there is 15% difference in their α- and β-chains (see Fig. 5.1). Such conservation of conformation is understandable in that the related proteins serve the same function in the different species, with very similar conformational requirements. The same protein fold can be defined by different amino acid sequences since not all positions are equally important for folding and structural integrity of the protein fold. In general, the sequences defining the main elements of secondary structure show higher conservation, and interior hydrophobic residues differ least frequently. Surface loops that connect β-strands or α-helices show the greatest dissimilarities in amino acid sequences (unless they fulfil some essential functional role). Deletions–insertions occur most frequently at surface loops and reverse turns connecting secondary structural elements, usually with little perturbation of the interior. Of course, in the case of orthologous proteins with the same function, the most highly conserved residues include those involved directly in the functional properties of the protein (see Figs 5.1 and 5.2). In general, the three-dimensional structures of orthologous proteins are usually remarkably conserved during their evolution. Changes in their primary structures accumulate during evolutionary divergence, but only those substitutions that are compatible with the folded conformation and function are retained.

Knowledge of the three-dimensional structures of evolutionarily related proteins may be very useful for evolutionary analyses since structurally equivalent residues are likely to be those that are of common ancestry. As discussed in Chapter 4, alignment of sequences of distantly related proteins is usually uncertain, especially when there have been numerous deletions and insertions. The proper positioning of the gaps may be facilitated by the knowledge of the three-dimensional structure (of any member of the family), since surface loops and reverse turns connecting secondary structural elements are the preferred locations of gaps.

5.7.1 Prediction of secondary structure of proteins

The great variation in amino acid sequences that can occur in essentially identical secondary structural elements indicates that there is significant flexibility in the rules relating primary to secondary to tertiary structures (see

Figs 5.1 and 5.2). In most cases it is not a specific amino acid residue that is conserved in a structurally critical position of a given protein family, but only the general type of side chain (e.g. aromatic, nonpolar, hydrogen bonding or ionic). Nevertheless, based on a survey of a large number of proteins the relative tendencies of the various residues to be involved in α-helices, β-strands and reverse turns (their conformational preferences P_α, P_β and P_t) can be defined and the observed preferences can be exploited for the prediction of secondary structure elements.

Since structural elements are formed as a result of the interaction of several residues, the propensities to form α-helices, β-sheets and reverse turns are determined by a number of adjacent residues. The best-known prediction scheme is that of Chou and Fasman (1978), which predicts an α-helix if four out of six adjacent residues are helix-favouring and if the average value of P_α is greater than 1.0 and greater than P_β; the helix is extended along the sequence until either Pro or a run of four sequential residues with an average value of $P_\alpha < 1.0$ is reached. A β-strand is predicted if three out of five residues are sheet-favouring and if the average value of P_β is greater than 1.0 and greater than P_α; the strand is extended along the sequence until a run of four residues with an average value of $P_\beta < 1.0$ is reached. A reverse turn is predicted if sequences of four residues characteristic of reverse turns are found. The reliability of prediction schemes can be further improved if the residue preferences in different positions of the secondary structure elements are also considered. For example, Asp and Glu residues predominate at the N-terminus, and basic Lys, Arg and His residues predominate at the C-terminus of α-helices, presumably as a result of favourable interactions with the partial charges of the helix dipole. The position factor is of even greater importance in the case of reverse turns, different types of turns following different rules (see section 2.5.3).

Based on such empirical schemes, predictions of α-helical, β-strand or nonregular conformations are usually about 60–70% correct. Current methods are based on similar principles but they use neural networks trained on protein sequences with known structure to predict secondary structure elements of proteins.

The principle that homologous proteins are likely to have homologous secondary structure elements in homologous positions can be exploited to increase the reliability of secondary structure prediction (e.g. Rost & Sander, 1994). If multiple homologous sequences are available, the secondary structures are predicted for each of these sequences and the predictions are averaged. The rationale of such approaches is that if, for example, a β-strand is predicted in the same position for all members of a family of homologous sequences we may be more confident that this prediction is correct than in the case of single-sequence predictions. The use of multiple sequence alignments of homologous sequences thus significantly enhanced the performance of protein secondary structure prediction tools.

5.7.2 Prediction of the three-dimensional structure of proteins

Protein structure prediction aims to determine the three-dimensional structure of proteins from their primary amino acid sequences. There are two major types of approaches of protein structure prediction: *ab initio* modelling and comparative modelling. ***Ab initio* protein-modelling** methods attempt to build three-dimensional protein models based on physical principles without reference to known protein structures. This is still a challenging task since the number of possible structures that a protein may assume is extremely large.

Comparative protein modelling that uses previously solved structures as a starting point is much more effective since the number of tertiary structural topologies appears to limited (cf. Chapter 2), therefore the new protein sequence is likely to adopt one of the known three-dimensional structural folds. Comparative protein modelling methods belong to two major types. **Protein-threading** methods align and assess the compatibility of the sequence of an unknown structure with known structures: these methods can assign the sequence to a known fold-family even in the absence of significant sequence similarity. In these approaches the query sequence (with unknown homology and unknown fold) is aligned in all possible ways on the known three-dimensional structures of proteins. For each alignment the probability is calculated for each of its amino acid residues that it would occur in such an environment in a folded conformation, based on the preferences of the amino acid residues observed in all known protein structures. The value calculated gives the probability that the query sequence would adopt that particular three-dimensional structure.

Homology modelling exploits the principle that homologous proteins have similar three-dimensional structures. Homology modelling is quite accurate for sequences that are closely related to (and can be unambiguously aligned with, cf. Chapter 4) the sequence of a protein whose three-dimensional structure is known. A model of the unknown protein structure can then be constructed using its amino acid sequence and the backbone conformation of the known structure of its homolog. For example, the Swiss-Model Automated Protein Modelling Service (Peitsch, 1995) first searches the Brookhaven Protein Data Bank for proteins that show a significant sequence similarity to the target protein. The framework structure of the model is produced by aligning the target sequence with the selected template sequences using a combination of sequence alignment tools and three-dimensional superposition. Gaps in the framework structure are then filled by structural similarity searches through the Brookhaven Data Bank. Obviously, the accuracy of the model depends on the degree of similarity between the two primary structures, especially the number of insertions and deletions of residues. If the sequences of homologous proteins diverged significantly, the difficulty lies both in the detection of the homology by sequence comparison (see below) and in the reliability of the prediction.

5.8 Detecting sequence homology of protein-coding genes

As discussed above, purifying selection dominates the evolution of orthologous proteins that fulfil similar functions in different species. Since functional constraints result in significant conservation of residues essential for the integrity of the three-dimensional structure and biological activity, sequence similarity may decay at a slow rate. As a consequence, orthologous proteins of species that diverged billions of years ago may retain sufficient sequence similarity to betray their common ancestry.

Detection of homology by detection of sequence similarity may be thus based on the principle that if two sequences are significantly more similar than expected by chance, then the most likely explanation for their similarity is that they arose from a common ancestor. For example, in the case of significant sequence similarities (e.g. >80% amino acid or nucleotide sequence identity) of two protein-coding genes there may be little doubt of common ancestry.

With greater evolutionary divergence the sequence similarity of orthologous protein-coding genes may become almost insignificant (<10% of amino acid sequence identity), even though they still have the same three-dimensional structure and fulfil the same biological function. Since randomly chosen, unrelated protein sequences (given the observed frequencies of the 20 amino acids) would be expected to be identical in 5–6% of their residues, chance similarity may not be readily distinguished from similarity stemming from common ancestry. The situation is even worse when comparing nucleic acid sequences of protein-coding genes: here the expected degree of sequence identity of unrelated sequences (for 50% GC content) is ~25%. A practical consequence of this difference between nucleic acid and protein sequences is that distant homologies are more likely to be detected by comparison of protein sequences.

In all types of homology searches the 'query' sequence is first compared with all other sequences in a database in order to find sequences that are most similar to (give the highest **alignment score** with) the query sequence (see section 4.1). In these comparisons an E-value is assigned to each comparison; the **E value (Expectation value)** gives the number of different alignments with scores equivalent to or better than the given score that are expected to occur in a database search by chance. The lower the E value, the more significant the match and the more confident one can be that the sequence similarity is due to common ancestry. When comparing distantly related protein sequences it is helpful if we use a scoring system that takes into account the different probabilities of amino acid substitutions (substitution matrices; see section 3.3). In other words, the sensitivity of detection of homologies can be significantly increased if we use the mutation probability matrices to score similarity of aligned sequences. The most frequently used scoring matrices are the PAM matrices and the BLOSUM matrices (see section 3.3).

In the case of very distantly related homologs their overall sequence similarity may be so weak that special scoring systems, search protocols and statistical tests may be required to detect the 'signals' of common ancestry. Since this is most common among paralogous proteins with drastically divergent functions, the special techniques used for the detection of distant homologies will be discussed in Chapter 7.

References

Ashburner, M., Ball, C.A., Blake, J.A. et al. (2000) Gene ontology: tool for the unification of biology. The Gene Ontology Consortium. *Nature Genetics* **25**, 25–9.

Blundell, T.L. & Wood, S.P. (1975) Is the evolution of insulin Darwinian or due to selectively neutral mutations? *Nature* **257**, 197–203.

Chothia, C. & Lesk, A.M. (1986) The relation between the divergence of sequence and structure in proteins. *EMBO Journal* **5**, 823–6.

Chou, P.Y. & Fasman, G.D. (1978) Empirical predictions of protein conformation. *Annual Review of Biochemistry* **47**, 251–76.

Hendriks, W., Leunissen, J., Nevo, E. et al. (1987) The lens protein α-crystallin of the blind mole rat, *Spalax ehrenbergi*: evolutionary change and functional constraints. *Proceedings of the National Academy of Sciences of the USA* **84**, 5320–4.

Jolles, J., Jolles, J.P., Bowman, B.H. et al. (1989) Episodic evolution in the stomach lysozymes of ruminants. *Journal of Molecular Evolution* **28**, 528–35.

Kleinschmidt, T., Marz, J., Jurgens, K.D. et al. (1986) Interaction of allosteric effectors with α-globin chains and high altitude respiration of mammals. The primary structure of two tylapoda hemoglobins with high oxygen affinity: vicuna and alpaca. *Biological Chemistry Hoppe-Seyler* **367**, 153–60.

Kornegay, J.R., Schilling, J.W. & Wilson, A.C. (1994) Molecular adaptation of a leaf-eating bird: stomach lysozyme of the hoatzin. *Molecular Biology and Evolution* **11**, 921–8.

Messier, W. & Stewart, C.B. (1997) Episodic adaptive evolution of primate lysozymes. *Nature* **385**, 151–4.

Peitsch, M.C. (1995) Protein modelling by e-mail. *Bio/Technology* **13**, 658–60.

Perutz, M.F. (1983) Species adaptation in a protein molecule. *Molecular Biology and Evolution* **1**, 1–28.

Piccinini, M.T., Kleinschmidt, T., Jurgens, K.D. et al. (1990) Primary structure and oxygen-binding properties of the hemoglobin from Guanaco. *Biological Chemistry Hoppe-Seyler* **371**, 641–8.

Rost, B. & Sander, C. (1994) Combining evolutionary information and neural networks to predict protein secondary structure. *Proteins* **19**, 55–72.

Stewart, C.B. & Wilson, A.C. (1987) Sequence convergence and functional adaptation of stomach lysozymes from foregut fermenters. *Cold Spring Harbor Laboratory. Symposia on Quantitative Biology* **52**, 891–9.

Tomarev, S.I. & Piatigorsky, J. (1996) Lens crystallins of vertebrates. Diversity and recruitment from detoxification enzymes and novel proteins. *European Journal of Biochemistry* **235**, 449–65.

Useful internet resources

Orthology relationship between genes/proteins

InParanoid is a database of pairwise orthologs. Inparanoid is a program that automatically detects orthologs (or groups of orthologs) from two species (http://abi.marseille.inserm.fr/cgi-bin/karine/inparanoid-para).

InParanoid: Eukaryotic Ortholog Groups (http://inparanoid.cgb.ki.se/).

Orthology Search (http://genome.imim.es/~talioto/OrthologySearch/).

Roundup Orthology Database is a large-scale database of orthology covering over 250 publicly available genomes (https://rodeo.med.harvard.edu/tools/roundup/).

Homology-based prediction of the function of proteins

Gene Ontology provides a controlled vocabulary to describe gene and gene product attributes in any organism (http://www.geneontology.org/).

Blast2GO uses BLAST to find homologous sequences to fasta formatted input sequences. The program extracts GO terms to each obtained hit by mapping to existent annotation associations (http://www.blast2go.de/).

GOblet is a platform for Gene Ontology annotation of anonymous sequence data (http://goblet.molgen.mpg.de/).

Gotcha uses sequence similarity searches to associate GO terms with the query sequence (http://www.compbio.dundee.ac.uk/gotcha/gotcha.php).

Prediction of the secondary structural elements of proteins, single-sequence methods

Prediction of β turns

BTPRED (the Beta-TurnPrediction Server) is a method for predicting the location and type of β turns in protein sequences (http://www.biochem.ucl.ac.uk/bsm/btpred/index.html).

Prediction of α-helices and β-strands

The **Chou–Fasman secondary structure prediction method** (http://fasta.bioch.virginia.edu/fasta_www2/fasta_www.cgi?rm=misc1).

The **GOR** (Garnier–Osguthorpe–Robson) secondary structure prediction method (http://fasta.bioch.virginia.edu/fasta_www2/fasta_www.cgi?rm=misc1).

Prediction of the secondary structural elements of proteins, multiple-sequence methods

Jpred is a consensus secondary structure prediction server. It predicts protein secondary structure from multiple-sequence alignments (http://www.compbio.dundee.ac.uk/~www-jpred/).

PSIPRED protein structure prediction server (http://bioinf.cs.ucl.ac.uk/psipred/)

PredictProtein is a service for protein sequence analysis, structure and function prediction, including protein secondary structure prediction (http://www.predictprotein.org).

PROF is a secondary structure prediction system (http://www.aber.ac.uk/%7Ephiwww/prof/).

NNPREDICT is a protein secondary structure prediction system (http://www.cmpharm.ucsf.edu/%7Enomi/nnpredict.html).

Sspro/Scratch is a server for protein secondary structure prediction (http://www.igb.uci.edu/tools/scratch/).

Prediction of the tertiary structure of proteins by comparative modelling

WHAT IF is a versatile molecular modelling package
(www.cmbi.kun.nl/whatif).

SWISS-MODEL is a fully automated protein structure homology modelling server.
SwissModel follows the standard protocol of homolog identification, sequence
alignment, determining the core backbone and modelling loops and side chains
(http://www.expasy.org/swissmod/).

CPHmodels 2.0 is a homology modelling server
(http://www.cbs.dtu.dk/services/CPHmodels/).

3D-JIGSAW is an automated system to build three-dimensional models for proteins
based on homologs of known structure (http://www.bmm.icnet.uk/~3djigsaw/).

Geno3D is a web server for homology or comparative protein structure modelling
(http://geno3d-pbil.ibcp.fr/cgi-bin/geno3d_automat.pl?page=/GENO3D/geno3d_
home.html).

PredictProtein is a service for sequence analysis, structure and function
prediction (http://www.predictprotein.org/).

The **PSIPRED** protein structure prediction server
(http://bioinf.cs.ucl.ac.uk/psipred/psiform.html).

ESyPred3D is an automated homology modelling program
(http://www.fundp.ac.be/urbm/bioinfo/esypred/).

MODELLER is a server for homology or comparative modelling of
protein three-dimensional structures (http://salilab.org/modeller/).

MODBASE is a database of comparative protein structure models
(http://modbase.compbio.ucsf.edu/modbase-cgi-new/search_form.cgi).

Fold recognition, threading

GenTHREADER is a fast fold recognition method of the PSIPRED protein
structure prediction server (http://bioinf.cs.ucl.ac.uk/psipred/index.html).

Fold Recognition is the UCLA-DOE FOLD server (http://www.doe-mbi.
ucla.edu/Services/FOLD/).

3D-PSSM is a web-based method for protein fold recognition using 1D and 3D
sequence profiles coupled with secondary structure and solvation potential
information (http://www.sbg.bio.ic.ac.uk/~3dpssm/index2.html).

The **Meta-predict-protein** server runs a variety of prediction methods, including
secondary structure, transmembrane regions, threading and homology modelling.
(http://www.predictprotein.org/newwebsite/meta/submit3.php).

Protein and nucleic acid sequence databases

GenBank is the NIH genetic sequence database, an annotated collection of all publicly
available DNA sequences. GenBank is the American member of the International
Nucleotide Sequence Database Collaboration DDBJ/EMBL/GenBank together with
DDBJ (Japan) and the EMBL nucleotide sequence database (United Kingdom)
(http://www.ncbi.nlm.nih.gov/Genbank/).

UniProt is a universal protein resource (http://www.expasy.uniprot.org/).

The **Entrez Nucleotides database** is a collection of nucleotide sequences
(http://www.ncbi.nlm.nih.gov/entrez/query.fcgi?db=Nucleotide).

The **Entrez** protein entries have been compiled from SwissProt, PIR, PRF, PDB
and translations from annotated coding regions in GenBank and RefSeq
(http://www.ncbi.nlm.nih.gov/entrez/query.fcgi?db=Protein).

The **Swiss-Prot** protein sequence knowledge base is a universal annotated
   protein sequence database covering proteins from many different species
   (http://www.ebi.ac.uk/swissprot/).

TrEMBL protein sequence database, a supplement to Swiss-Prot, is a universal
   computer-annotated protein sequence database covering proteins from many
   different species (http://www.ebi.ac.uk/swissprot/).

dbEST Expressed Sequence Tags database is a database containing sequence
   data and mapping information on expressed sequence tags (ESTs) from various
   organisms (http://www.ncbi.nlm.nih.gov/dbEST/).

The **EMBL Nucleotide Sequence Database** is the European member of the
   International Nucleotide Sequence Database Collaboration DDBJ/EMBL/
   GenBank together with DDBJ (Japan) and GenBank (United States)
   (http://www.ebi.ac.uk/embl/).

The **DNA Databank of Japan (DDBJ)** is a member of the International
   Nucleotide Sequence Database Collaboration DDBJ/EMBL/GenBank, together
   with GenBank (United States) and the EMBL nucleotide sequence database
   (http://www.cib.nig.ac.jp).

The **Protein Information Resource** (PIR) is a universal annotated protein sequence
   database covering proteins from many different species (http://pir.georgetown.edu/).

Sequence similarity searches

The **Basic Local Alignment Search Tool (BLAST)** finds regions of local similarity
   between sequences. BLAST is a suite of programs facilitating rapid searching of
   nucleic acid and protein databases (http://www.ncbi.nlm.nih.gov/BLAST/).

Fasta Protein provides sequence similarity searching against protein databases using
   the Fasta programs (http://www.ebi.ac.uk/fasta33/).

Fasta Nucleotide provides sequence similarity searching against nucleic acid databases
   using the Fasta programs (http://www.ebi.ac.uk/fasta33/nucleotide.html).

MPsrch is a biological sequence comparison tool that implements the Smith
   and Waterman algorithm. MPsrch is the most sensitive sequence comparison
   method available. The method is only implemented for protein sequences
   (http://www.ebi.ac.uk/MPsrch/).

Chapter 6: Formation of novel protein-coding genes

6.1 *De novo* formation of novel protein-coding genes

The chance of *de novo* formation of protein-coding genes from noncoding DNA appears to be highest for short proteins that lack any stable secondary or tertiary structure (e.g. Levine et al., 2006). *De novo* creation of simple structural elements of proteins (e.g. α-helices, β-sheets, reverse turns) also seems to be rather trivial, primarily since there are so many alternative ways of forming such structures. Obviously, such basic structural units were 'invented' independently several times during evolution. Similarly, although transmembrane segments or signal peptides are similar in chemical composition and physicochemical character they are not all derived from a single ancestor.

For the very same reason, those proteins that consist of just a single secondary structural element had the greatest chance of *de novo* formation. Although a short secondary structure is unstable, it is a relatively simple process to increase the stability by repeating these elements: proteins with repetitive sequences have the greatest chance to arise *de novo*. Moreover, repetitive oligonucleotide sequences may rapidly expand and the resulting repetitive genes may encode periodic amino acid sequences, which in turn are likely to define periodic protein structures (e.g. fibrous proteins like collagens, coiled coil proteins, fibroins and leucine-rich-repeat proteins; see sections 2.7.2 and 2.7.3). It has been frequently suggested (although it is hard to prove) that collagen-like, coiled coil and leucine-rich-repeat (LRR) proteins arose several times during evolution. The increased chance of *de novo* formation of proteins by this mechanism is illustrated by the fact that the first (and so far the only) unambiguous example for convergent *de novo* evolution of similar protein structures involves such a mechanism: the evolution of antifreeze proteins in Arctic and Antarctic fishes.

In the waters of the Antarctic and Arctic, fish live at temperatures as low as −1.9°C. Nevertheless, they do not freeze since they have serum antifreeze glycoproteins (AFGPs) that interfere with the growth of ice crystals and lower the freezing temperature of the fish. It was shown (Chen, DeVries & Cheng, 1997a,b) that fish from the opposite ends of the earth have evolved very similar AFGPs independently. The antifreeze protein of Antarctic fish, made up of a simple sequence of repeating tripeptide units, has evolved from the pancreatic enzyme trypsinogen: it arose by recruitment of the 5′ and 3′ ends of an ancestral trypsinogen gene (which provided the secretory signal and the 3′ untranslated regions) and by *de novo* amplification of a 9bp Thr-Ala-Ala element. Arctic cod also have a similar Thr-Ala-Ala tripeptide repeat-based AFGP, but this has no relationship with the trypsinogen gene. The threonines of the Thr-Ala-Ala repeats of the AFGPs in both fishes are O-linked to galactosyl-*N*-acetylgalactosamine. The regular tripeptide repeating structure ensures the proper positioning of the disaccharides for ice binding: the periodicity in the antifreeze glycoprotein matches the periodicity of water molecules in the ice lattice. In other words, fish in the Arctic and Antarctic adapted to the low temperatures independently, and, through convergent evolution, arrived at the same sequence of antifreeze proteins.

The probability of *de novo* creation of genes encoding more complex, globular proteins is inversely proportional with their complexity. Those that consist of just a single supersecondary structural element (like the TIM barrel proteins) have a higher probability of independent *de novo* creation. Indeed, it seems possible that this structure was invented several times independently (see below).

Although there is evidence that folded proteins occur frequently in random amino acid sequences (Davidson & Sauer, 1994) and functional proteins may be found in random-sequence libraries (Keefe & Szostak, 2001) it seems much easier to remodel replicas of old protein folds than to invent them from scratch. According to this view, creation of the first folded proteins was probably a rate-limiting step in the appearance of protein-based life, and all extant proteins have arisen from a limited number of ancestral protein folds by divergence (see section 2.7.1). Consequently, if there is significant structural similarity between two proteins (significantly greater than expected by chance) then this similarity reflects their common ancestry. The other explanation that significant structural similarities of two complex proteins arose by convergence is much less likely and much more difficult to prove.

It must be emphasized that here we are discussing convergence to similar primary, secondary and tertiary structures (**structural convergence**). Unlike structural convergence, **functional convergence** and **mechanistic convergence** are relatively common. For example, there are several types of proteinases (sulphydryl proteinases, metalloproteinases, aspartyl proteinases, serine proteinases) that have similar function (they cleave proteins) but they have different structures, employ different catalytic mechanisms and have evolved independently. There are also cases of **mechanistic convergence**, when a similar catalytic mechanism evolved independently in different proteins. The textbook examples of mechanistic convergence are the serine proteinases of the subtilisin and trypsin families. These two families of proteinases have similar active sites and catalytic mechanisms but have no sequence or conformational homology. Their difference is underlined by the fact that the equivalent catalytic site residues (His, Asp, Ser) occur in different order in their primary structures.

In contrast with the abundance of examples of functional and mechanistic convergence, there are no clear-cut cases in which similarities in the conformations of complex globular proteins can be shown to have arisen by convergence. The primary problem is that it is very difficult to prove convergence. The folded conformations of proteins derived from a common ancestor may be conserved even when there is practically no detectable similarity of their primary structure. In such cases it may be impossible to decide whether similarity of the protein folds simply reflects the general principles of protein conformation (convergence) or reflects common ancestry despite the extreme sequence divergence. The most plausible candidates for evolutionary convergence are the $(\beta/\alpha)8$ barrel (TIM barrel) proteins: this protein fold has been found in at least 16 proteins with no detectable amino acid sequence homology. However, since most proteins with this structure also have significant functional similarities (they have their catalytic sites in the same position of the barrel), it may be argued that they all diverged from a common ancestor (Farber, 1993).

6.2 Gene duplications

The evolutionary significance of gene duplication is that it gives rise to a redundant duplicate of a gene that may acquire divergent mutations and eventually emerge as a new gene (Ohno, 1972). Studies in the last decades clearly show that gene duplication is the predominant and most important mechanism by which new genes can arise. Genes derived by duplication events within a species are said to be **paralogous** (they are found in different loci of the chromosome), as opposed to orthologous genes derived by speciation events (they are found in corresponding loci of the different species).

An increase in the number of copies of a DNA segment can be brought about by several types of DNA duplications within a species. DNA duplications involved may be classified according to the extent of the genomic region.

1 Partial, intragenic or internal gene duplication. In this case only an internal segment of a protein-coding gene is duplicated.
2 Complete gene duplication. In this case a DNA segment containing an entire protein-coding gene (including those flanking regions that are required for its expression) is duplicated.
3 Partial chromosomal duplication. In this case several adjacent genes of a chromosome are duplicated.
4 Chromosomal duplication (aneuploidy). A complete chromosome is duplicated.
5 Genome duplication (polyploidy). The entire haploid set of chromosomes is duplicated.

DNA duplication has long been known as an important factor in the evolution of genome size. In particular, the duplication of the entire genome or a major part of it may result in a sudden substantial increase in genome size. Genome duplication (polyploidization) events resulting in genome enlargement have been observed in various groups of organisms, primarily in some lineages of plants, bony fishes and amphibians. Of greatest evolutionary significance are those whole genome duplications in the mainstream of the vertebrate lineage that contributed significantly to their increased complexity (see section 9.6.3).

As opposed to intraspecies DNA duplication that leads to the formation of paralogs, **lateral gene transfer** (or **horizontal gene transfer**) between species can lead to the acquisition of novel genes (that have no homologs in the recipient) or to the acquisition of a 'foreign' homolog – a **xenolog** – of existing genes.

There are several diagnostic signs of lateral gene transfer that indicate that its evolutionary history differs from those of vertically transmitted genes (for examples see Chapter 9):

1 Unrecognized **xenology** can cause taxon phylogenies that are in conflict with current evolutionary knowledge. In fact aberrant phylogeny is one of the major manifestations of lateral gene transfer. For example, the acquisition of chloroplasts by a eukaryote was a xenologous (endosymbiotic) event and if we construct a phylogenetic tree for plant chloroplast genes and their nuclear and prokaryotic homologs, the tree will reveal a closer relationship of plants with cyanobacteria than with eukaryotes.

2 Since lateral gene transfer involves the introduction of a gene into a single lineage, the acquired gene (and the associated traits) will be limited to the descendents of the recipient strain and absent from closely related taxa producing a **scattered phylogenetic distribution**.

3 Sequences that were introduced through lateral transfer may retain the **sequence characteristics** of the donor genome and may be distinguished from DNA of the recipient. For example, bacterial species display a wide degree of variation in their overall G+C content, but the genes in a particular species' genome are fairly similar with respect to their base compositions, patterns of codon usage and frequencies of di- and trinucleotides. Consequently, sequences that are 'foreign' to a bacterial genome can be distinguished using these parameters.

6.2.1 Mechanisms of gene duplication

The mechanism that usually underlies very short intragenic duplications is disengagement – for various reasons – of the DNA polymerase from the strand that is being copied and its subsequent reattachment at the wrong point (see section 3.1.2). The major mechanism for duplication of longer segments of genes, entire genes or segments of chromosomes involves **unequal crossing over**. A large number of cases are now known that indicate the frequent involvement of unequal crossing over in these processes, and analysis of these cases has illuminated the factors that facilitate its occurrence. Extensive amplification of protein-coding genes may also occur via multiple replications of a single replicon. This is the major mechanism for somatic **amplification** of genes.

Unequal crossing over involves a **mistaken pairing** and recombination between homologous chromosomes. Recombination in such misaligned regions gives rise to two different recombinants: one in which the region that lies between the given site and the misaligned site is duplicated; another in which the same region is deleted. If the two sites involved in recombination are within the same gene, unequal crossing over gives rise to intragenic duplication; if they are located outside the 5′ and 3′ boundaries of a gene, the duplication of the complete gene will ensue. If the two sites involved in recombination are separated by several genes, duplication of an entire chromosome segment occurs. The chances of misalignment by mistaken pairing are greatly increased if the misaligned site has significant homology with the given site. Since such a homology may be the result of an earlier tandem duplication, once a duplication has occurred (and has been fixed in the population), the chances of further duplications are increased, since there will be more opportunities for mistaken pairing of homologous sequences.

This has two distinct consequences. First, since duplications increase the chances of more unequal crossing-over events (and more duplications/deletions), this implies a saltatory effect, which could permit rapid expansion of repeats within genes (see examples in Chapters 3 and 8) as well as expansion of gene families. Since each unequal crossing over results in asymmetrical products (deletion or duplication of the region involved), repeated regions undergo both shrinkage and expansion.

Second, the frequent misalignment and unequal crossing over in homologous regions may facilitate the 'homogenization' of their sequences and thus may slow down their divergence (see 'Concerted evolution'; section 6.2.2). A duplicated gene has an increased chance of starting its independent career if it is **translocated** to an unrelated chromosome where the chances of misalignment with its relatives are diminished.

The types of problems that arise from frequent misalignment of tandem duplicated genes (duplicons) may be illustrated by the human visual pigment genes. As will be discussed below, the human X chromosome has two closely linked, closely related genes for colour pigment, which are sensitive to red and green light, respectively. Due to their high similarity, the duplicated regions are frequently involved in unequal crossing over, resulting in further gene duplications/deletions (if the recombination point is outside the boundaries of the misaligned genes) or the formation of chimeric genes (if the recombination point is within the misaligned gene). As a result of unequal crossing over some individuals lack the green gene (deletion), other individuals may have two or three tandemly arrayed green genes (duplications). There are forms of colour blindness where the affected individuals have an altered pigment with a shifted absorption spectrum since they contain a colour pigment gene that is a chimera of the green and red pigment genes.

Another example is provided by the hybrid genes that may form within the β-globin gene cluster. As mentioned in Chapter 3 (see p. 49), in the case of the variant human haemoglobin Lepore and anti-Lepore genes, hybrid haemoglobin chains are formed through recombination of the β-chain and δ-chain genes.

The chances of misalignment by mistaken pairing may be increased even if the two sites share just a segment that is homologous. For example, partial homology between otherwise unrelated DNA regions may exist if there are related middle repetitious sequence elements at the two sites. There is now convincing evidence that such elements greatly facilitate unequal crossing over. The role of such elements in intronic recombination, intragenic duplications and exon shuffling will be discussed in Chapter 8. Here we will give just one example to illustrate how repetitive elements facilitate gene duplications.

The bacteriolytic activity of lysozyme, which is usually exploited by nonspecific host defence mechanisms (and is usually expressed in myeloid cells), has been recruited to a digestive function not only in ruminants but also in the mouse. However, in contrast to the case of ruminants, where the need for high levels of lysozyme in the digestive tract was accompanied by an increase in the number of lysozyme genes to ≈10 and a decrease in the levels of enzyme in extraintestinal tissues (Irwin & Wilson, 1989), in the mouse high expression in the small intestine has been achieved without loss of the extraintestinal enzyme. The predominant form of lysozyme mRNA in the mouse small intestine is encoded by the lysozyme P gene, while the lysozyme M gene accounts for the vast majority of lysozyme mRNA detectable in other tissues. Cross and Renkawitz (1990) have shown that the mouse M and P lysozymes are the products of two closely linked genes that arose by gene duplication about 30–50 million years ago, and have defined precisely the point of recombination. They have shown that the original gene duplication involved homologous recombination between Alu-like B2 middle repetitive elements that flanked the ancestral gene (Fig. 6.1). The resulting downstream copy of the gene has retained the myeloid specificity of expression, while the upstream copy is expressed in the small intestine.

Gene duplication by whole-genome duplication is a consequence of the lack of disjunction of daughter chromosomes following DNA replication (this will be discussed in Chapter 9).

Retrosequences

Duplicates of protein-coding genes may also be produced by **duplicative transposition**, a process in which a copy of genetic material is moved from one chromosomal location to another (one copy of the genetic information remains at the original site). The information of RNA-coding or protein-coding genes may be transposed through RNA. In this case the DNA is first

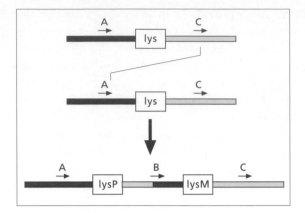

Fig. 6.1 Lysozyme gene duplication by unequal homologous crossing over between the B2 repeats A and C. One of the product chromosomes carries two copies of the gene separated by a hybrid B2 repeat (B). (From Cross, M. & Renkawitz, R. (1990) Repetitive sequence involvement in the duplication and divergence of mouse lysozyme genes. *EMBO Journal* **9**, 1283–8.) EMBO Journal by Cross, M. & Renkawitz, R. Copyright 2007 by Nature Publishing Group (Permissions). Reproduced with permission of Nature Publishing Group (Permissions) in the format Textbook via Copyright Clearance Center.

transcribed into RNA, which is then **reverse transcribed** into cDNA, therefore this mode of transposition is usually referred to as **retroposition**. When a reverse transcribed transposable element (cDNA) is inserted into the genome, a small segment of the host DNA at the insertion site (usually 4–12 bp) is duplicated. Since these duplicated repeats are in the same orientation they are called **direct repeats**. The presence of such direct repeats is diagnostic of insertion by transposition.

The genomic sequences that have been derived through the reverse transcription of an RNA transcript of a gene and the subsequent integration of the cDNA into the genome are usually called **retrosequences**. Since they originated from RNA transcripts, they show the diagnostic signs of RNA processing, therefore they are also referred to as **processed sequences**.

The most significant diagnostic features of retrosequences include:

1 they lack introns (where the 'parent' gene would have introns);
2 the boundaries of the retrotranscript coincide with or are within the transcribed regions of genes (lack upstream promoter regions of their parent);
3 they contain stretches of poly(A) at the 3' end;
4 there are short direct repeats at both ends;
5 their chromosomal position is not closely linked to the locus of the original gene from which the RNA was transcribed, indicating that they were produced by retroposition.

Depending on whether the copy is functional or not, we may distinguish **processed genes (retrogenes)** and **processed pseudogenes (retropseudogenes)**.

A processed protein-coding gene is a functional retrosequence producing a protein that is related to that produced by the gene from which the retrogene has been derived. There are several reasons why it is unlikely that a reverse-transcribed gene will be functional. First, the process of reverse transcription is much more inaccurate than DNA replication, and many differences (mutations) between the RNA template and the cDNA can occur. Second, since it contains only regions that are normally transcribed it usually does not contain the necessary regulatory sequences that reside in the untranscribed promoter regions, hence it is unlikely to be properly transcribed. Third, many processed pseudogenes are truncated at their 5′ ends. A major source of 5′ truncation is that reverse transcriptase fails to complete the reverse transcription to the 3′ end of the RNA molecule (i.e. the 5′ end of the cDNA). Finally, a processed gene may be inserted at a genomic location that may not be adequate for its expression. As a result of all these adverse effects, in the vast majority of cases a processed gene is already nonfunctional at the time of its formation (i.e. it is a processed pseudogene). These factors, combined with the fact that duplicate copies of genes are more likely to be inactivated than retained, make it very unlikely that processed genes contribute to the formation of novel genes. As soon as a processed pseudogene is established within the genome it is nonfunctional and free from all selective constraints. Because of the lack of function, pseudogenes accumulate point mutations and deletions very rapidly, and these mutations will eventually obliterate the sequence similarity between the pseudogene and its functional progenitor.

Nevertheless, some processed functional genes have been found, although they seem to be very rare (Vinckenbosch, Dupanloup & Kaessmann, 2006). For example, the human phosphoglycerate kinase (PGK) multigene family consists of an active X-linked gene, a processed X-linked pseudogene and an additional autosomal gene. The X-linked gene contains 11 exons and 10 introns; its autosomal homolog, on the other hand, is unusual in that it has no introns and is flanked on its 3′ end by remnants of a poly(A) tail, strongly suggesting a reverse transcription process involving mRNA. Interestingly, the autosomal PGK gene is expressed almost exclusively in the testes. Thus, the reverse-transcribed PGK gene has not only maintained an intact reading frame and the ability to transcribe and produce a functional polypeptide, but also has acquired a novel tissue specificity. The possible reasons for the survival of such processed genes are discussed in Chapter 7.

The rat and mouse preproinsulin I gene represents a rare instance of a **semiprocessed retrogene**. This gene contains a single intron in the 5′ untranslated region, whereas its homolog, preproinsulin II, contains the same intron plus an additional intron within the coding region. This seems to be the 'original' exon–intron organization since all preproinsulin genes from other mammals, including other rodents, contain these two introns. The preproinsulin I gene is flanked by short repeats, and the polyadenylation signal is followed by a short poly(A) tract, suggesting that it arose by retroposition of a partially processed preproinsulin II pre-mRNA. Preproinsulin I appears to have been derived from an aberrant pre-mRNA transcript that initiated 500 base pairs upstream of the normal cap site. Because the aberrant transcript contained 5′ regulatory sequences not normally transcribed this retrogene inherited the capability of being transcribed following its integration into a new genomic location.

In the case of some highly expressed RNA genes the number of processed pseudogenes can exceed the number of their functional counterparts by orders of magnitude. The best-known example is the Alu family of processed pseudogenes (the name comes from the fact that the sequence contains a characteristic restriction site for the *Alu*1 endonuclease). Alu sequences are processed pseudogenes of the RNA gene specifying 7SL RNA, a constituent of the signal recognition particle that is essential in the cutting of signal sequences of secreted proteins. Alu sequences are about 300 bp long and they belong to a family of repeated sequences that appear more than 1,300,000 times in the human genome, constituting about 10% of the genome. The human Alu sequences have been derived from a functional 7SL sequence by a series of steps, involving a duplication, two deletions and many nucleotide substitutions. Most human Alu sequences have a dimeric structure, whereas the rodent Alu equivalent, the B1 family, is almost exclusively monomeric. Due to the ubiquity of reverse transcription, the genomes of mammals are bombarded with copies of reverse-transcribed sequences such as the Alu repeats. The generation of such repeated sequences in huge copy numbers has a great impact on genomic plasticity and such highly repetitive elements also play a significant role in protein evolution since they facilitate misalignment and unequal crossing over, therefore they facilitate gene duplication (see Fig. 6.1) and exon shuffling (see section 8.1).

In the case of protein-coding genes, processed pseudogenes are preferentially produced from genes of housekeeping enzymes that are highly expressed in germline cells (Piechaczyk et al., 1984).

Lateral gene transfer

In lateral gene transfer genetic material from one species is transferred to another species. As will be discussed in Chapter 9, lateral gene transfer has played a major role in the evolution of Eubacteria and Archaea, but it also occurs in eukaryotes (Allers & Mevarech, 2005; Andersson, 2005; Ochman, Lawrence & Groisman, 2000).

There are three major mechanisms for lateral gene transfer into bacteria (Chen, Christie & Dubnau, 2005). **Transformation** occurs through the uptake of naked DNA; since this mechanism is independent from the source of DNA it can mediate transfer of DNA from very distantly related organisms (including higher eukaryotes). The second major mechanism is **conjugation** that involves physical contact between donor and recipient cells. New genetic material can also be introduced into a bacterium by **generalized transduction**, i.e. by bacteriophages that have packaged random DNA fragments of the donor microorganism in which they have replicated.

In the case of eukaryotes the above mechanisms are less widespread. The fact that lateral gene transfer is most frequent in phagotrophic single-cell eukaryotes suggests that **ingestion of foreign DNA** is the main mechanism for acquisition of foreign DNA.

Endosymbiosis may be considered as a special case of a lateral gene transfer event; in this case entire genomes (mitochondrial genome, chloroplast genome, plastid genome, nuclear genome) were transferred from one group of organisms to another through the enslavement of these organisms (see Chapter 9).

6.2.2 Fate of duplicated genes

All types of gene duplication, apart from partial gene duplications, produce two identical copies of the 'original' gene(s). The fate of duplicated genes is determined by the functional consequences of having extra copies of the same gene and increased amounts of the same protein.

Gene duplications, just like any other mutation event, can be advantageous, deleterious or neutral. For example, if an organism is exposed to a toxic environment, there may be selective advantage in overproducing those proteins that decrease toxicity. Similarly, the need for a greater supply of lysozyme in foregut fermenters may have favoured the duplication of lysozyme genes (Irwin & Wilson, 1989). Disadvantage will result if increased amounts of the gene product upset some fine regulatory balance and cause confusion within the cell. Initially, the majority of gene duplications are likely to be neutral; their fate is determined by natural selection and random genetic drift in much the same way as discussed for point mutations (see section 3.3).

Studies of the genetic basis of insecticide resistance suggest that spontaneous gene duplications occur at a high rate in insects (Devonshire & Field, 1991). Similarly, studies on increased resistance of mammalian cells and untreated human tumours to cytotoxic drugs indicate that tandem duplications occur frequently in mammals (Brodeur et al., 1984; Schimke, 1986). However, in the absence of such positive selection for advantageous gene duplications, the duplicated gene is unlikely to be fixed unless it acquires a novel and useful function.

Accordingly, one of the most typical routes of duplicated genes is that one of the copies is inactivated by deleterious mutations and becomes a pseudogene. This is the most likely fate for deleterious duplications and, to a lesser extent, for neutral duplications.

In the case of advantageous duplications, the duplicons may retain their original function, enabling the organism to produce a larger quantity of the encoded proteins.

The greatest evolutionary significance of gene duplications lies in the fact that they may lead to the emergence of genes with novel or modified functions. This may happen if one of the duplicated genes retains its original function while the other accumulates molecular changes that adapt it to perform a task different to that of the ancestral gene. Such a change in function – **neofunctionalization** – can be achieved through changes in amino acid sequence (e.g. leading to the development of novel type of activity) or through changes in the gene's expression pattern (e.g. expression of a gene in a tissue where the ancestral gene was not previously expressed). The mode of gene duplication can lead to spontaneous neofunctionalization through incomplete duplication of the regulatory region of a gene or when a retrogene is transposed in a novel chromosomal environment. As an example of neofunctionalization we may refer to the lysozyme P gene of mouse that is expressed in the intestine and fulfils a novel, digestive function, whereas the lysozyme M gene is expressed in myeloid tissues and fulfils a role in defence against bacterial infections (cf. section 6.2.1).

In many cases both duplicates acquire functions that differ from that of their common ancestor inasmuch as they specialize in different subfunctions of their ancestor. In the **subfunctionalization** pathway the two daughter genes accumulate changes resulting in division of the ancestral function, e.g. division of the ancestral expression pattern or division of the ancestral spectrum of molecular functions (cf. the genes for trypsin, chymotrypsin, elastase and visual pigments; see section 7.2.1). In complex organisms sub-functionalization can perhaps most readily be detected if it causes changes in the **expression profiles** of genes. In the long run, this type of subfunc-tionalization should lead to a decrease in expression breadth and the develop-ment of tissue-specific genes, as has been demonstrated in some previous studies on individual genes where subfunctionalization has occurred (Prince & Pickett, 2002).

Sometimes duplicated genes acquire drastically novel functions that show no apparent similarity to the function of the common ancestor (e.g. hepatocyte growth factor vs. plasminogen; see section 7.2.2).

Nonfunctionalization of duplicated genes

Even if the gene duplication *per se* is selectively neutral, the redundant copy of a gene is more likely to become nonfunctional than to evolve into a new gene, simply because deleterious mutations occur far more frequently than advantageous ones. Substitutions that inactivate one copy of the duplic-ated gene are selectively neutral, since the other copy is there to perform the original function. Following the first mutation that is the direct cause of gene inactivation, the inactivated duplicon, a **pseudogene**, accumulates multiple defects, frameshifts, stop codons, etc. since it is not protected by natural selection.

Pseudogenes produced from duplicons created by unequal crossing over or amplification resemble the original gene in retaining their introns, therefore they are usually called **unprocessed pseudogenes** to distinguish them from the processed pseudogenes (see above).

Formation of gene families

In many cases repeated gene duplications give rise to large families of repeated genes. The paralogous genes in a genome that have descended from a common ancestral gene by gene duplications are usually referred to as a gene family or multigene family. Closely related members of a gene family frequently reside in close proximity on the same chromosome, reflecting the fact that they arose by tandem gene duplications relatively recently. More distantly related members may be located on other chromosomes as a result of chromosome or whole-genome duplications (see section 9.4) or as a result of genomic rearrangements that occurred following tandem duplication.

There are cases when different members of a multigene family still fulfil the same biological function; the genes are practically invariant, and their sequences are identical or nearly identical. Such **invariant repeated genes** are common among protein-coding genes whose product is required in large quantities for the normal function of the organism. The best-known examples are the genes for histones, which constitute the chief protein component of chromosomes, and therefore must be synthesized at a high rate during a well-defined short period of cell division and DNA replication. The highly repetitive histone genes are very similar to each other. Purifying selection (reflecting the high functional constraints on these vital proteins) is not the only factor responsible for the homogeneity of their genes since their conservation often extends to regions devoid of any functional significance. The maintenance of homogeneity is due to mechanisms that homogenize their sequences within the same species (see below).

In other cases different members of a multigene family fulfil essentially the same molecular function but differ from each other in tissue specificity, developmental regulation and in some of their biochemical properties. The duplicated genes coding for different **isozymes** of lactate dehydrogenase, aldolase, creatine kinase, pyruvate kinase, etc. catalyse the same biochemical reaction but are expressed in different tissues at different developmental stages, and differ in some of their catalytic properties. Another instructive example of gene duplication giving rise to proteins with slightly different functions is the case of visual pigment genes of Old World primates: they encode pigment proteins that are sensitive to blue, green or red light.

Frequently, members of a gene family perform more drastically different functions. For example, both trypsin – a proteinase that cleaves proteins in the digestive tract – and haptoglobin – a globin-binding protein that lacks proteolytic activity – belong to the gene family of serine proteinases.

Similarly, lysozymes, which dissolve bacteria by cleaving the polysaccharide component of their cell walls, and lactalbumin, a carbohydrate-binding protein devoid of catalytic activity, arose by duplication of a common ancestral gene.

Concerted evolution in multigene families

It has been frequently observed that paralogous members of some multigene families are very similar to each other within one species, although orthologous members of the same family from even closely related species may differ greatly from each other. In other words, the genes in each species evolve together, although they diverge between species.

Such observations emphasize that the within-species conservation cannot be explained by functional constraints and suggest that a 'horizontal' correction mechanism operates within the species. As a consequence of this correction mechanism, the family evolves together as a unit, in a concerted fashion. Unequal crossing over and gene conversion are the two most important mechanisms responsible for the **concerted evolution** of multigene families, ensuring horizontal transfer of mutations among the members of the family. The term **molecular drive** has been proposed for the process of concerted evolution of multigene families under the effect of random genetic drift.

The existence of concerted evolution highlights the importance of interactions between duplicated genes – their evolution may not be independent. As long as the two copies of the genes are closely linked, and as long as their sequences are significantly similar, the two copies may evolve in a concerted fashion. In this early period of the existence of duplicated genes their divergence is slowed down, reducing their chances for the acquisition of new functions but also protecting redundant copies from rapid nonfunctionalization.

The existence of correction mechanisms also has important implications for the rate of evolution of duplicated genes. Assuming that the rate of mutations of duplicated genes is the same and that the functional constraints are constant for the duplicons the sequences of duplicated genes would be expected to diverge as a linear function of time. However, as a consequence of concerted evolution (correction of one duplicon by the other) their divergence may not be a monotonic function of time: duplicates may become more similar than they were before. Furthermore, since the efficiency of the correction mechanism decreases with the divergence of the duplicated genes, divergence may accelerate as the correction mechanism becomes less and less efficient. Such complications may cause significant confusions in the estimation of the time of the gene duplication event. In large multigene families where gene correction events occur frequently it may be very difficult to trace the true evolutionary relationships among family members.

6.2.3 Fate of genes acquired by lateral gene transfer

A special aspect of lateral transfer of genetic material is that the codon usage (and sometimes even the genetic code) of the species involved may not be compatible and this incompatibility may be an important barrier to the fixation of the transferred gene(s) (Medrano-Soto et al., 2004). Another important aspect is that the transferred genetic material may contain just one of several genes defining a complex **functional module** (e.g. pathway etc.) that is absent from the recipient organism. In this case, only the transfer of all the genes of the functional module would be meaningful. Accordingly, successful transfer of complex functional traits would require the physical clustering of the corresponding genes so that all necessary genes could be transferred in a single step. It is therefore expected that lateral gene transfer would select for gene clusters and operons that can be expressed in the recipient cells.

There are numerous examples that illustrate that these difficulties can be overcome provided that acquisition of the new gene(s) confers significant selective advantage on the recipient organism. Lateral gene transfer has been shown to play a crucial role in the spread of **antibiotic resistance** since the acquisition of antibiotic resistance genes allows a microorganism to proliferate in the presence of these compounds. Lateral gene transfer has also been shown to be involved in the adoption of a pathogenic lifestyle in different groups of bacteria: **bacterial virulence** results from the acquisition of genes that are absent from avirulent forms. Recent studies have discovered that horizontally acquired '**pathogenicity islands**' are major contributors to the virulent nature of many pathogenic bacteria. These chromosomally encoded pathogenicity islands contain large clusters of virulence genes and can transform a benign organism into a pathogen. Lateral gene transfer has also played a significant role in the acquisition of genes that allowed recipient organisms to explore new environments. Whole-genome comparisons have identified sets of genes that are restricted to organisms that have independently adapted to a common lifestyle, such as Archaeal and Bacterial hyperthermophiles or the intracellular pathogens *Rickettsia* and *Chlamydia*. The presence of such genes in phylogenetically divergent but ecologically similar microorganisms supports the notion that they were acquired by lateral gene transfer.

6.2.4 Dating gene duplications

Assuming that duplicated genes diverge at a constant rate, we can estimate the date of a gene duplication, T_D, that gave rise to two paralogous genes (A and B) if we have sequences of these paralogs from two different species (species 1 and 2) and we know the time of speciation, T_S, that led to the divergence of these species. (Of course, if both species 1 and 2 have orthologs of A and B, then the gene duplication preceded speciation of 1 and 2, i.e. $T_D > T_S$.)

If the genes evolved at a constant rate then the average number of substitutions per site ($[K_A K_B]/2$) in the two ortholog comparisons (A1 vs. A2, B1 vs. B2) is proportional to the species divergence time, T_S, and the average number of substitutions per site, K_{AB}, in the four paralogous comparisons (A1 vs. B1, A2 vs. B2, A1 vs. B2 and A2 vs. B1) is proportional to the time, T_D, that passed since the gene duplication. In this case the following equation holds:

$$T_D/T_S = 2K_{AB}/[K_A + K_B]$$

The above calculation assumes rate constancy in all lineages, i.e. that both paralogs evolve at the same rate in both species, an assumption that is rarely valid (e.g. accelerated evolution during adaptive radiation). The validity of this assumption can be tested by checking whether an approximate equality holds among the appropriate pairwise comparisons (see section 4.4.3). As mentioned above, problems due to concerted evolutionary events may lead to serious underestimation of T_D.

An independent way of checking the validity of estimates of T_D values is by considering whether or not the phylogenetic distribution of duplicated genes is consistent with or contradicts the known divergence dates of species (which may be known from palaeontological data). For example, the fact that only one X-linked colour pigment gene is present in New World monkeys but there are two closely related X-linked genes in Old World primates is consistent with a recent gene duplication that occurred in the Old World primate lineage. Similarly, the fact that apolipoprotein(a) is present only in Old World primates, but is absent from all other species (including New World monkeys) is consistent with estimates dating the gene duplication event that gave rise to apolipoprotein(a) to about 40 million years ago (McLean et al., 1987; Lawn, Schwarz & Patthy, 1997).

As another example, we may refer to the globin family. All vertebrates have myoglobin in their muscle and haemoglobin in their blood. Myoglobin differs from both the α and β subunits of haemoglobin more than they differ from each other, indicating that myoglobin diverged ($T_D \approx 600-800$ million years ago) before the α and β genes arose ($T_D \approx 500$ million years ago). Mammals, reptiles, birds, amphibians and bony fish all have distinct α and β subunits, whereas the most primitive vertebrates, the Agnatha (jawless fish), contain only one type of haemoglobin subunit. This pattern of distribution is consistent with the estimated dates of the gene duplications: myoglobin and haemoglobin diverged prior to the separation of agnathans and jawed vertebrates, whereas the duplication giving rise to α and β genes occurred in the ancestor of all jawed vertebrates following its divergence from the agnathans.

The evolutionary history and linkage pattern of the α- and β-globin gene clusters of humans are also instructive. In humans, the gene cluster of the α-globin family (present on chromosome 16) consists of four functional genes (ζ, α_1, α_2, θ_1) and three unprocessed pseudogenes (ψζ, ψα_1, ψα_2). The gene cluster of the β-globin family (present on chromosome 11) contains five functional genes (ϵ, G_γ, A_γ, β and δ) and one unprocessed pseudogene ($\psi\beta$). (ψ is used to distinguish pseudogenes from functional genes.) Within the α-globin family, the embryonic type ζ is the most divergent, with an estimated T_D of more than 300 million years ago. Somewhat less divergent is θ_1, with an estimated T_D of about 260 million years ago. The genes α_1 and α_2 produce an identical polypeptide (and have almost identical DNA sequences) suggesting a recent divergence time.

Within the β-globin family there are two subfamilies. The adult types (β and δ) and the nonadult types (ϵ and γ) diverged about 155–200 million years ago. The ancestor of the two γ genes diverged from the ϵ gene about 100–140 million years ago. Based on the distribution of the γ genes, it is clear that the duplication responsible for creating G_γ and A_γ occurred following the separation of the human lineage from the New World monkey lineage, about 35 million years ago. In the lineage of adult type β-chains, the divergence between the δ and β genes occurred about 80 million years ago.

It is noteworthy that the α and β clusters ($T_D \approx 500$ million years ago) are on separate chromosomes and that within both α- and β-globin gene clusters the arrangement of the genes correlates with the time of divergence. This is a clear reflection of the mechanism whereby they arose: tandem duplication by unequal crossing over. In general, within gene clusters there seems to be a correlation between the time of duplication and the distance between duplicated genes. In fact, molecular clocks may be based on the expansion of gene families (Trusov & Dear, 1996) since there seems to be a linear relationship between intergene distance and time of divergence (Fig. 6.2).

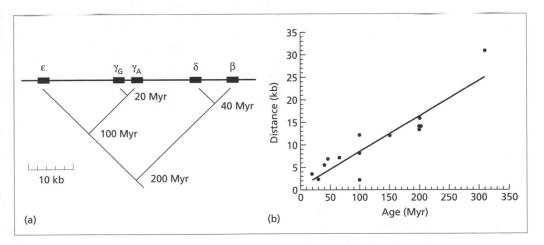

(a)   (b)

Fig. 6.2 (a) The human β-like globin gene cluster. The physical map is aligned with the evolutionary tree of this gene family. The approximate ages of the duplication events are indicated. (b) Relationship between intergene distance and age since duplication for the globin, haptoglobin and kallikrein gene clusters. (From Trusov, Y.A. & Dear, P.H. (1996) A molecular clock based on the expansion of gene families. *Nucleic Acids Research* **24**, 995–9, reprinted by permission of Oxford University Press.)

The globin superfamily may also serve to illustrate the possible evolutionary routes that are open for duplicated genes: some duplicates have become nonfunctional ($\psi\zeta$, $\psi\alpha_1$, $\psi\alpha_2$, $\psi\beta$), some have retained the same function, and others acquired new functional, regulatory and expression characteristics (e.g. myoglobin vs. haemoglobins, embryonic vs. foetal vs. adult forms). Myoglobin has adapted to be the oxygen-storage protein of muscle, whereas haemoglobin has become the oxygen carrier in blood. As dictated by its specific roles, myoglobin evolved a higher affinity for oxygen than haemoglobin, while the functions of the various haemoglobins have become much more refined according to their special needs. The embryonic, foetal and adult haemoglobins fulfil slightly different functional roles, therefore differences in oxygen-binding affinity have evolved among these globins. The foetal haemoglobin has a higher oxygen affinity than adult haemoglobins and can thus function more efficiently in the low-oxygen environment of the foetus.

References

Allers, T. & Mevarech, M. (2005) Archaeal genetics – the third way. *Nature Reviews in Genetics* **6**, 58–73.

Andersson, J.O. (2005) Lateral gene transfer in eukaryotes. *Cellular and Molecular Life Sciences* **62**, 1182–97.

Brodeur, G.M., Seeger, R.C., Schwab, M. et al. (1984) Amplification of N-*myc* in untreated human neuroblastomas correlates with advanced disease stage. *Science* **224**, 1121–4.

Chen, L., DeVries, A.L. & Cheng, C.H.C. (1997a) Evolution of antifreeze glycoprotein gene from a trypsinogen gene in Antarctic notothenioid fish. *Proceedings of the National Academy of Sciences of the USA* **94**, 3811–16.

Chen, L., DeVries, A.L. & Cheng, C.H.C. (1997b) Convergent evolution of antifreeze glycoproteins in Antarctic notothenioid fish and Arctic cod. *Proceedings of the National Academy of Sciences of the USA* **94**, 3817–22.

Chen, I., Christie, P.J. & Dubnau, D. (2005) The ins and outs of DNA transfer in bacteria. *Science* **310**, 1456–60.

Cross, M. & Renkawitz, R. (1990) Repetitive sequence involvement in the duplication and divergence of mouse lysozyme genes. *EMBO Journal* **9**, 1283–8.

Davidson, A.R. & Sauer, R.T. (1994) Folded proteins occur frequently in libraries of random amino acid sequences. *Proceedings of the National Academy of Sciences of the USA* **91**, 2146–50.

Devonshire, A.L. & Field, L.M. (1991) Gene amplification and insecticide resistance. *Annual Review of Entomology* **36**, 1–23.

Farber, G.K. (1993) An β/α-barrel full of evolutionary trouble. *Current Opinion in Structural Biology* **3**, 409–12.

Irwin, D. & Wilson, A.C. (1989) Multiple cDNA sequences and the evolution of bovine stomach lysozyme. *Journal of Biological Chemistry* **264**, 11387–93.

Keefe, A.D. & Szostak, J.W. (2001) Functional proteins from a random-sequence library. *Nature* **410**, 715–18.

Lawn, R.M., Schwarz, K. & Patthy, L. (1997) Convergent evolution of apolipoprotein(a) in primates and hedgehog. *Proceedings of the National Academy of Sciences of the USA* **94**, 11992–7.

Levine, M.T., Jones, C.D., Kern, A.D., Lindfors, H.A. & Begun, D.J. (2006) Novel genes derived from noncoding DNA in *Drosophila melanogaster* are frequently X-linked and exhibit testis-biased expression. *Proceedings of the National Academy of Sciences of the USA* **103**, 9935–9.

McLean, J.W., Tomlinson, J.E., Kuang, W.J. et al. (1987) cDNA sequence of human apolipoprotein(a) is homologous to plasminogen. *Nature* **300**, 132–7.

Medrano-Soto, A., Moreno-Hagelsieb, G., Vinuesa, P., Christen, J.A. & Collado-Vides, J. (2004) Successful lateral transfer requires codon usage compatibility between foreign genes and recipient genomes. *Molecular Biology and Evolution* **21**, 1884–94.

Ochman, H., Lawrence, J.G. & Groisman, E.A. (2000) Lateral gene transfer and the nature of bacterial innovation. *Nature* **405**, 299–304.

Ohno, S. (1972) *Evolution by Gene Duplication*. New York: Springer.

Prince, V.E. & Pickett, F.B. (2002) Splitting pairs: the diverging fates of duplicated genes. *Natural Reviews in Genetics* **3**, 827–37.

Picchaczyk, M., Blanchard, J.M., Sabouty, S.R.E. et al. (1984) Unusual abundance of glyceraldehyde 3-phosphate dehydrogenase pseudogenes in vertebrate genomes. *Nature* **312**, 469–71.

Schimke, R.T. (1986) Methotrexate resistance and gene amplification. Mechanisms and implications. *Cancer* **57**, 1912–17.

Trusov, Y.A. & Dear, P.H. (1996) A molecular clock based on the expansion of gene families. *Nucleic Acids Research* **24**, 995–9.

Vinckenbosch, N., Dupanloup, I. & Kaessmann, H. (2006) Evolutionary fate of retroposed gene copies in the human genome. *Proceedings of the National Academy of Sciences of the USA* **103**, 3220–5.

Useful internet resources

Paralogy
Paradb is a database created to predict and map paralogous regions in vertebrate genomes (http://abi.marseille.inserm.fr/paradb/).

Lateral gene transfer
Horizontal Gene Transfer Database (HGT-DB) is a genomic database that includes statistical parameters such as G+C content, codon and amino acid usage, as well as information about which genes deviate in these parameters for prokaryotic complete genomes (http://www.tinet.org/~debb/HGT/).

Pseudogene
Hoppsigen is a nucleic acid database of homologous processed pseudogenes (http://pbil.univ-lyon1.fr/databases/hoppsigen.html).

The **Pseudogene.org** site contains a comprehensive database of identified pseudogenes, utilities used to find pseudogenes, various publication data sets and a pseudogene knowledge base (http://www.pseudogene.org/).

The **UI Pseudogenes** website serves as a repository for all pseudogenes in the human genome (http://genome.uiowa.edu/pseudogenes/).

LINEs
L1Resources provides detection and annotation of full-length, intact L1 elements (http://line1.molgen.mpg.de/).

Chapter 7: Evolution of paralogous proteins

As mentioned in the previous chapter, the fate of duplicated genes (and xenologous genes) is determined by whether the duplication is advantageous, deleterious or neutral.

Duplication of a complete protein-coding gene (by unequal crossing over or amplification, or by chromosome or genome duplication) results in two identical duplicons that encode the same protein as the original gene and express it in the same way as the original gene (if the promoter regions are also duplicated). Accordingly, the duplicated genes may produce twice as much of the protein as previously. The duplication will be advantageous if an increased supply of the gene product is advantageous (e.g. in the case of histones and ribosomal proteins).

Duplication of a protein-coding gene by retroposition results in a processed duplicon that encodes the same protein as the original gene but whose expression characteristics are likely to be different from that of the progenitor gene, since it is placed in a novel chromosome environment and usually lacks the original promoter region. Consequently, the retrogene is likely to have unique expression characteristics (tissue specificity, developmental stage, etc.). The duplication will be advantageous if such a change in expression pattern has biological advantage (e.g. testis-specific retrogenes; section 7.1.2).

Since the majority of gene duplications are selectively neutral and since for a neutral allele the fixation probability equals its frequency in the population, gene duplications have a significant fixation probability even if they are selectively neutral. This may be illustrated by the general observation that functional redundancy among genes is frequently observed in a wide range of processes, where the selective value of such redundancy is not immediately obvious. Although **fully redundant functions** (neutral duplications) are expected to be evolutionarily unstable and may be lost by mutational drift they can survive long enough to have a chance to acquire a useful function that would make them advantageous.

In this chapter we will survey some selectively advantageous duplications, but the major emphasis will be on how mutations of duplicons created by selectively neutral gene duplications can endow them with advantageous novel functions. Mutation here is used in its broadest sense, including not just point mutations and short insertions/deletions but also insertion, deletion, duplication and transposition of entire domains as well as fusion of genes and chimera formation. The examples to be discussed will illustrate the maxim that new proteins with novel functions arise mainly by the modulation of existing ones: the germ of the function of novel proteins is usually present in its ancestor(s).

7.1 Advantageous duplications

7.1.1 Unprocessed genes

Duplications are advantageous and are maintained by positive selection if having multiple genes performing the same function ensures enhanced efficiency for an organism. Textbook examples of advantageous duplications are the histone genes, which are present in multiple identical copies in eukaryotes. The selection pressure that maintains the 'redundancy' of histone genes is that multiple copies of histone genes are needed to supply large amounts of histones at a certain point of the cell cycle. In birds and mammals the repetition frequency of histone genes is 10–40, in *Drosophila melanogaster* it is about 100, whereas in some sea urchin species there may be 300–600 copies of each histone gene.

A striking example of positive selection for advantageous duplications is provided by the chorion genes of *Drosophila*. Two clusters of chorion genes are specifically amplified in ovarian follicle cells when large amounts of their gene products are required. Positive selection for 'advantageous' gene duplications can also be observed in insects exposed to insecticides, in tumours exposed to cytotoxic drugs or in protozoa exposed to drugs (Ouellette et al., 1991).

The stomach lysozymes of ruminants may also illustrate the benefits of having multiple copies of the same gene with the same function. Irwin and Wilson (1989) have identified seven types of mRNA sequences that encode various stomach isozymes of bovine lysozyme. The sequences are closely related, they differ by only 1–3 amino acid replacements, and represent the products of around 10 cow lysozyme genes. Evolutionary analysis has shown that the multiple lysozyme genes expressed in the cow stomach are the result of gene duplications that occurred during ruminant evolution. The recruitment of lysozyme as a major enzyme of the stomach thus has involved a 4–7-fold increase in expression thanks to gene duplication. Thus it appears that in addition to a change in the site of lysozyme expression, an increase in the number of lysozyme genes has also occurred in ruminants. It seems clear that serial duplication of the lysozyme gene was an adaptive response to increase the expression of lysozyme.

7.1.2 Processed genes

Due to the unique mode of their creation, processed genes differ from duplicons produced by unequal crossing over in several ways. Whereas at the time of their birth duplicated genes have exactly the same structure and are located in the same genomic environment (and are likely to produce increased amounts of the same protein in exactly the same expression pattern), processed genes are transposed to a different genomic environment (which may have different expression characteristics), and their 5′ regulatory regions are likely to differ from those of their progenitors (either because they were created by using an abnormal promoter or because they acquired a new promoter at the site of their insertion). As a consequence, a processed gene is likely to have altered regulatory features from the beginning: in this case it may not be competing with its progenitor, and its fate is decided in its own right.

Since reverse transcription is of lower fidelity, the processed gene may have mutations right from the beginning. Since they are most likely to be deleterious, processed genes have relatively little chance of surviving: they are most likely to be converted to processed pseudogenes. However, if they do survive, their function will usually differ from that of the progenitor, primarily in the expression-characteristic (developmental stage, tissue distribution). The role of positive selection in increasing the chance of survival of processed genes may be illustrated by two testis-specific processed genes.

The phosphoglycerate kinase retrogene

Phosphoglycerate kinase (PGK) is a metabolic enzyme of the pathway that converts glucose to pyruvate. There are two functional loci for the production of PGK in the mammalian genome: *PGK-1* is an X-linked gene expressed constitutively in all somatic cells; *PGK-2* is a functional autosomal gene expressed in a tissue-specific manner exclusively in the late stages of spermatogenesis (McCarrey & Thomas, 1987; McCarrey, 1990). The human *PGK-1* gene consists of 11 exons and 10 introns, whereas *PGK-2* is a processed gene: it completely lacks introns and contains characteristics of a processed gene, including the remnants of a poly(A) tail.

The conservation of function in the processed *PGK-2* gene and its tissue-specific expression in spermatogenesis are best explained as a compensatory response to the inactivation of the X-linked gene in spermatogenic cells before meiosis. Since mature spermatozoa require significant amounts of phosphoglycerate kinase to participate in the metabolism of fructose present in semen, the inactivation of the single X chromosome in spermatogenic cells before meiosis called for a functional autosomal PGK locus. The random creation of processed genes could give rise to the requisite autosomal PGK gene. (Note that tandem gene duplication by unequal crossing over could not have solved the problem created by X inactivation.)

It seems that the processed gene survived since it had an immediate advantage from the beginning: it permitted the expression of a gene in a tissue, where the X-linked gene is inactivated. There is also evidence for adaptive evolution in this case. First, a comparison of the human and mouse PGK genes indicates that *PGK-2* has evolved more rapidly than *PGK-1*. Second, the *PGK-2* retrogene initially included a copy of the *PGK-1* housekeeping promoter and subsequently evolved a testis-specific promoter. Selection for high-level, cell type-specific expression in spermatogenic cells thus had a primary influence on the survival and evolution of the *PGK-2* retrogene.

A similar explanation seems to hold for the evolution of another processed gene, the testis-specific form of pyruvate dehydrogenase E1α subunit.

Pyruvate dehydrogenase E1α retrogene

The pyruvate dehydrogenase (PDH) complex catalyses an essential and rate-limiting step in aerobic glucose metabolism. The gene for the PDH E1α subunit of this complex contains 10 introns and is located on the X chromosome and is expressed in somatic tissues. The human genome contains another PDH E1α gene (on chromosome 4) that is testis-specific, and is expressed in postmeiotic spermatogenic cells (Dahl et al., 1990). The coding sequence of this autosomal PDH E1α gene is strikingly similar to that of the X chromosome-linked gene: at the nucleotide level they are 84% identical. The autosomal human gene, however, completely lacks introns, and possesses all other characteristics of a functional processed gene. In addition to the lack of introns, there is a remnant of a poly(A) tract following a poly(A) addition signal in the 3′ untranslated region. Immediately downstream of this poly(A) tract and upstream from its coding region there are 10-bp direct repeats.

Energy production in spermatozoa is almost entirely by aerobic carbohydrate metabolism, for which they may utilize lactate and fructose as well as glucose as energy sources. After the last meiotic division the spermatids go through a maturation process and throughout the period of maturation, storage and release, the haploid sperm is dependent on energy generated from pyruvate via the PDH complex, thus a PDH E1α subunit is indispensable. However, the X chromosome is inactivated in postmeiotic spermatogenic cells (not to mention that only half of the spermatozoa contain an X chromosome). Similarly to PGK, the solution to this problem was the evolution of an alternative gene that is not X-linked. The retroposition of a processed gene to an autosome has permitted its expression during spermatogenesis, given the acquisition of regulatory features that ensure its testis-specific expression: PDH E1α is switched on at the onset of spermatogenesis in spermatogenic cells.

Thus, to circumvent the problem of X chromosome inactivation or the absence of an X chromosome in haploid spermatogenic cells, autosomal 'back-up' genes have evolved for both PGK and PDH. It is of interest to note that in both cases the novel gene evolved via a reverse transcriptase-mediated step. The explanation for this seems to be obvious: since the transfer of the gene to another chromosome was essential, (retro)position – rather than tandem gene duplication – was the favoured mechanism.

Recent studies confirm the hypothesis that X-chromosome inactivation may be a major driving force for the generation of autosomal retrocopies of

X-linked genes. Vinckenbosch, Dupanloup and Kaessmann (2006) have identified more than 1000 transcribed retrocopies in the human genome, of which at least ~120 have evolved into bona fide genes. Among these, ~50 retrogenes have evolved functions in testes, more than half of which were recruited as functional autosomal counterparts of X-linked genes during spermatogenesis.

7.2 Neutral duplications

If there is no immediate selective advantage in having duplicated genes (overproduction, novel expression characteristics) then functional constraints protecting the gene(s) from deleterious mutations may be completely relaxed. In this case mutations preventing the expression of one of the copies will be accepted since this will merely restore the situation that existed before the gene duplication. Since most frequently overproduction is selectively neutral, this is a likely fate of duplicons: one of the copies will be converted to a pseudogene. A wealth of data indicates that clusters of duplicated genes usually contain pseudogenes (e.g. the pseudogenes of the globin gene cluster). Duplicated genes may avoid inactivation if soon after duplication they acquire some advantageous mutations. In other words, there are two evolutionary paths for a duplicated gene: either it becomes a pseudogene or it fixes an advantageous mutation that permits its survival. In the latter case it still has a chance of evolving a new function. Since the ratio of the probabilities of advantageous to deleterious mutations is very low, a duplicated gene has little chance of survival without the help of positive selection.

Advantageous mutations may rescue duplicons from inactivation if one of the copies begins to adapt to some functional needs of the organism by slightly modifying the original function of the protein. This pathway of the evolution of paralogous genes leads to diversification of the original function of the protein (e.g. changes in the substrate specificity of enzymes) and is the major mechanism for the evolution of proteins/enzymes with modified specificity, and for the evolution of biochemical pathways and regulatory cascades.

If one of the diverging paralogs accepts mutations in regions that control its expression, such regulatory mutations may have immediate effect on its developmental or tissue expression pattern. Such mutations may give rise to forms expressed at a unique developmental stage (e.g. foetal, embryonic and adult forms of globins in mammals) or to tissue-specific forms. In both cases, the proteins are challenged by altered functional needs and they may have to adapt to altered environments. During such adaptations, there is positive selection for advantageous mutations that make the proteins better adjusted to their modified biological roles. Since acceptance of adaptive mutations is favoured, there will be an acceleration of nonsynonymous mutations relative to synonymous mutations. Once the protein has adapted to its modified function, the new functional constraints will slow down the non-synonymous rate.

In general, the emergence of a new, advantageous function in a protein sequence occurs by advantageous mutations. As we have seen in Chapter 5 (e.g. haemoglobin, lysozyme), and will illustrate below (visual pigments, trypsin, chymotrypsin), advantageous modification of a function may involve relatively few critical residues, or a few bases in regulatory regions that alter the gene's expression pattern (e.g. PGK) so the chances of the acquisition of novel functions may be quite high.

In more complex cases, when there is a major change of function, several critical sites may be involved and the new function might not be fully manifested until several sites had adapted to a novel function. In such cases, many early mutational steps might have been selectively nearly neutral, serving as the initial steps towards the evolution of a new function (preadaptive mutations). Even if some early mutations of duplicated genes may be selectively neutral, advantageous mutations must occur, otherwise the duplicated gene may be eliminated. Positive selection helps the advantageous mutations to become established in the population; and there will be accelerated evolution of the new function. Once the duplicated gene has established its novel function the significance of positive selection is diminished and purifying selection becomes dominant; there will be a slowdown in the rate of evolution.

A diagnostic sign of positive selection for advantageous point mutations is that the rate of nonsynonymous mutations is significantly higher than in the case of related proteins with established functions (e.g. lysozyme; see section 5.2). Another manifestation of positive selection is that the ratio of nonsynonymous/synonymous substitutions is 'abnormally' high since selection for advantageous substitutions favours amino acid-changing mutations over silent mutations. Thanks to positive selection, new functions may arise by combining the minor contributions of several advantageous mutations. Continuous modification of an original function may eventually lead to a drastically different, novel function. According to this view emergence of a new function by point mutations is expected to lead through a continuum of functions.

Unlike a point mutation, which usually can lead only to a minor improvement of function, domain duplication, domain-shuffling mutation or gene fusion may have a major and immediate selective advantage, increasing the chances of survival and acquisition of a new function by a duplicated gene. As will be discussed in Chapter 8, domain duplication and domain shuffling have played an important role in protein evolution and organismal evolution. Since module shuffling is a much more rare genetic event than point mutations, the unusual frequency of module insertions in some proteins proves that there was significant positive selection for such 'mutations'.

7.2.1 Modification of function by point mutations

Visual pigment proteins

Old World primates possess three colour-sensitive pigment proteins, which are sensitive to blue, green and red light. The green- and red-absorbing photoreceptor proteins are encoded by two closely related (96% sequence identity), closely linked genes on the X chromosome, whereas the blue pigment is encoded by an autosomal gene. The close linkage and high homology between the red and green pigments point to a very recent gene duplication. The fact that New World monkeys have only one X-linked pigment gene indicates that the duplication occurred in the ancestor of Old World primates after their divergence from the New World monkeys (about 35–40 million years ago). As a consequence of duplication of the X-linked gene, humans, apes and Old World monkeys can distinguish three colours, whereas New World monkeys can distinguish only two colours. Yokoyama and Yokoyama (1990) have reconstructed the phylogeny of these receptors, and have shown that, prior to the emergence of the three colour pigments, the rate of nonsynonymous substitution exceeded the synonymous rate, suggesting that there was positive selection for three-colour vision. It is noteworthy that a shift in the absorption spectrum of the duplicated genes could be achieved by very few (15 of 348) amino acid substitutions.

Although loss of any one of the visual pigment genes results in vision that is virtually normal, colour discrimination is impaired. The observed positive selection for three-colour vision thus proves that it must have had significant selective value in primates (e.g. more efficient detection of predators, fruits, etc.).

Serine proteinases and their inhibitors

The family of trypsin-like serine proteinases provides many illustrative examples of how duplicated genes may acquire novel functions. Catalytically active members of this family all have a His-57, an Asp-102 and a Ser-195 residue at the active site, which together form the catalytic triad. All trypsin-like serine proteinases employ similar catalytic mechanisms to cleave peptide bonds of proteins. In higher vertebrates this family includes various proteinases of the digestive tract (trypsins, chymotrypsins, elastases, enterokinase), proteinases of lymphocytes (e.g. granzymes) and a variety of plasma proteinases involved in blood coagulation, fibrinolysis and complement activation cascades.

The pancreatic proteinases (trypsin, chymotrypsin, elastase) involved in the digestion of proteins present in foodstuffs may serve to illustrate the divergence of paralogs with relatively minor modifications of their function. The three-dimensional structures of chymotrypsin, trypsin and elastase are strikingly similar, e.g. the conformations of trypsin and chymotrypsin can be superimposed so that on average the atoms of their polypeptide back-bone differ in relative positions by only 0.75 Å. Despite their close structural similarity, these proteinases differ markedly in their primary substrate speci-ficity, inasmuch as their sequence specificity is different. Elastase cleaves in the vicinity of amino acids with small nonpolar side chains (Ala, Val, Ser, etc.), chymotrypsin cleaves at bulky hydrophobic residues (Trp-X, Phe-X, Tyr-X bonds, etc.), whereas trypsin cleaves only at Arg-X or Lys-X bonds of proteins. The advantage of having multiple digestive proteinases with different sequence specificity is clear: their combined activities ensure more efficient degradation (and utilization) of proteins in foodstuffs, compared to the digestive capabilities of any one of them alone. Duplicons that arose by duplication of the common ancestor of digestive proteinases could survive since they acquired advantageous mutations that diversified their function.

The molecular basis of the differences in the sequence specificities of these serine proteinases may be rationalized from their three-dimensional structures and a deeper understanding of the catalytic mechanism of these proteinases. The specificity of trypsins for Arg and Lys residues is due primarily to two factors: (i) the substrate-binding site of trypsins is deep, so it can accommodate the bulky Arg/Lys side chains); and (ii) there is an Asp-189 residue at the bottom of the binding pocket that neutralizes the charge of the Arg or Lys residue of the substrate (Fig. 7.1(a)). Similarly, the specificity of chymotrypsin for bulky aromatic residues is explained by its large hydrophobic substrate-binding pocket. In chymotrypsin, position 189 at the bottom of the pocket is occupied by a small, neutral (usually Ser) residue (Fig. 7.1(b)). Conversely, the specificity of elastase for small nonpolar side chains is due to the fact that its binding site is shallower than that of trypsin or chymotrypsin. For example, whereas in the case of trypsin and chymotrypsin, residues 216 and 226 – present in the binding pocket – have small side chains (Gly, Ala), in the case of elastases these residues are usually replaced by bulkier residues (Val, Thr, Ser) (Fig. 7.1(c)). These larger side chains at the bottom of the binding pocket prevent the binding of larger side chains of substrates.

An important conclusion from structure–function studies on pancreatic proteinases is that substitution of a few key residues may alter the sequence specificity without eliminating enzyme activity. For example, replacement of Asp-189 of trypsin by a Ser residue (to mimic chymotrypsin) greatly dimin-ishes the catalytic activity of trypsin towards Lys and Arg substrates, and increases activity towards hydrophobic substrates 10- to 50-fold. Nevertheless, the fact that this increase is much less than expected if this single mutation could convert it to 'chymotrypsin', indicates that several other readjustments (at other positions) had to occur during step-by-step divergence of trypsin and chymotrypsin from a common ancestor. The example of pancreatic pro-teinases seems to support the notion that a new function may emerge by continual 'improvement' of function, with positive selection for substitutions

```
                            EEE    EEEEEE EEEEEEEEE    EEEE        EEEE           EEEEEEEEE                    EEEEEEE
tryl_salsa   IVGGYECKAYSQTHQVSLNSGYHFCGGSLVNENWVVSAAHCYKSRVEVRLGEHNIKVTEGSEQFISSSRVIRHPNYSSYNIDNDIMLIKLSKPATLNTYV
try2_salsa   IVGGYECKAYSQPHQVSLNSGYHFCGGSLVNENWVVSAAHCYQSRVEVRLGEHNIQVTEGSEQFISSSRVIRHPNYSSYNIDNDIMLIKLSKPATLNTYV
trya_rat     IVGGYTCQEHSVPYQVSLNAGSHICGGSLITDQWVLSAAHCYHPQLQVRLGEHNIYEIEGAEQFIDAAKMILHPDYDKWTVDNDIMLIKLKSPATLNSKV
tryb_rat     IVGGYTCQEHSVPYQVSLNAGSHICGGSLITDQWVLSAAHCYHPQLQVRLGEHNIYEIEGAEQFIDAAKMILHPDYDKWTVDNDIMLIKLKSPATLNSKV
tryl_human   IVGGYNCEENSVPYQVSLNSGYHFCGGSLINEQWVVSAGHCYKSRIQVRLGEHNIEVLEGNEQFINAAKIIRHPQYDRKTLNNDIMLIKLSSRAVINARV
try2_human   IVGGYICEENSVPYQVSLNSGYHFCGGSLISEQWVVSAGHCYKSRIQVRLGEHNIEVLEGNEQFINAAKIIRHPKYNSRTLDNDILLIKLSSPAVINSRV
tryp_bovin   IVGGYTCGANTVPYQVSLNSGYHFCGGSLINSQWVVSAAHCYKSGIQVRLGEDNINVVEGNEQFISASKSIVHPSYNSNTLNNDIMLIKLKSAASLNSRV
Consensus    IVGGY-C------QVSLN-G-H-CGGSL----WV-SA-HCY-----VRLGE-NI----EG-EQFI-----I-HP-Y-----NDI-LIKL--A--N--V
                     *                                 *

                 EEEEEE       EEEEEEEEHHHHHHHH        EEEEE                          EEEEE          EEEEE      EEEEEEE
tryl_salsa   QPVALPTSCAPAGTMCTVSGWGNTMSSTADS.NKLQCLNIPILSYDCNNSYPGMITNAMFCAGYLEGGKDSCQGDSGGPVVCNGELQGVVSWGYGCAEP
try2_salsa   QPVALPTSCAPAGTMCTVSGWGNTMSSTADK.NKLQCLNIPILSYDCNNSYPGMITNAMFCAGYLEGGKDSCQGDSGGPVVCNGELQGVVSWGYGCAEP
trya_rat     STIPLPQYCPTAGTECLVSGWG.VLKFGFESPSVLQCLDAPVLSDSVCHKAYPRQITNNMFCLGFLEGGKDSCQYDSGGPVVCNGEVQGIVSWGDGCALE
tryb_rat     STIPLPQYCPTAGTECLVSGWG.VLKFGFESPSVLQCLDAPVLSDSVCHKAYPRQITNNMFCLGFLEGGKDSCQYDSGGPVVCNGEVQGIVSWGDGCALE
tryl_human   STISLPTAPPATGTKCLISGWGNTASSGADYPDELQCLDAPVLSQAKCEASYPGKITSNMFCVGFLEGGKDSCQGDSGGPVVSNGELQGIVSWGYGCAQK
try2_human   SAISLPTAPPAGTESLISGWGNTLSSGADYPDELQCLDAPVLSQAKCEASYPGKITSNMFCVGFLEGGKDSCQGDSGGPVVSNGELQGIVSWGYGCAQK
tryp_bovin   ASISLPTSCASAGTQCLISGWGNTKSSGTSYPDVLKCLKAPILSDSSCKSAYPGQITSNMFCAGYLEGGKDSCQGDSGGPVVCSGKLQGIVSWGSGCAQK
Consensus    ---LP-----GT-----SGWG-----------L-CL--P-LS----C---YP--IT--MFC-G-LEGGKDSCQ-DSGGPVV--G--QG-VSWG-GCA--
                                                                                              *

                 EEEEE  HHHHHHHHHHH
tryl_salsa   GNPGVYAKVCIFNDWLTSTMASY~
try2_salsa   GNPGVYAKVCIFNDWLTSTMATY~
trya_rat     GKPGVYTKVCNYLNWIHQTIAEN~
tryb_rat     GKPGVYTKVCNYLNWIQQTVAAN~
tryl_human   NKPGVYTKVNYVKWIKNTIAANS
try2_human   NRPGVYTKVYNYVDWIKDTIAANS
tryp_bovin   NKPGVYTKVCNYVSWIKQTIASN~
Consensus    --PGVY-KV----W----T-A--~
```

(a)

Fig. 7.1 (a) Multiple alignment of the amino acid sequences of the catalytic chains of human, bovine, rat and salmon (*Salmo salar*) trypsins. The consensus sequence shows residues identical in all sequences. The asterisks indicate the residues of the catalytic triad and the arrows indicate residues involved in defining the substrate specificity of trypsins. The residues involved in β-strands or α-helices of bovine trypsin are indicated by E and H, respectively. The ribbon diagram of bovine trypsin is also shown. (*Continued*)

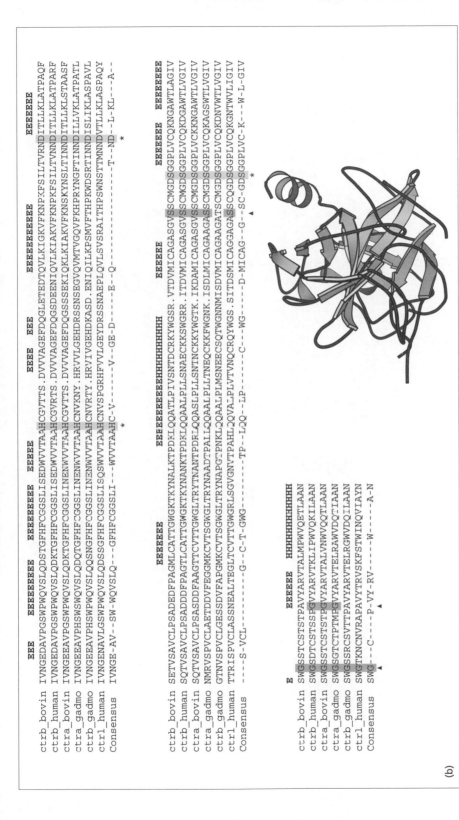

(b)

Fig. 7.1 (*Continued*) (b) Multiple alignment of the amino acid sequences of the catalytic chains of human, bovine and Atlantic cod (*Gadus morhua*) chymotrypsins. The consensus sequence shows residues identical in all sequences. The asterisks indicate the residues of the catalytic triad and arrows indicate the residues involved in defining the substrate specificity of chymotrypsins. The residues involved in β-strands or α-helices of bovine chymotrypsin *a* are indicated by E and H, respectively. The ribbon diagram of bovine α chymotrypsin is also shown. (*Continued*)

```
              EEE    EEEEEEEEE  EEEEEEEEEE   EEE            EEEEE         EEE    EEEEEEEEEEE    EEEE
el3a_human  VVHGEDAVPYSWPWQVSLQYEKSGSFYHTCGGSLIAPDWVVTAGHCISRDLTYQVVLGEYNLAVKEGPEQVIPINSEELFVHPLWNRSCVACGNDIALIK
el3b_human  VVNGEDAVPYSWPWQVSLQYEKSGSFYHTCGGSLIAPDWVVTAGHCISSSRTYQVVLGEYDRAVKEGPEQVIPINSGDLFVHPLWNRSCVACGNDIALIK
el1_pig     VVGGTEAQRNSWPSQISLQYRSGSSWAHTCGGTLIRQNWVMTAAHCVDRELTFRVVVGEHNLNQNDGTEQVYVGVQ..KIVVHPYWNTDDVAAGYDIALLR
el1_rat     VVGGAEARRNSWPSQISLQYLSGGSWYHTCGGTLIRRNWVMTAAHCVSSQMTFRVVVGDHNLSQNDGTEQYVSVQ..KIMVHP`WNSNNVAAGYDIALLR
el2_mouse   VVGGQEATPNTWPWQVSLQYVSSGRWRHNCGGSLVANNVLTAAHCLSNYQTYRVLLGAHSLSNPGAGSAAVQVS..KLVVHQRWNSQNVGNGYDIALIK
el2_rat     VVGGQEASPNSWPWQVSLQYLSSGKWHHTCGGSLVANNWLTAAHCISNSRTYRVLLGRHSLSTSESGSLAVQVS..KLVVHEKWNAQKLSNGNDIALVK
el2_pig     VVGGEDARPNSWPWQVSLQYDSSGQWRHTCGGTLVDQSWVLTAAHCISSSRTYRVLGRHSLSTNEPGSLAVKVS..KLVVHQDWNSNQLSNGNDIALLK
Consensus   ---G--A---WP-Q-SLQ------H-CGG-L---W-TA-HC------V--G------------VH--WN------G-DIAL--
                                                     *                                              *
```

```
            EE       EEEEEE        EEEEEEEEEEHHHHHH              EEEEE        EEEEEEE
el3a_human  LSRSAQLGDAVQLASLPPAGDILPNKTPCYITGWGRLYTNGPLPDKLQQARLPVVDYKHCSRWNWGSTVKKTMVCAGGY.IRSGCNGDSGGPLNCPTED
el3b_human  LSRSAQLGDAVQLASLPPAGDILPNETPCYITGWGRLYTNGPLPDKLQEALLPVVDYEHCSRWNWGSSVKKTMVCAGGD.IRSGCNGDSGGPLNCPTED
el1_pig     LAQSVTLNSYVQLGVLPRAGTILANNSPCYITGWGLTRTNGQLAQTLQQAYLPTVDYAICSSSSYWGSTVKNSMVCAGGDGVRSGCQGDSGGPLHCLV.N
el1_rat     LAQSVTLNNYVQLAVLPQEGTILANNNPCYITGWGRTRTNGQLSQTLQQAYLPSVDYSICSSSSYWGSTVKTTMVCAGGDGVRSGCQGDSGGPLHCLV.N
el2_mouse   LASPVTLSKNIQTACLPPAGTILPRNYPCYVTGWGLLQTNGNSPDTLRQGRLLVVDYATCSSASWWGSSVKSSMVCAGDGVTSCNGDSGGPLNCRASN
el2_rat     LASPVALTSKIQTACLPAGTILPNNYPCYVTGWGRLQTNGATPDVLQQGRLLVVDYATCSSASWWGSSVKTNMVCAGDGVTSCNGDSGGPLNCQASN
el2_pig     LASPVSLTDKIQLGCLPAAGTILPNNYVCYVTGWGRLQTNGASPDILQQGQLLVVDYATCSKPGWWGSTVKTNMICAGGDGIISSCNGDSGGPLNCQGAN
Consensus   L----L---Q---LP--G-IL-----CY-TGWG--TNG-----L----L---VDY--CS--WGS-VK--M-CAGG---S-C-GDSGGPL-C---
                 *                                                                                  *
```

```
            EEEEEEEEE                 EEEEEE  HHHHHHHHHHHH
el3a_human  GGWQVHGVTSFVSGFGCNFIWKPTVFTRVSAFIDWIEETIASH
el3b_human  GGWQVHGVTSFVSAFGCNTRRKPTVFTRVSAFIDWIEETIASH
el1_pig     GQYAVHGVTSFVSRLGCNVTRKPTVFTRVSAYISWINNVIASN
el1_rat     GQYSVHGVTSFVSSMGCNVSKKPTVFTRVSAYISWMNNVIAYT
el2_mouse   GQWQVHGIVSFGSSLGCNYPRKPSVFTRVSNYIDWINSVMARN
el2_rat     GQWQVHGIVSFGSTLGCNYPRKPSVFTRVSNYIDWINSVIAKN
el2_pig     GQWQVHGIVSFGSSLGCNYYHKPSVFTRVSNYIDWINSVIANN
Consensus   G---VHG--S--S--GCN---KP--FTRVS----W----A--
```

Fig. 7.1 (*Continued*) (c) Multiple alignment of the amino acid sequences of the catalytic chains of human, porcine, rat and mouse elastases. The consensus sequence shows residues identical in all sequences. The asterisks indicate the residues of the catalytic triad and arrows indicate the residues involved in defining the substrate specificity of elastases. The residues involved in β-strands or α-helices of elastase are indicated by E and H, respectively. The ribbon diagram of porcine elastase is also shown.

(c)

giving selective advantage. The sequence specificities of some members of the trypsin family have been narrowed to such an extent that their activity is restricted to a single peptide bond of just one or a few proteins. Such extreme specificities are usually employed in the regulation of processes such as the blood coagulation cascade (e.g. thrombin), fibrinolytic cascades (e.g. plasminogen activators), complement activation cascade and the activation of pancreatic proteinases (e.g. enterokinase). In most cases each member of the cascade can cleave and activate only the zymogen of the downstream member of the cascade. As an example of such specificity we can refer to plasminogen activators (t-PA and u-PA), which have only a single substrate, namely plasminogen. Both enzymes activate plasminogen by cleaving a single Arg–Val bond of this zymogen.

Functional diversification of proteinases calls for the functional adaptation of proteins that control proteinase activity. Following the appearance of a proteinase with novel specificity there will be positive selection for the emergence of a proteinase inhibitor with a corresponding specificity. Indeed, there is evidence for an accelerated evolution in the reactive centre regions of serine proteinase inhibitors when challenged with novel proteinases of invading parasites (Hill & Hastie, 1987; Borriello & Krauter, 1991; Ray, Gao & Ray, 1994). In the case of the serpin family of serine proteinase inhibitors an unusually high substitution rate was observed in a narrow region surrounding their reactive sites, suggesting positive selection for new reactive site sequences that can cope with an increasing number of attacking proteinases.

The porcine elafin family of elastase inhibitors (Tamechika et al., 1996) provides another illustrative example for accelerated evolution in region(s) interacting with the proteinase active site. The porcine genome contains three closely related genes of the elafin family that diverged by accelerated evolution as a result of positive selection for advantageous mutations. Evolutionary analyses indicate that the duplications of the elafin gene and their diversification occurred in the porcine lineage, following divergence from the human lineage. Since porcine elafins are abundantly expressed in the trachea and intestine, the most likely selective forces for the accelerated evolution are extrinsic proteinases produced by invasive microorganisms (trachea and intestine are exposed to a variety of infectious agents such as bacteria and parasites). A striking aspect of these closely related elafin genes is that although their intron sequences are related to each other with sequence similarities of 93–98%, the exon 2 sequences (encoding the proteinase inhibitor domain of these proteins) exhibit only 60–77% similarities. This is just the opposite of what generally holds for protein-coding genes (introns are usually less conserved than exons; see section 4.3). The extreme divergence in the exon 2 sequences compared to the highly conserved intron sequences was generated by accelerated mutations in the functionally most important regions of these proteins, the highest divergence being in the active centre region. Quantitative analysis of substitution data confirmed the existence of positive selection: in the regions that define inhibitory activity, the K_A values exceed significantly the K_S values, and the ratio K_A/K_S is well above 1.

Phospholipases and phospholipase inhibitors of snake venoms

Crotalinae snake venom gland phospholipase A2 (PLA2) isozymes have evolved via accelerated evolution to gain new physiological activities that were critical for the toxicity of the venom (Nakashima et al., 1993, 1995; Nobuhisa et al., 1996). A striking sign of positive selection for nonsynonymous substitutions is that the intron sequences and untranslated regions of exons of the snake venom phospholipase A2 isozyme gene families are more conserved than the regions coding for the hypervariable regions.

The serum of crotaline snakes also contains phospholipase A2 inhibitors (PLIs) that provide the snakes with an effective defence against toxicity of their own venom gland phospholipase A2 enzymes (PLA2s). Clearly, there must have been strong selective pressure on PLIs to evolve against venom PLA2 isozymes. Studies on two PLI genes of the habu snake, *gPLI A* and *gPLI B*, have provided evidence for positive selection for amino acid-changing mutations (Nobuhisa et al., 1997). Comparison of the sequences of *gPLI A* and *gPLI B* genes showed that these genes are highly homologous, including introns, except that exon 3 is rich in nonsynonymous substitutions, which are almost four times as frequent as synonymous substitutions: the K_A/K_S ratio is 3.61 for exon 3. This variable region of PLI genes seems to be crucial for their specificity to interact with particular PLA2s.

UDP-glucuronosyltransferases

UDP-glucuronosyltransferases (UDPGTs) make up a family of enzymes that detoxify hundreds of compounds by their conjugation to glucuronic acid (Brierly & Burchell, 1993). In this glucuronidation pathway, the sugar moiety of UDP-glucuronic acid is covalently linked to a toxic xenobiotic or endobiotic compound, rendering them harmless and more water soluble, thereby facilitating their removal from the body in the urine or bile. The evolutionary diversification of UDPGTs reflects the constant selective pressure to get rid of a large variety of toxic compounds. One major source of toxic compounds for plant-eating animals is ingested plant phenolic metabolites. The appearance of a novel type of detoxifying enzyme (with a novel substrate specificity) could confer a selective advantage on an organism: it could enable the organism to utilize additional food sources (despite their toxic contents) or it could allow the organism to inhabit 'toxic' environments. In mammals bilirubin is one of the most critical compounds that must undergo detoxification by conjugation with glucuronic acid. Since this toxic metabolite of haem is produced in large quantities from the normal turnover of haemoglobin and other haemoproteins it represents a powerful selective pressure to evolve an efficient detoxifying system. In harmony with the multiplicity of toxic compounds (phenolic compounds, steroids, bilirubin, etc.), glucuronidation is catalysed by a large family of UDPGT enzymes, which have different but overlapping substrate specificities. It must be mentioned that, due to their rather broad substrate specificity, UDPGTs can lower the level not only of toxic xenobiotics, but also of a variety of hormones (steroid hormones, tyrosine-derived hormones, etc.), therefore UDPGT induction can perturb hormone levels and can interfere with normal homeostasis. Consequently, overproduction (which necessarily occurs as a result of whole-gene duplication) could have an immediate deleterious effect on the organism.

Importantly, the UDPGTs are proteins with distinct domains serving different subfunctions. They have an N-terminal globular domain that is responsible for binding the toxic substrate, and a C-terminal globular domain that is involved in UDP–glucuronic acid binding and catalysis. Diversification of the substrate specificity of UDPGTs occurred through a process of amino acid substitutions in restricted regions of the substrate-binding domains.

In mammals there are two subfamilies of UDPGTs, which differ markedly in the evolutionary strategy used for diversification of functions. The UDPGT2B gene subfamily has evolved by a series of classical whole-gene duplications, resulting in the emergence of several isoforms. In humans these UDPGT2B genes are clustered on chromosome 4. The primary substrates for these enzymes are 4-hydroxysterone, hyodeoxycholic acid of bile and oestriol.

In contrast with this 'classical' gene-duplication strategy, the single UDPGT1 gene complex, localized to human chromosome 2, has diversified by duplication of only exon 1, which specifies the N-terminal, substrate-binding domain (Ritter et al., 1992). The human UDPGT1 gene complex contains six closely related exon 1 variants; each of them encodes a variant of the N-terminal, substrate-binding domain. The UDPGT1 gene complex contains only a single set of the four exons that encode the C-terminal parts of UDP-glucuronosyltransferases. The mRNAs of the various isoforms of UDPGT1s are produced by differential splicing of one of the exon 1 variants to the four-exon block encoding the 'constant domain' of UDPGTs. In other words, evolutionary expansion and diversification within this UDPGT1 gene complex has occurred only by duplication of a part of the gene, i.e. exon 1. The UDPGT1 isozymes produced by alternative splicing differ only in their N-terminal domains but have identical carboxyl terminal domains. The primary substrates for the various UDPGT1 isozymes are bilirubin and a variety of phenolic compounds.

There is an interesting major difference between these two diversification strategies: in the case of whole-gene duplications, duplication leads to an initial overproduction of the gene product, whereas in the case of the exon-duplication/differential splicing strategy overproduction could be avoided. Overproduction may be deleterious in the case of isoforms that mediate removal of hormones (since it may cause some disturbance of homeostasis), but could be advantageous in the case of isoforms that eliminate toxic compounds. It seems likely that these factors have played a role in the choice of the two evolutionary strategies.

7.2.2 Major change of function by point mutations

Some members of the serine proteinase family (e.g. haptoglobin, hepatocyte growth factor, macrophage stimulating protein, azurocidin) have lost their capacity to act as proteinases: they lack one or all of the residues of the catalytic triad. Nevertheless they have some very important biological functions that apparently have nothing to do with the proteinase function of their active ancestors. For example, haptoglobin serves to bind globin released from lysed erythrocytes; hepatocyte growth factor and macrophage stimulating protein act through specific receptor tyrosine kinases to stimulate the growth of certain cell types; and azurocidin has bactericidal activity. Such drastic changes in function are most intriguing since they pose the possibility of major functional jumps. Careful analysis, however, sometimes indicates plausible pathways for a rather smooth transition from one function to another. This may be best illustrated in the case of azurocidins.

Azurophil granules, the specialized lysosomes of neutrophils, contain several proteins implicated in the killing of microorganisms. Among these are serine proteinases (termed serprocidins) that may also cause degradation of connective tissues: cathepsin G, neutrophil elastase and proteinase 3. An additional member of this group is azurocidin, which is highly homologous to neutrophil elastase, proteinase 3 and cathepsin G; however, it lacks proteolytic activity (Almeida et al., 1991). In azurocidin a Gly replaced the active site Ser-195, and a Ser replaced the His-57 of the catalytic triad (Fig. 7.2).

The potent antibiotic, bactericidal activity of azurocidin (also known as heparin binding protein) is mediated by its tight binding to the anionic lipopolysaccharide (LPS), a component of the Gram-negative bacterial envelope. It thus appears that the serine proteinase fold has been used as a scaffold for an endotoxin-binding capacity (Rasmussen et al., 1997). Considering the fact that azurocidins share the most recent common ancestor with proteinases that have antibacterial activity (cathepsin G, neutrophil elastase, proteinase 3), it seems clear that this was a common function of this ancestor. In the branch leading to azurocidins the new function (based on binding LPS of bacteria) emerged probably before the original function (based on proteolytic activity) was lost.

Haptoglobin is another member of the serine proteinase family that has lost its ability to cleave proteins. The closest relatives of haptoglobin are the serine proteinases of the C1r/C1s branch of the complement cascade. This conclusion is supported not only by the similarity of their amino acid sequences (Fig. 7.3), but also by the organization of their genes: unlike other

```
              EEE     EEEEEE     EEEEEEEE    EEEE     EEEE      EEE    EEEEEEEE         EEEEE

catg_human    IIGGRESRPHSRPYMAYLIQISPAGQSRCGGFLVREDFVLTAAHCWGS....NINVTLGAHNIQRRE.NTQQHITARRAIRHPQYNQRTIQNDIMLLQLS
catg_mouse    IIGGREARPHSYPYMAFLLIQSPEGLSACGGFIVREDFVLTAAHCLGS...SINVTLGAHNIQMRE.RTQQLITVLRA.RHPDYNPQNIRNDIMLLQLR
elne_human    IVGGRRARPHAWPFMVSLQLR...GGHFCGATLIAPNFVMSAAHCVANVNRAVRVVLGAHNLSRREP.TRQVFAVQRIFEN.GYDPVNLLNDIVILQLN
elne_mouse    IVGGRPARPHAWPFMASLQRR...GGHFCGATLIARNFVMSAVHCVNGLNFRSVQVVLGAHDLRRQER.TRQTFSVQGIFEN.GFDPSQLLNDIVIIQLN
prn3_human    IVGGHEAQPHSRPYMASLQMRGNPGSHFCGGTLIHPSFVLTAAHCLRDIPQRLVNVLGAHNVRTQEP.TQQHFSVAQVFLN.NYDAENKLNDILLIQLS
prn3_mouse    IVGGHEARPHSRPYVASLQLSRFPGSHFCGGTLIHPRFVLTAAHCLQDISWQLVTVVLGAHDLLSSEP.EQQKFTISQVFQN.NYNPEENLNDVLLLQLN
cap7_human    IVGGRKARPPQFPFLASIQ...NQGRHFCGGALIEARFVMTAASCFQSQNPGVSTVVLGAYDLRRREQRSRQTFSISSMSEN.GYDPQQNLNDLMLLQLD
cap7_pig      IVGGRRAQPQEFPFLASIQ...KQGRPFCAGALVHPRFV.TAASCFRGKNSGSASVVLGAYDLRQQE.QSRQTFSIRSISQN.GYDPRQNLNDVLLLQLD
Consensus     I-GG----P---P-------G----C----L----FV--A--C---------------V-LGA-------B-----Q-----------ND----QL-
                                                                                                *

              EEEEEEEE EEEE  EEEEEEEEEEEEEE                                           EEE    EEEEE     EEEE

catg_human    RVRVRNRNVNPVALPRAQEGLRPGTLCTVAGWGRVSMRGI.DTLREVQLRVQRDRQCLRIFGSYDPRRQICVGDRRERKAAFKGDSGGPLLCNNVAHGI
catg_mouse    RRARRSGSVKPVALPQASKKLQPGDLCTVAGWGRVSQSRGI.NVLQEVQLRVQMDQMCANRFQFYNSQTQICVGNPRERKSAFRGDSGGPLVCSNVAQGI
elne_human    GSATINANVQVAQLPAQGRRLGNGVQCLAMGWGLLGRNGLIASVLQELNVTVV.TSLC.R......RSNVCTLVRGRQAGVCFGDSGGPLVCNGLLHGI
elne_mouse    GSATINANVQVAQLPAQGQVGDRTPCLAMGWGRLGTNFSPSVLQELNVTVV.TNMCPR......RVNVCTLVPRRQAGICFGDSGGPLVCNNLVQGI
prn3_human    SPANLSASVTSVQLPQQDQPVPHGTQCLAMGWGRVGAHDFPAQVLQELNVTVV.TFFC.R......PHNICTFVPRRKAGICFGDSGGPLICDGIIQGI
prn3_mouse    RTASLGKEVAVASLPQQDQTLSQGTQCLAMGWGRLGTQAFTPRVLQELNVTVV.TFLC.R......EHNVCTLVPRRAAGICFGDSGGPLICNGILHGV
cap7_human    REANLTSSVTILPLPLQNATVEAGTRCQVAGWGSQRSGGELSRFPRFVNVTVTPEDQC.R......PNNVCTGVLTRRGGICNGDGGTPLVCEGLAHGV
cap7_pig      REARLTPSVALVPLPPQNATVEAGTNCQVAGWGTQRLRRLF3RFPRVLNVTVT.SNPC.L......PRDMCIGVFSRRGRISQGDRGTPLVCNGLAQGV
Consensus     ----------V---LP------C---GWG--------------V----C------V-----GD-G-FL-C----G-
                                                                                                *

              EEEE     EEEEE HHHHHHHH

catg_human    VSY...GKSSGVPPEVFTRVSSFLPWIRTTMRSFKLLDQMETPL~~~~~~~
catg_mouse    VSY...GSNNGNPPAVFTKIQSFMPWIKRTMRRF..APRYQPPANSLSQAQT
elne_human    ASFVRGGCASGLYPDAFAPVAQFVNWIDSIIQRSEDNPCPHPRDPPDASRTH
elne_mouse    DSFIRGGCGSGLYPDAFAPVGEFVDWINSIIRRHDHLLTHPK..DRRGRTN
prn3_human    DSFVIWGCATRLFPDFFTRVALYVDWIRSTLRRVEAKGRP~~~~~~~~~~
prn3_mouse    DSFVIRECASLQFPDFFARVSMYVDWIQNVLRGAEP~~~~~~~~~~~
cap7_human    ASFSLGPCGRG..PDFFTRVALFRDWIDGVLNNPGFGPA~~~~~~~~~~
cap7_pig      ASFLRRFRRS..SGFFTRVALFRNWIDSVLNNP.F~~~~~~~~~~
Consensus     -S----------------F-----WI--------------G-
```

Fig. 7.2 Multiple alignment of the amino acid sequences of the catalytic chains of serprocidins neutrophil elastase, proteinase 3, cathepsin G and azurocidins (cap7). The consensus sequence shows residues identical in all sequences. The asterisks indicate the residues of the catalytic triad. Note that in azurocidins the active site residues, His-57 and Ser-195, are replaced. The residues involved in β-strands or α-helices of leukocyte elastase are indicated by E and H, respectively. The ribbon diagram of human neutrophil elastase is also shown.

```
hpt2_human  ILGGHLDAK.GSFPWQAKMVSHHNLTTGATLINEQWLLTTAKNLF.LNHSENATAK.D..IAPT..LTLYVGKKQLVEIEKVLHPNYSQV..........
hpt1_human  ILGGHLDAK.GSFPWQAKMVSHHNLTTGATLINEQWLLTTAKNLF.LNHSENATAK.D..IAPT..LTLYVGKKQLVEIEKVLHPNYSQV..........
hptr_human  ILGGHLDAK.GSFPWQAKMVSHHNLTTGATLINEQWLLTTAKNLF.LNHSENATAK.D..IAPT..LTLYVGKKQLVEIEKVLHPNYHQV..........
c1r_human   IIGGQ.KAKMGNFPWQV.FTNIHGRGGGA.LLGDRWILTAAHTLYPKEHEAQSNASLDVFLGHTNVE.ELMKLGNHPIRRVSVHPDYR...QDESYNF
c1s_human   IIGGS.DADIKNFPWQV.FFD.NPWAGGA.LINEYWVLTAAHVV......EGNREPTMYVGSTSVQTSRLAKSKMLTPEHVFIHPGWKLLEVPEGRTNF
Consensus   I-GG---A----FPWQ--------GA-L------W-LT-A-----------------T----K-----V--HP-----
                                                     *
```

```
hpt2_human  ..DIGLIKLKQKVSVNERVMPICLP..SKDY.AEVGRVGYVSGWGRNANFKFTDHLKYVMLPVADQDQCIRHYEGSTVPEKKTPKSPVGVQP.ILNEHTF
hpt1_human  ..DIGLIKLKQKVSVNERVMPICLP..SKDY.AEVGRVGYVSGWGRNANFKFTDHLKYVMLPVADQDQCIRHYEGSTVPEKKTPKSPVGVQP.ILNEHTF
hptr_human  ..DIGLIKLKQKVLVNERVMPICLP..SKNY.AEVGRVGYVSGWGQSDNFKLTDHLKYVMLPVADQYDCITHYEGSTCPKWKAPKSPVGVQP.ILNEHTF
c1r_human   EGDIALLELENSVTLGPNLLpPICLPDNDTFYDL..GLMGYVSGFGVMEEK.IAHDLRFVRLPVANPQAC........ENWLRGKNRMD...VFSQNMF
c1s_human   DNDIALVRLKDPVKMGPTVSPICLPGTSSDYNLMDGDLGLISGWGRTEKRDRAVRLKAARLPVAPLRKC.......KEVKVEKPTADAEAVVFTPNMI
Consensus   --DI-L-L--V------PICLP------Y----G--G--SG-G-----------L---LPVA-----------C----------K----
              *
```

```
hpt2_human  CAGMSKYQEDTCYGDAGSAFAVHD.LEEDTWYATGILSFDKSCAVAEYGVYVKVTSIQDWVQKTIAEN~~~~~~
hpt1_human  CAGMSKYQEDTCYGDAGSAFAVHD.LEEDTWYATGILSFDKSCAVAEYGVYVKVTSIQDWVQKTIAEN~~~~~~
hptr_human  CVGMSKYQEDTCYGDAGSAFAVHD.LEEDTWYAAGILSFDKSCAVAEYGVYVKVTSIQDWVQKTIAEN~~~~~~
c1r_human   CAGHPSLKQDACQGDSGGVFAVRDPN.TDRWVATGIVSWGIGCS.RGYGFYTKVLNVVDWIKKEMEEED~~~~~
c1s_human   CAGGEK.GMDSCKGDSGGAFAVQDPNDKTKFYAAGLVSWGPQCG.T.YGLYTRVKNYVDWIMKTMQENSTPRED
Consensus   C-G-----D-C-GD-G--FAV-D------A-G--S----C----YG-Y--V----DW--K--E-------
                                  *
```

Fig. 7.3 Multiple alignment of the amino acid sequences of the catalytic chains of complement proteases C1r, C1s and haptoglobins. The consensus sequence shows residues identical in all sequences. The asterisks indicate the positions corresponding to residues of the catalytic triad. Note that in haptoglobins active sites His-57 and Ser-195 are replaced.

members of the serine proteinase family, they have genes where the serine proteinase domain is intronless. The fact that C1r and C1s are the closest relatives of haptoglobin also has interesting implications for the evolution of novel functions. Complement proteinases C1r and C1s are part of the complement cascade whose role includes lysis and elimination of aged erythrocytes and other types of target cells. The main function of haptoglobin is the scavenging of haemoglobin α/β heterodimers released from lysed erythrocytes. A distant functional link between haptoglobin and complement proteinases C1r and C1s is thus apparent: they may be descendants of an ancestral protein system involved in elimination of target cells. Haptoglobin shifted in specificity from binding other components of the complement cascade (e.g. C1q) to binding haemoglobin released from target cells.

The evolutionary lineage of plasminogen (fibrinolytic branch of serine proteinases) gave rise to several proteins that are devoid of proteolytic activity. Primate apolipoprotein(a), a close relative of plasminogen with a large number of kringle domains, arose relatively recently, in the lineage of Old World primates (McLean et al., 1987; Lawn, Schwarz & Patthy, 1997). No adaptive value has yet been established for this molecule, but since it forms a complex with lipoprotein B-100, it is generally thought to be involved in delivering cholesterol to the sites of tissue injury for proliferating fibroblasts. The fact that a similar apolipoprotein(a) molecule (with a large number of kringles) has emerged independently in the insectivore lineage (Lawn et al., 1997) suggests positive selection for the emergence of apolipoprotein(a). The fact that plasminogen is also capable of binding lipoprotein B-100 suggests a possible pathway for the functional transition.

There are two other plasminogen-like molecules, hepatocyte growth factor and macrophage stimulating protein, that have lost catalytic activity because of replacement of the residues in their active centre (Fig. 7.4). These proteins acquired the capacity to stimulate the growth of different cell lines by binding to receptor tyrosine kinases of the MET family. At present, there is no clear explanation of how this functional transition (from proteinase to growth factor) occurred. In plasminogen and hepatocyte growth factor/ macrophage stimulating proteins we see the final results of a change of function but there are no hints as to the possible intermediate stages of the functional transition.

7.2.3 Major change of function by domain acquisitions

Serine proteinases

In contrast with the zymogens of simple proteinases of the trypsin family (pancreatic proteinases, glandular kallikreins, serprocidins, etc., in which only a signal peptide is attached to the amino-terminal end of the catalytic region), in the proteinases involved in regulation of blood coagulation,

```
                EEEEEE        EEEEEEEE       EEEE          EEE                    EEEEEE              EEEEEE
plmn_human   VVGGCVAHPSWPWQVSLRTRF.GMHFCGGTLISPEWVLTAAHCLEKSPRP.SSYKVILGAHQE.V.NLEPHVQEIEVSRLFLEPTRKDIALLKLSSPAV
plmn_mouse   VVGGCVANPHSWPWQISLRTRFTGQHFCGGTLIAPEWVLTAAHCLEKSSRP.EFYKVILGAHEEYIRGLD..VQEISVAKLILEPNNRDIALLKLSRPAT
hgf_mouse    VVNGIPTQTTV.GWMVSLKYR..NKHICGGSLIKESWVLTARQCFPARNKDLKDYEAWLGIHDVHERGEEKRKQILNISQLVVGPEGSDLVLLKLARPAI
hgf_human    VVNGIPTRTNI.GWMVSLRYR..NKHICGGSLIKESWVLTARQCFPSR..DLKDYEAWLGIHDVHGRGDEKCKQVLNVSQLVVGPEGSDLVLMKLARPAV
hgf_xenopus  VVNGIPAQTRK.VWMVSVRYR..NAHKCGGTLIKENWVLTARQCFLSGDIDLKYYEAWLGVHNIYST.TEKHKQILNISQLVHGPKGSNLVLLKLSRPAT
msp_human    VVGGHPGNSP...WTVSLRNR.QGQHFCGGSLVKEQWILTARQCFSSCHMPLTGYEVWLGTLFQNPQHGEPSLQRVPVAKMVCGPSGSQLVLLKLERSVT
Consensus    VV-G-------W--S----R---H-CGG-L---W-LTA--C-------Y---LG--------Q------P----L-KL-----
                                                            *                                     *

                        EEEE                   EEEEE                          EEE                 EEE
plmn_human   ITDKVIPACLPSPNYVVADRTECFITGWGETQGTFGA.GLLKEAQLPVIENKVCNRYEFLNGRV..QSTELCAGHLAGGTDSCQGDSGGPLVCFEKDKYI
plmn_mouse   ITDKVIPACLPSPNYMVADRTICYITGWGETQGTFGA.GRLKEAQLPVIENKVCNRVEYLNNRV..KSTELCAGQLAGGVDSCQGDSGGPLVCFEKDKYI
hgf_mouse    LDNFVSTIDLPSYGCTIPEKTTCSIYGWGYT.GLINADGLLRVAHLYIMGNEKCSQHH..QGKVTLNESELCAGAEKIGSGPCEGDYGGPLICEQHKMRM
hgf_human    LDDFVSTIDLPNYGCTIPEKTSCSVYGWGYT.GLINVDGLLRVAHLYIMGNEKCSQHH..RGKVTLNESEICAGAEKIGSGPCEGDYGGPLVCEQHKMRM
hgf_xenopus  LNAYVDRIKLPNYGCTIPEKTTCSVYGWGHT.GTNDYDGQLQEGTLHIVGNEKCNENH..KGKITVNESEICAIGETANIGPCERDYGGPLICEENRTHL
msp_human    LNQRVALICLPPEWYVVPPGTKCEIAGWGETKGTGN.DTVLNVALLNVISNQECNIKH..RGRV.RESEMCTEGLLAPVGACEGDYGGPLACFTHNCWV
Consensus    ----V----LP------T-C---GWG-T-G-----L---L--N--C------------E-C------C--D-GGPL-C-----
                                                            ▲              *

                EEEEE                 EEEE
plmn_human   LQGVTSWGLGCARPNKPGVYVRVSRFVTWIEGVM....RNN
plmn_mouse   LQGVTSWGLGCARPNKPGVYVRVSRFVDWIEREM....RNN
hgf_mouse    VLGVIVPGRGCAIPNRPGIFVRVAYYAKWIHKV ILTYKL...
hgf_human    VLGVIVPGRGCAIPNRPGIFVRVAYYAKWIHKIILTYKVPQS
hgf_xenopus  VQGVIIPGRGCAIQKRPVIFVRVAYYAKWIHKIMLTYKAP..
msp_human    LEGIIIPNRVCARSRWPAVFTRVSVFVDWIHKVM...RL..G
Consensus    --G-----CA----P----RV----WI------
```

Fig. 7.4 Multiple alignment of the amino acid sequences of the catalytic chains of plasmins (human, mouse and the toad *Xenopus laevis*) and macrophage stimulating factor (human). The consensus sequence shows residues identical in all sequences. The asterisks indicate the positions corresponding to residues of the catalytic triad and the arrow indicates the residue defining the primary specificity of plasmins. Note that in macrophage stimulating protein and hepatocyte growth factors the His-57, Asp-102 and Ser-195 residues of the catalytic triad as well as Asp-189 of the primary specificity site are replaced. The residues predicted to be in β-strands or α-helices by homology modelling are indicated by E.

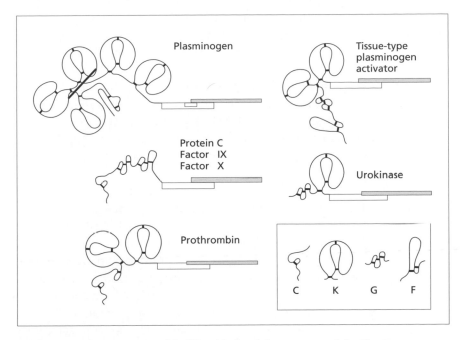

Fig. 7.5 Modular structure and building blocks of the proteases of the blood coagulation and fibrinolytic cascades. The hatched bars represent the trypsin-homolog serine protease domain. The box shows the schemes of vitamin K-dependent calcium-binding module (C), kringle module (K), growth factor module (G) and finger module (F). (From Patthy, L. (1985) Evolution of the proteases of blood coagulation and fibrinolysis by assembly from modules. *Cell* **41**, 657–63 © Cell Press.) Reprinted from Cell, 41, Patthy, L., Evolution of the proteases of blood coagulation and fibrinolysis by assembly from modules, 657–663, Copyright 1985, with permission from Elsevier.

fibrinolysis and complement activation very large segments are joined to the trypsin-homolog region (Fig. 7.5). These 'nonproteinase' parts of plasma proteinases consist of multiple structural–functional domains that were introduced into ancestral serine proteinases (between their signal-peptide and serine-proteinase domains) by exon shuffling (see section 8.1.3). In other words, the function of the proteinases was modified not only by point mutations but also by domain insertions and duplications (Patthy, 1985). In the majority of cases, the serine proteinase domains of the resulting multidomain proteinases retained proteolytic activity, and point mutations in these domains have led to altered (usually narrower) sequence specificity as dictated by their specific regulatory roles.

The selective value of the domains that were joined to the serine-proteinase domain may be illustrated by the fact they are usually involved in interactions with cofactors, substrates and inhibitors that are crucial for the regulation of the activity and activation of these enzymes. For example, the vitamin K-dependent calcium-binding domains of prothrombin, and of coagulation factors VII, IX, X and protein C, anchor these proteinases and their zymogens to phospholipid membranes, thereby ensuring proper regulation of the coagulation cascade. Similarly, the kringle domains of plasmin (and plasminogen) are critical for the binding of this proteinase to its primary substrate, fibrin. In the case of plasminogen the functional import- ance of the nonproteinase part is especially obvious. The serine-proteinase domain of plasmin has proteinase specificity very similar to that of trypsin: it cleaves all kinds of proteins at Lys-X and (at a lower rate) at Arg-X bonds. Its fibrin specificity is due primarily to the fact that its kringle domains have specific fibrin-binding sites that target the enzyme to fibrin.

The value of domain-acquisition mutations is that they can endow the recipient proteinase with novel binding specificities and can lead to dramatic changes in its regulation and targeting. Furthermore, in such modular proteins numerous distinct binding specificities may coexist, making such proteins ideal members of regulatory networks where multiple interactions are critical. Moreover, and perhaps most significantly, the 'domain-acquisition mutation' has a great chance of conferring immediate selective advantage, thereby increasing the chance of survival of the fusion mutant.

The power of domain acquisitions is also illustrated by protein-engineering experiments, in which the same principle has been exploited to change the regulation, specificity and targeting of some serine proteinases (Patthy, 1993). Research on fibrinolytic proteinases (plasminogen, and urokinase-type and tissue-type plasminogen activators) has extensively employed such domain- shuffling experiments to create chimeric proteins, variants with module deletions, duplications and rearrangements, to engineer novel fibrinolytic enzymes. The general success of this approach provides convincing evidence that domain acquisitions can bring about sudden advantageous changes in the properties of enzymes. It thus appears that acquisition of modules is a major source of evolutionary novelty. The fact that the overwhelming majority of the regulatory proteinases were constructed from domains borrowed from other proteins underlines the extreme importance of this evolutionary mechanism. What made this mechanism so popular is that acquisition of a domain possessing a given specificity can immediately endow the 'host protein' with a novel specificity, without the loss of its earlier binding specificities. In contrast, changes in specificity by point muta- tions usually may be achieved only at the expense of the loss of another specificity. (For example, a proteinase with a trypsin-like sequence specificity can be converted to an enzyme with chymotrypsin-like specificity only at the expense of losing the trypsin-like specificity.)

Zinc metalloproteinases

The matrixin family of zinc metalloproteinases consists of a number of enzymes (interstitial collagenases, stromelysin, proteoglycanase, gelatinases, etc.) involved in the degradation of various constituents of the extracellular matrix. Gelatinases are unique among matrixins in having high affinity for denatured collagens, this affinity being critical for their substrate specificity. These enzymes arose from an ancestral collagenase-like enzyme as a result of the acquisition of a domain closely related to the gelatin-binding domain of fibronectin (for further details see Chapter 8). Acquisition of a gelatin-binding domain could result in a sudden change in the substrate specificity of the ancestral collagenase-like enzyme, illustrating the immediate advantage of borrowing a module with established binding specificity.

7.3 Similarities and differences in the evolution of paralogous and orthologous proteins

As we have discussed in Chapter 5, the three-dimensional structure of proteins is highly conserved in divergent protein families. The common protein fold is thus preserved in both paralogs and orthologs, and the structural elements important for the integrity of the protein fold usually accept mutations at similar rates in orthologous and paralogous proteins. One general difference between orthologous and paralogous proteins is, however, that orthologous proteins are likely to fulfil very similar functions in different species (except for some modifications as discussed in Chapter 5), whereas paralogous proteins are more likely to have diversified in function (except for cases when numerous paralogous genes are needed for the production of large quantities of the same protein). Consequently, in comparing orthologous proteins, the residues that are critical for structure, function and specificity are equally likely to be conserved; whereas in comparing paralogous proteins that fulfil very different functions, only the residues essential for the preservation of structure may be conserved.

A practical consequence of this is that when orthologous proteins fulfilling the same function in different species are compared, all those residues that are critical for maintaining the three-dimensional structure or for the performance of the specific biological function will be identified as conserved residues. For example, if we compare orthologs of trypsin from a variety of species (see Fig. 7.1(a)), the conserved residues will include the residues critical for maintaining the trypsin fold (defining its topology), those critical for proteinase activity (spatial arrangement of the catalytic triad: His-57, Asp-102 and Ser-195), and those essential for defining the cleavage specificity of trypsin (Asp-189, Gly-216 and Gly-226). Similarly, if we compare orthologs of chymotrypsin (see Fig. 7.1(b)) or orthologs of elastase (see Fig. 7.1(c)) from different species, the residues essential for maintaining the serine-proteinase fold, those critical for proteinase activity (spatial arrangement of the catalytic triad: His-57, Asp-102, Ser-195), and those essential for defining their distinct cleavage specificities will be conserved in these ortholog comparisons.

If we compare proteolytically active paralogs of the trypsin family that differ only in their cleavage specificities (e.g. human trypsin vs. human chymotrypsin vs. human elastase), residues involved in defining the different

```
ctrb_human  IVNGEDAVPGSWPWQVSLQ.DKTG..FHFCGGSLISEDWVTAAHCGVR..TSDVVVAGEFDQGSDEENIQVLKI..AKVFKNPKF..SILTVNNDITLL
el3a_human  VVHGEDAVPYSWPWQVSLQYEKSGSFYHTCGGSLIAPDWVVTAGHCISRDLTYQVVL.GEYNLAVKEGPEQVIPINSEELFVHPLWNRSCVACGNDIALI
try1_human  IVGGYNCEENSVPYQVSL...NSG..YHFCGGSLINEQWVVSAGHCYKSRI..QVRL.GEHNIEVLEGNEQF..INAAKIIRHPQYDRK..TLNNDIMLI
Consensus   -V-G-----S-P-QVSL----G--H-CGGSLI--WVV-A-HC------V----GE-----E---Q-----I-----P------NDI-L-
                                       *                                                              *

ctrb_human  KLATPARFSQTVSAVCLPSADDDFPA.GTLCATTGWGKTKYNANKTPDKLQQAALPLLSNAECKK..SWGRRITDVMICAG..ASGVSSCMGDSGGPLVC
el3a_human  KLSRSAQLGDAVQLASLPPAGDILPN.KTPCYITGWGRL.YTNGPLPDKLQQARLPVVDYKHCSRWNWWGSTVKKTMVCAG..GYIRSGCNGDSGGPLNC
try1_human  KLSSRAVINARVSTISLPTAP...PATGTKCLISGWGNTASSGADYPDELQCLDAPVLSQAKCE..ASYPGKITSNMFCVGFLEGGKDSCQGDSGGPVVC
Consensus   KL---A----V----LP-A----P---T-C---GWG-------PD-LQ-----P-----C-------M-G------C-GDSGGP--C
                                                                                     ▲              *

ctrb_human  .QKDGAWTLVGIVSWGSD.TCS.TSSPGVYARVTKLIPWVQKILAAN~
el3a_human  PTEDGGWQVHGVTSFVSGFGCNFIWKPTVFTRVSAFIDWIEETIASH~
try1_human  ...NG..QLQGVVSWGDG..CAQKNKPGVYTKVYNYVKWIKNTIAANS
Consensus   ----G--S-----G---S-----C---P-V---V----W----A--~
                           ▲                 ▲
```

Fig. 7.6 Multiple alignment of the amino acid sequences of the catalytic chains of human trypsin, chymotrypsin and elastase. The consensus sequence shows residues identical in all sequences. The asterisks indicate the residues of the catalytic triad and arrows indicate the residues involved in defining the substrate specificity of these enzymes. Note that the residues of the catalytic triad are conserved, but the residues controlling substrate specificity are not.

cleavage specificities of trypsin, chymotrypsin, elastase, etc. will no longer show up among the residues that are conserved, whereas the residues essential for correct topology and catalytic activity (His-57, Asp-102, Ser-195) are still conserved (Fig. 7.6).

If we compare a group of paralogs in which members devoid of proteo-lytic activity (haptoglobin, hepatocyte growth factor, macrophage stimulating protein, some azurocidins, etc.) are included then the residues conserved in such comparisons may no longer include residues critical for catalytic activ-ity but still include those that are essential for the maintenance of the three-dimensional structure (see Figs 7.2–7.4).

The three types of analyses listed above thus permit the systematic identification of residues essential for structural integrity, catalytic activity and substrate specificity in this protein family. Such analyses are very useful for the prediction of structure–function aspects of paralogous proteins. In general, residues critical for the three-dimensional structure of a protein fold are likely to show detectable conservation in all members of the family (since they all have similar protein folds), whereas residues critical for a specific function will show detectable conservation only in those members of the family that share that function.

We may also illustrate this principle with the kringle domains of plasma proteinases, which fulfil diverse binding functions in these proteins. Despite their functional diversity, the sequences of the five kringle domains of human plasminogen, the two kringles of human prothrombin, the two kringles of human tissue plasminogen activator (t-PA), the kringle of factor XII and the kringle of human urokinase are closely related (Fig. 7.7(a)) and their three-dimensional structures are also strikingly similar. In the multiple alignment of these paralogous sequences the segments involved in the maintenance and stability of the kringle fold (β-structures, disulphide bonds) are conserved, whereas flexible loops between conserved structural elements coincide with variable regions and gaps of the multiple alignment.

Kringle 1 and kringle 4 of plasminogens are known to have related lysine-binding sites that are critical for the affinity of plasmin for the fibrin substrate. These kringles were shown to recognize a pair of NH2– and COOH– groups in ligands (such as lysine) provided that the distance between these posit-ively and negatively charged groups is the same as in lysine. Since this lysine affinity is shared by kringle 1 and kringle 4 domains of plasminogens from various vertebrate species, the residues involved in binding are expected to be conserved in all kringle 1 and kringle 4 domains. Comparison of kringle 1 and kringle 4 sequences identifies some regions uniquely conserved in these kringles (Fig. 7.7(b), cf. Fig. 7.7(a)), suggesting that these regions might be important for lysine binding. Studies on kringles have indeed identified the conserved Asp-56 and Arg-70 as primary determinants of lysine binding: they provide complementary negative and positive charges that interact with the charged groups of the ligand (Trexler, Vali & Patthy, 1983).

```
                EEEE   EEEE                                              EEEEEEE  EEEEEEEE
k1_tpa      CYEDQGISYRGTWSTAESGAECTNWNSSALAQKPYSGRRPDAIRLGLGNHNYCRNPDRDSK.PWCYVFKAGKYSSEFCSTPAC
k2_tpa      CYFGNGSAYRGTHSLTESGASCLPWNSMILIGKVYTAQNPSAQALGLGKHNYCRNPDGDAK.PWCHVLKNRRLTWEYCDVPSC
k_uk        CYEGNGHFYRGKASTDTMGRPCLPWNSATVLQQTYHAHRSDALQLGLGKHNYCRNPDNRRR.PWCYQVGLKPLVQECMVHDC
k_XII       CYDGRGLSYRGLARTTLSGAPCQPWASEA.TYRNVTAEQ..ARNWGLGGHAFCRNPDNDIR.PWCFVLNRDRLSWEYCDLAQC
k1_plg      CKTGNGKNYRGTMSKTKNGITCQKWSSTS.PHRPR.FSPATHPSEGL.EENYCRNPDNDPQGPWCYTTD.PEKRYDYCDILEC
k2_plg      CMHCSGENYDGKISKTMSGLECQAWDSQS.PHAHG.YIPSKFPNKNL.KKNYCRNPDRELR.PWCFTTD.PNKRWELCDIPRC
k3_plg      CLKGTGENYRGNVAVTVSGHTCQHWSAQT.PHTHN.RTPENFPCKNL.DENYCRNPDGKRA.PWCHTTN.SQVRWEYCKIPSC
k4_plg      CYHGDGQSYRGTSSTTTTGKKCQSWSSMT.PHRHQ.KTPENYPNAGL.TMNYCRNPDADKG.PWCFTTD.PSVRWEYCNLKKC
k5_plg      CMFGNGKGYRGKRATTVTGTPCQDWAAQE.PHRHSIFTPETNPRAGL.EKNYCRNPDGDVGGPWCYTTN.PRKLYDYCDVPQC
k1_pt       CAEGLGTNYRGHVNITRSGIECQLWRSRY.PHKPE.INSTTHPGADL.QENFCRNPDSSTTGPWCYTTD.PTVRRQECSIPVC
k2_pt       CVPDRGQQYQGRLAVTTHGLPCLAWASAQ.AKALS.KHQDFNSAVQL.VENFCRNPDGDEEGVWCYVAG.KPGDFGYCDLNYC
Consensus   C----G--Y-G-------G--C--W--------------------L-----CRNPD------WC-----------C----C
```

(a)

```
k1_bovin    CKTGNGQTYRGTTAETKSGVTCQKWSATSPHVPKFSPEKFPLAGLEENYCRNPDNDENGPWCYTTDPDKRYDYCDIPEC
k1_hedgehog CKVGNGKYYRGTVSKTKTGLTCQKWSAETPHKPRFSPDENPSEGLDQNYCRNPDNDPKGPWCYTMDPEVRYEYCEIIQC
k1_human    CKTGNGKNYRGTMSKTKNGITCQKWSSTSPHRPRFSPATHPSEGLEENYCRNPDNDPQGPWCYTTDPEKRYDYCDILEC
k1_mouse    CKTGIGNGYRGTMSRTKSGVACQKWGATFPHVPNYSPSTHPNEGLEENYCRNPDNDEQGPWCYTTDPDKRYDYCNIPEC
k1_pig      CKTGNGKNYRGTTSKTKSGVICQKWSVSSPHIPKYSPEKFPLAGLEENYCRNPDNDEKGPWCYTTDPETRFDYCDIPEC
k1_rhesus   CKTGNGKNYRGTMSKTRTGITCQKWSSTSPHRPTFSPATHPSEGLEENYCRNPDNDGQGPWCYTTDPEERFDYCDIPEC
k4_bovin    CYHGNGQSYRGTSSTTITGRKCQSWSSMTPHRHLKTPENYPNAGLTMNYCRNPDAD.KSPWCYTTDPPRVRWEFCNLKKC
k4_chicken  CYQGNGVSYRGTASFTITGKKCQAWNSMSPHRHNKTESHFPNADLRQNYCRNPDAD.RSPWCYTTDPSVRWEYCNLKRC
k4_hedgehog CYQGNGQSYRGTSSTTITGKKCQPWTSMRPHRHSKTPENYPDADLTMNYCRNPDGD.KGPWCYTTDPSVRWEFCNLKKC
k4_human    CYHGDGQSYRGTSSTTTTGKKCQSWSSMTPHRHQKTPENYPNAGLTMNYCRNPDAD.KGPWCFTTDPSVRWEYCNLKKC
k4_mouse    CYQSDGQSYRGTSSTTITGKKCQSWAAMFPHRHSKTPENFPDAGLEMNYCRNPDGD.KGPWCYTTDPSVRWEYCNLKRC
k4_pig      CYRGNGESYRGTSSTTITGKKCQSWVSMTPHRHEKTPGNFPNAGLTMNYCRNPDAD.KSPWCYTTDPSVRWEYCNLKKC
k4_rhesus   CYHGDGQSYRGTSSTTTTGKKCQSWSSMTPHWHEKTPENFPNAGLTMNYCRNPDAD.KGPWCFTTDPSVRWEYCNLKKC
k4_wallaby  CYEGKGENYRGTTSTTISGKKCQAWSSMTPHQHKKTPDNFPNADLIRNYCRNPDGD.KSPWCYTMDPTVRWEFCNLEKC
Consensus   C----G--YRGT---T--G--CQ-W----PH---------P---L--NYCRNPD-D---PWC-T-DP--R---C----C
                                                                       *                *
```

(b)

Fig. 7.7 (a) Multiple alignment of the kringle domains of human plasma proteases (tpa, tissue plasminogen activator uk, urokinase XII, factor XII; plg, plasminogen; pt, prothrombin). The consensus sequence shows residues identical in all sequences. The residues involved in β-strands are indicated by E. (b) Multiple alignment of the sequences of the lysine-binding kringle 1 and kringle 4 domains of plasminogens from different vertebrate species. Note that residues specifically conserved in these lysine-binding kringles include Asp-56 and Arg-70 residues (asterisks), which are directly involved in binding the α-COOH and ε-NH$_2$ groups of lysine.

7.4 Predicting the function of proteins by homology

Since orthologous proteins are likely to fulfil very similar functions in different species, the function of a protein-coding gene can be predicted with great confidence if it can be shown (by evolutionary analyses) to be orthologous with genes of known function (cf. section 5.6).

Prediction of the function of paralogs is more difficult, since duplicated genes are more likely to acquire very different functions. Nevertheless, changes in the function of paralogs by point mutations usually occur gradually, some of the earlier functions of the common ancestor being retained. If point mutations reshape older functions gradually and if transition to a novel function leads through a continuum of functions then the function of a novel member of a protein family may be predicted by evolutionary analyses. Based on this principle, the function of a newly identified gene is most likely to be similar to its closest homologs. Prediction of the function of a novel gene thus must be based on analysis of the evolution of functions in that family. For example, if sequence comparison assigns a novel protein to the family of trypsin-like serine proteinases then this conclusion has the implication that the novel gene may also encode a proteinase. If phylogenetic analyses unambiguously place this novel gene in the trypsin family then the most plausible prediction is that the novel gene also encodes a proteinase with trypsin activity.

Some of the examples discussed above prompt caution that such prediction schemes may be misleading in the case of more drastic changes of function. For example, based on homology and phylogenetic analyses alone one could have been led to the (erroneous) conclusion that hepatocyte growth factor, macrophage stimulating protein or apolipoprotein may have enzymatic functions very similar to that of the fibrinolytic proteinase, plasmin.

Nevertheless, there are several cases that demonstrate the usefulness of homology-based function prediction even in the case of distant homologs, such as homologous domains of otherwise unrelated multidomain proteins. For example, the homology of the large extracellular protein fibronectin and the matrix metalloproteinase gelatinase A (cf. section 7.2.3) is restricted to the presence of fibronectin type II (FN2) domains in both proteins. Since the FN2 domains of fibronectin were shown to be involved in binding this protein to denatured collagen, it was plausible to predict that the FN2 domains of gelatinase A are responsible for the high affinity of this enzyme to gelatin, a prediction that was confirmed experimentally (Bányai & Patthy, 1991).

As another example, we may mention the case of RYK receptor tyrosine kinases. The RYK receptor tyrosine kinases form a distinct subfamily of tyrosine kinases with an unusually short extracellular (putative ligand-binding) region. Although RYK receptor tyrosine kinases were expected to function in the transduction of growth-regulatory information across the plasma membrane, nothing was known about the ligands of these orphan receptors. Using sensitive tools for the detection of distant homologies it has been shown that the extracellular domain of RYK receptors is related to the Wnt-binding domain of Wnt-inhibitory-factor-1 (WIF-1) and therefore it was predicted that the extracellular WIF-related domain of RYK receptors might serve to bind to Wnt proteins or related ligands (Patthy, 2000). This prediction has gained support from experiments that showed that RYKs function as Wnt coreceptors (Yoshikawa et al., 2003; Lu et al., 2004; Inoue et al., 2004).

7.5 Nonhomology-based methods for the prediction of the function of proteins

The 'function' of a protein has various aspects from molecular function to its involvement in a biological process through interaction with other proteins. Some aspects of the molecular function of proteins (e.g. presence of functional sites, signals for subcellular localization, post-translational modifications) can be predicted even if no functional information is available for its homologs.

Functional associations between proteins can frequently be inferred from genomic associations between their genes. For example, groups of genes that are required for the same function usually show similar phylogenetic distribution, are often located in close proximity on the genome (e.g. in operons) and tend to be involved in gene-fusion events. The large amount of completely sequenced genomes allows genomic context analyses to predict reliable functional associations between proteins. These non-homology-based comparative genomic methods include methods that rely on the fact that genes encoding physically interacting partners or members of shared metabolic pathways tend to be proximate on the genome (**genomic proximity methods**), tend to evolve in a correlated manner (**phylogenetic profiling methods**) and to be fused as a single sequence in some organisms (**domain fusion methods**).

7.6 Detecting distant homology of protein-coding genes

7.6.1 Detecting distant homology by consensus approaches

Diversification of function of paralogous protein-coding genes can eliminate all sequence similarity in regions involved in the distinct functions of the paralogs; sequence similarity may be preserved only in short regions that are critical for the maintenance of the given protein fold (Lesk & Fordham, 1996). Since similar three-dimensional structures may be built from very different amino acid sequences, in the case of distantly related paralogs it may be impossible to detect homology by global scoring systems. In this zone of very weak sequence similarities (the 'twilight zone') knowledge-based techniques are needed to decide whether a certain degree of similarity is due to common ancestry or merely reflects chance similarity of unrelated proteins.

Statistical tests are commonly used to assess whether two sequences are more similar than would be expected by chance. In such approaches, the similarity score of the actual comparison is compared with the distribution of the scores determined for pairs of a large number of random permutations of the two sequences, and the standard deviation of the comparison above the mean of the randomized comparisons is calculated. Although these tests readily identify very significant similarities that reflect genuine homology, the general experience is that distant relationships with lower scores remain buried in a background of irrelevant chance similarities. The lesson from such analyses is that a sharp separation of low-scoring related and unrelated sequences cannot be achieved by these statistical analyses (Pearson, 1991). A biologically significant similarity may not be significant statistically, and vice versa. For example, homology search protocols that use the same scoring system for all positions of the aligned sequences are unable to distinguish the lowest scoring related sequences from the highest scoring unrelated sequences

(Fig. 7.8). An unrelated sequence may give the same or an even higher similarity score to a test sequence than a genuinely homologous sequence.

To devise search protocols that can detect biologically significant similarities we must first analyse what distinguishes related and unrelated sequences that otherwise have identical scores with a query sequence. Since the sum of the scores is the same, the distinguishing feature can only be the **pattern of the scores**. As we have seen above, in multiple alignments of various families the conservation of residues is not randomly distributed but follows a pattern dictated by the structure of the protein fold. As a consequence, in the case of true homologs the scores come from positions that are characteristically conserved in the family of the query sequence, whereas in the case of false positives the scores come from positions that show little conservation during the evolution of that family. To eliminate false positives one must use search protocols that concentrate on scores coming from positions conserved in the family of the query sequence but suppress scores originating from variable positions. Another important point is that different positions of a sequence family differ not only with respect of the degree of conservation but also in their substitution patterns. There may be positions where substitution patterns conform to the mutation data matrix of the general data set, whereas in other positions a different substitution matrix may hold (compare different substitution patterns in transmembrane segments, reverse turns, α-helices, etc.).

A number of protocols permit the independent scoring of positions showing different degrees of conservation (and different substitution patterns); these are the **position-dependent scoring protocols** (Taylor, 1986; Patthy, 1987; Barton & Sternberg, 1990; Gribskov, Lüthy & Eisenberg, 1990; Taylor & Jones, 1991). A common feature of these approaches is that they all exploit the evolutionary information present in multiple alignments of homologous sequences. There are, however, significant differences in the way the information is extracted and utilized. Most approaches search for pairwise similarities to some type of **consensus sequence** (a profile, a template or a pattern) derived from a predefined multiple alignment. Once a biologically valid multiple alignment is constructed (see section 4.1), the evolutionary information characterizing the given protein family is condensed into a consensus sequence (pattern, template, profile, etc.). The various pattern-matching methods differ in their decision as to what features of the multiple alignment are emphasized and to what degree certain features of the related sequences are exaggerated. At one extreme we find procedures that retain all the **global information** present in the multiple alignment (Gribskov et al., 1990; Eddy, 1996), while the other extreme is represented by procedures (Henikoff & Henikoff, 1992; Bairoch & Bucher, 1994) that concentrate only on the most highly conserved **local similarities** (blocks, motifs, signatures).

The consensus sequence procedure (Patthy, 1987) occupies an intermediate position: it condenses global similarities of protein folds into consensus sequences and reduces the information to a character string in which low-

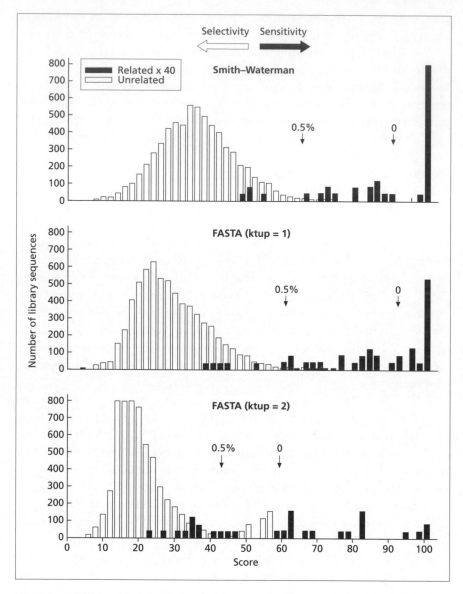

Fig. 7.8 Distribution of similarity scores for related and unrelated sequences in a protein sequence library. The query sequence was human calmodulin, which was compared with a protein sequence library using the Smith–Waterman algorithm or the FASTA program. The white bars indicate the number of unrelated sequences, the black bars indicate the number of related sequences. 0.5% and 0 mark cut-off values that exclude 99.5% and 100% of unrelated sequences. (From Pearson, W.R. (1991) Identifying distantly related protein sequences. *Current Opinion in Structural Biology* **1**, 321–6.) Reprinted from Current Opinion in Structural Biology, 1, Pearson, W.R., Identifying distantly related protein sequences, 321–326, Copyright 1991, with permission from Elsevier.

scoring positions (variable positions that are likely to have been randomized) are suppressed. The rationale of this approach is that in distantly related members of a protein family, all those regions that are involved in their vastly divergent functions may have been randomized, and only those regions carry remnants of common ancestry that were always required for the maintenance of the same overall three-dimensional structure. In multiple alignments of such distantly related sequences, regions essential for the structural integrity of the protein fold (e.g. regions that form regular secondary structures, cysteines forming disulphide bonds) are likely to be conserved, whereas regions that correspond to external loops connecting structural motifs are more tolerant to variation in sequence and to gap events. Consequently, each protein fold may be characterized by a unique pattern of accepted mutations, and this pattern may be formulated in some way to distinguish conserved positions, variable positions and gap positions.

The typical evolutionary 'behaviour' of the distribution of similarity scores along the consensus sequence of a protein family is shown in Fig. 7.9. As sequences diverge from perfect identity ($t^* = 100$, start of divergence) to lower values of t^*, scores are preferentially lost from 'variable' positions, while the scores in 'consensus' positions are much less affected. In the case of distantly related sequences ($t^* < 30$) similarity in the variable positions is indistinguishable from that expected by chance; the scores reflecting

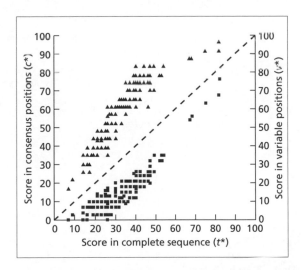

Fig. 7.9 Typical pattern of scores in a divergent protein family. The distribution of scores in pairwise comparisons of complement B modules was analysed with the consensus sequence characteristic of this family. Percent maximal scores in conserved positions (c*) and in variable positions (v*) are plotted as a function of the percent maximal score of the complete sequences (t*). Note that during divergence (as we move from (t* 100 to lower values) scores are preferentially lost from variable positions. (From Patthy, L. (1996) Consensus approaches in detection of distant homologies. *Methods in Enzymology* **266**, 184–98.) Reprinted from Methods in Enzymology, 266, Patthy, L., Consensus approaches in detection of distant homologies, 184–198, Copyright 1996, with permission from Elsevier.

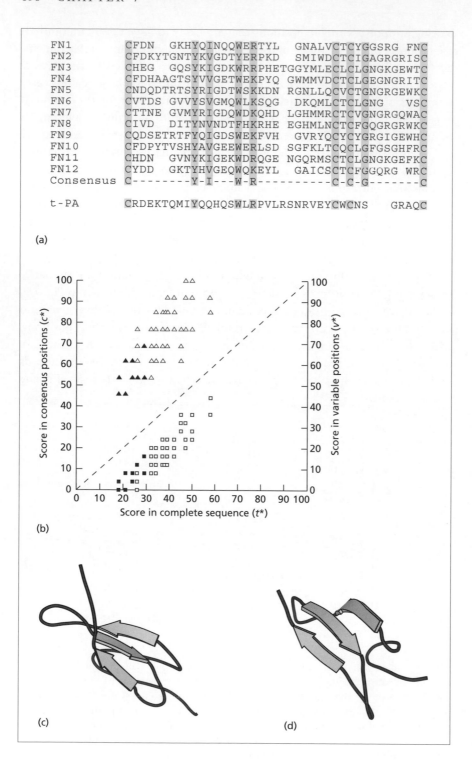

```
FN1        CFDN  GKHYQINQQWERTYL   GNALVCTCYGGSRG FNC
FN2        CFDKYTGNTYKVGDTYERPKD   SMIWDCTCIGAGRGRISC
FN3        CHEG  GQSYKIGDKWRRPHETGGYMLECLCLGNGKGEWTC
FN4        CFDHAAGTSYVVGETWEKPYQ   GWMMVDCTCLGEGNGRITC
FN5        CNDQDTRTSYRIGDTWSKKDN   RGNLLQCVCTGNGRGEWKC
FN6        CVTDS GVVYSVGMQWLKSQG   DKQMLCTCLGNG    VSC
FN7        CTTNE GVMYRIGDQWDKQHD   LGHMMRCTCVGNGRGQWAC
FN8        CIVD  DITYNVNDTFHKRHE   EGHMLNCTCFGQGRGRWKC
FN9        CQDSETRTFYQIGDSWEKFVH   GVRYQCYCYGRGIGEWHC
FN10       CFDPYTVSHYAVGEEWERLSD   SGFKLTCQCLGFGSGHFRC
FN11       CHDN  GVNYKIGEKWDRQGE   NGQRMSCTCLGNGKGEFKC
FN12       CYDD  GKTYHVGEQWQKEYL   GAICSCTCFGGQRG WRC
Consensus  C--------Y-I---W-R----------C-C-G-------C

t-PA       CRDEKTQMIYQQHQSWLRPVLRSNRVEYCWCNS    GRAQC
```

(a)

(b)

(c)   (d)

homology are concentrated only in the conserved positions. Accordingly, when distantly related proteins are compared, the conserved positions of the consensus sequence are the really sensitive indicators of homology, whereas similarity in variable positions is expected to be negligible. By disregarding similarity in variable positions the signal-to-noise ratio may be significantly improved, permitting the more sensitive detection of distantly related members of a protein family. When searching a protein sequence database with a consensus sequence, variable positions are masked to suppress scores that could lead to false positives.

Obviously, if we determine the pattern of scores for a protein family over a wide sequence similarity range, this pattern may be used to predict the probable distribution of scores for any new member of the family. Because the distribution of sequence similarity scores is unique for a protein fold, such analyses may be used to evaluate the biological significance of low similarity scores. In other words, if some sequence comparison procedure detects a marginal similarity of two proteins, the significance of this finding may be evaluated by determining whether the pattern of similarity scores agrees or conflicts with that expected on the basis of homology.

The consensus sequence procedure has proved to be very useful for the detection of distant homologies that were missed by conventional search procedures, and has been instrumental in the recognition of distant homologies of modules of mosaic proteins and has thus contributed significantly to the formulation of the principles of modular protein evolution by exon shuffling (Patthy, 1985). For example, the detection of the homology of domains of fibronectin and tissue plasminogen activator provided the first unequivocal example of modular protein evolution by exon shuffling (Bányai, Váradi & Patthy, 1983). Here we will illustrate the rationale of this approach with the finger and type III domains of fibronectin.

Consensus sequences were derived from multiple alignments of the finger domains of fibronectin (Fig. 7.10(a)) and were used to establish the

Fig. 7.10 (*Opposite*) (a) Multiple alignment of the 12 finger domains of fibronectin, FN1–FN12. The consensus sequence shows residues chemically conserved in all sequences. Chemically similar amino acids were grouped on the basis of the mutation probability matrix (see Table 3.1(a).): (Y,W,F), (I,L,V,M), (D,E,N,Q), (R,K), (S,T), H, P, G, A and C. Note that the majority of positions conserved in fibronectin fingers (highlighted with background tint) are also conserved in a segment of tissue plasminogen activator (t-PA). (b) Pattern of scores in the fibronectin type I (finger) module family analysed as in Fig. 7.9. Open symbols represent fibronectin/fibronectin comparisons, filled symbols represent fibronectin/t-PA comparisons. Note that distribution of scores in the two types of comparisons fit the same pattern. (From Patthy, L. (1996) Consensus approaches in detection of distant homologies. *Methods in Enzymology* **266**, 184–98.) (c) Ribbon diagram of the finger domain of t-PA. (d) Ribbon diagram of the fourth finger domain of fibronectin. Reprinted from Methods in Enzymology, 266, A,C,D, author original, B, Patthy, L., Consensus approaches in detection of distant homologies, 184–198, Copyright 1996, with permission from Elsevier.

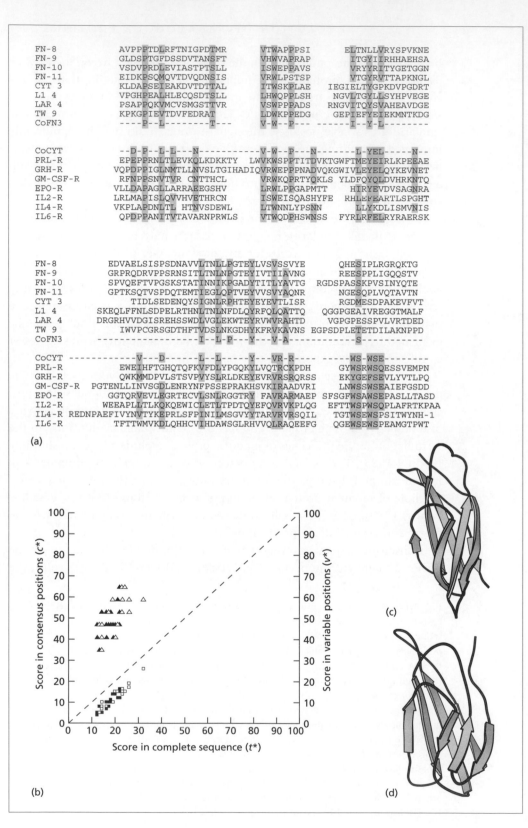

(a)

(b)

(c)

(d)

evolutionary behaviour of the distribution of scores in this small protein fold (Fig. 7.10(b)). Using the consensus sequence of fibronectin fingers we have found that a short segment of tissue plasminogen activator (t-PA) shows similarity with the finger domains of fibronectin (Bányai et al., 1983). Because of the low degree of overall sequence similarity (20–27% identity) of a very short segment, and because of gaps in the alignments, etc. this similarity is necessarily missed by conventional sequence comparison procedures as being statistically insignificant. Nevertheless, it is likely to reflect common ancestry since most of the residues conserved in all finger domains of fibronectin are also conserved in t-PA: the consensus sequence of fibronectin fingers is also valid for this short segment of t-PA (Fig. 7.10(a)). Furthermore, the distribution of scores in comparisons of this short segment of t-PA with the finger domains of fibronectin fits the pattern characteristic of fibronectin fingers (Fig. 7.10(b)); the pattern of similarity scores agrees with that expected on the basis of homology of fibronectin and t-PA. The implication of this conclusion was that the finger domains of fibronectin and this short segment of t-PA have a common evolutionary origin, i.e. they must have similar three-dimensional structures. Subsequent studies on the three-dimensional structures of finger domains of tissue plasminogen activator (Fig. 7.10(c)) and fibronectin (Fig. 7.10(d)) have confirmed this prediction: their protein fold is strikingly similar (Baron et al., 1990; Downing et al., 1992).

Such analyses may also show that two protein families are distantly related, i.e. they belong to the same superfamily. This point may be illustrated with the homology of members of the cytokine receptor family and members of the fibronectin type III domain family. Consensus sequences derived from a multiple alignment of the type III domains of fibronectin, tenascin, twitchin, etc. were used to establish the evolutionary behaviour of this protein fold (Fig. 7.11(a)). Using the consensus sequence of such fibronectin type III domains it was found (Patthy, 1990) that a segment of the extracellular portion of growth hormone receptor and other members of the cytokine receptor family conforms to this consensus sequence (although they show

Fig. 7.11 (*Opposite*) (a) Multiple alignment of type III domains (upper part) and cytokine receptors (lower part). In both cases the consensus sequence shows residues chemically conserved in the majority (>80%) of sequences (highlighted with background tint). (b) Pattern of scores in the fibronectin type III module family analysed with the type III consensus sequence as in Fig. 7.9. Open symbols denote comparisons of known type III modules, filled symbols denote comparisons of type III modules with cytokine receptors. Note that the distribution of scores in the two types of comparisons fit the same pattern. (From Patthy, L. (1996) Consensus approaches in detection of distant homologies. *Methods in Enzymology* **266**, 184–98.) (c) Ribbon diagram of the type III domain of growth hormone receptor. (d) Ribbon diagram of the tenth type III domain of fibronectin. Reprinted from Methods in Enzymology, 266, A,C,D, author original, B, Patthy, L., Consensus approaches in detection of distant homologies, 184–198, Copyright 1996, with permission from Elsevier.

only 14–23% sequence identity with type III modules). Despite this weak sequence similarity, the finding reflects common ancestry since most of the residues conserved in known type III domains are also conserved in all cytokine receptors: the 'consensus sequences' of the two groups are strikingly similar. Analysis of the distribution of scores in comparisons of type III domains with cytokine receptors using the consensus sequence revealed that this pattern of scores is superimposable with that observed in comparisons of type III domains (Fig. 7.11(b)). In other words, the consensus sequence characteristic of the known fibronectin type III domains is valid for a segment of the extracellular part of the cytokine receptors. The implication of this conclusion is that some modules of fibronectin and cytokine receptors have a common evolutionary origin, i.e. they must have similar three-dimensional structures. Subsequent studies on the structures of growth hormone receptor (Fig. 7.11(c)) and type III domains of fibronectin (Fig. 7.11(d)) and tenascin have indeed revealed that their protein fold is strikingly similar (Baron et al., 1992; Leahy et al., 1992).

Similar approaches were able to detect the distant homology of the NTR domains of netrins, secreted frizzled related proteins, complement components C3, C4, C5 and type I procollagen C-proteinase enhancer protein with the N-terminal domains of tissue inhibitors of metalloproteases (Bányai & Patthy, 1999). The validity of this homology is supported by the NMR structure of the NTR domain of human type I procollagen C-proteinase enhancer: a search of the Protein Data Bank for structurally related proteins with the program DALI yielded the N-terminal domain of TIMP-2 (and the laminin-binding domain of agrin) as the proteins with the highest similairity (Liepinsh et al., 2003).

As another example we may mention the distant homology of the N-terminal domains of plasminogen and hepatocyte growth factor with the 'apple' domains of the prekallikrein family, several invertebrate proteins and some proteins of apicomplexan parasites (Tordai, Bányai & Patthy, 1999). The justification for the inclusion of these diverse domains in a single (PAN-domain) family has been confirmed by the finding that the structure of an apple domain of the microneme protein of the parasite *Eimeria tenella* is similar to the N-terminal module of hepatocyte growth factor (Brown et al., 2003).

Similar analyses may also lead to the conclusion that a marginal sequence similarity is due to chance and not to homology. This point may be illustrated by the sequence similarity of surfactant protein B (SP-B) with the kringle domain of the blood coagulation factor XII. SP-B was found to exhibit 26% sequence identity with the kringle of factor XII (Fig. 7.12(a)) and it has been suggested that kringles and surfactant protein B might be related. To decide whether the pattern of scores is consistent with this assumption, the distributions of scores may be analysed with consensus sequences determined for both the saposin module family, to which SP-B belongs (Patthy, 1991), and the kringle module family, to which the kringle of factor XII belongs (see Fig. 7.7(a)). Comparison of SP-B with kringles (using kringle con-

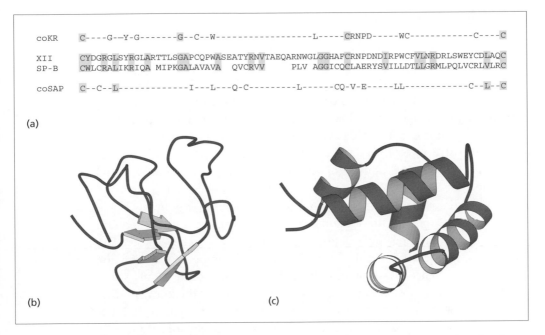

(a)

```
coKR    C----G--Y-G-------G--C--W-----------------L-----CRNPD-----WC------------C----C

XII     CYDGRGLSYRGLARTTLSGAPCQPWASEATYRNVTAEQARNWGLGGHAFCRNPDNDIRPWCFVLNRDRLSWEYCDLAQC
SP-B    CWLCRALIKRIQA MIPKGALAVAVA  QVCRVV    PLV AGGICQCLAERYSVILLDTLLGRMLPQLVCRLVLRC

coSAP   C--C--L-------------I---L---Q-C---------L------CQ-V-E-----LL------------C--L--C
```

(b)   (c)

Fig. 7.12 (a) Alignment of the sequences of surfactant protein B and the kringle of factor XII with consensus sequences of kringles (coKR) and saposins (coSAP). Note that the similarities in the SP-B/ FXII alignment do not conform to either the kringle or the saposin consensuses. (b) Ribbon diagram of kringle 4 of human plasminogen. (c) Ribbon diagram of the saposin fold of porcine NK-lysin.

sensus sequences) or the kringle of factor XII with members of the saposin family (using saposin consensus sequences) gave distributions of scores that were inconsistent with the assumption that the two families are related. Since neither the kringle consensus sequence nor the saposin consensus sequence is valid for the distribution of scores in the SP-B/factor XII comparison (Fig. 7.12(a)) their sequence similarity does not reflect homology. The implication of this conclusion is that the three-dimensional structure of the saposin domain is unlikely to be related to the characteristic all-β structure of kringles (Fig. 7.12(b)). Studies on the saposin domain (Liepinsh et al., 1997) have indeed shown that its 'four-helix bundle' structure assigns it to the all-α group of proteins (Fig. 7.12(c)).

7.6.2 Detecting distant homology by comparing three-dimensional structures

It is obvious from some of the above examples that proteins that have evolved from a common ancestor may still fold into homologous structures in spite of retaining little sequence similarity. The significance of this may be best illustrated by the fact that in a database of structurally aligned remote homologs (Holm et al., 1993; Holm & Sander, 1994) thousands of homologs show less than 25% pairwise sequence identity. Since homology of structure is better conserved than sequence, in principle distant homology may be more reliably detected by comparison of three-dimensional structures of proteins. However, information about three-dimensional structures is available for only a small fraction of known proteins, therefore the direct comparison of three-dimensional structures is rarely possible. Nevertheless, there are many cases where this type of information could be used to decide whether marginal sequence similarity is due to chance or reflects common evolutionary origin. For example, a DALI database search has revealed that the three-dimensional structure of the laminin-binding domain of agrin is homologous with the N-terminal domain of TIMPs (Stetefeld et al., 2001). It should be pointed out that this homology was not detected by sequence comparison alone, since sequence similarity between these domains is very low (<15%).

Another way of detecting distant homology is based on fold recognition by sequence **threading** (Bowie, Luthy & Eisenberg, 1991; Fischer & Eisenberg, 1996; Rost, Schneider & Sander, 1997). In these techniques one takes the amino acid sequence of a protein (with unknown three-dimensional structure) and evaluates how well this sequence fits into one of the known three-dimensional protein structures (see also section 5.7.2). In some techniques we first predict the secondary structure and solvent accessibility for each residue and then thread the resulting one-dimensional (1D) profile of predicted structure assignments into each of the known three-dimensional (3D) structures. Such 1D–3D threading methods essentially capture the fitness of an amino acid sequence for a particular 1D succession of secondary structure segments and residue solvent accessibility (Rost et al., 1997). The chances of finding the correct structure are quite good, especially since the number of different protein folds is limited (Chothia, 1992).

7.6.3 Detecting distant homology by comparing exon–intron structures

Insertion/removal of introns into/from protein-coding genes is a much rarer genetic event than point mutations, therefore protein-coding genes that have evolved from a common ancestor may have very similar exon–intron structures although they have retained little sequence similarity. Accordingly, the detection of distant homology of protein-coding genes may be based on the detection of the similarity of exon–intron organization of the genes (Patthy, 1988; Brown et al., 1995). The exon–intron organization of two homologous protein-coding genes is identical if they have the same set of homologous introns. As mentioned in Chapter 4, introns are considered to be homologous (irrespective of their size or sequence) if they are found in homologous positions of aligned nucleic acid and amino acid sequences, and if in translated regions they split the reading frame in the same phase.

The utility of this approach may be illustrated by the fact that a number of module homologies were detected by exploiting this technique (Patthy,

1988). Nevertheless, it must be emphasized that homologous protein-coding genes may have very different exon–intron structures due to intron insertion and removal (see Chapters 8 and 9). Accordingly, lack of homology of genomic organization cannot be taken as evidence against homology of the genes.

References

Almeida, R.P., Melchior, M., Campanelli, D. et al. (1991) Complementary DNA sequence of human neutrophil azurocidin, an antibiotic with extensive homology to serine proteases. *Biochemistry and Biophysics Research Communications* **177**, 683–95.

Bairoch, A. & Bucher, P. (1994) PROSITE: recent developments. *Nucleic Acids Research* **22**, 3583–98.

Bányai, L. & Patthy, L. (1991) Evidence for the involvement of type II domains in collagen binding by 72 kDa type IV procollagenase. *FEBS Letters* **282**, 23–5.

Bányai, L., Váradi, A. & Patthy, L. (1983) Common evolutionary origin of the fibrin-binding structures of fibronectin and tissue-type plasminogen activator. *FEBS Letters* **163**, 37–41.

Bányai, L. & Patthy, L. (1999) The NTR module: domains of netrins, secreted frizzled related proteins, and type I procollagen C-proteinase enhancer protein are homologous with tissue inhibitors of metalloproteases. *Protein Science* **8**, 1636–42.

Baron, M., Norman, D., Willis, A. et al. (1990) Structure of the fibronectin type 1 module. *Nature* **345**, 642–6.

Baron, M., Main, A.L., Driscoll, P.C., Mardon, H.J., Boyd, J. & Campbell, I.D. (1992) 1H NMR assignment and secondary structure of the cell adhesion type III module of fibronectin. *Biochemistry* **31**, 2068–73.

Barton, G.J. & Sternberg, M.J.E. (1990) Flexible protein sequence patterns. A sensitive method to detect weak structural similarities. *Journal of Molecular Biology* **212**, 389–402.

Borriello, F. & Krauter, K.S. (1991) Multiple murine alpha 1-proteinase inhibitor genes show unusual evolutionary divergence. *Proceedings of the National Academy of Sciences of the USA* **88**, 9417–21.

Bowie, J.U., Luthy, R. & Eisenberg, D. (1991) A method to identify protein sequences that fold into a known three-dimensional structure. *Science* **253**, 164–9.

Brierly, C.H. & Burchell, B. (1993) Human UDP-glucuronosyltransferases: chemical defence, jaundice and gene therapy. *Bioessays* **15**, 749–54.

Brown, N.P., Whittaker, A.J., Newell, W.R. et al. (1995) Identification and analysis of multigene families by comparison of exon fingerprints. *Journal of Molecular Biology* **249**, 342–59.

Brown, P.J., Mulvey, D., Potts, J.R., Tomley, F.M. & Campbell, I.D. (2003) Solution structure of a PAN module from the apicomplexan parasite *Eimeria tenella*. *Journal of Structural and Functional Genomics* **4**, 227–34.

Chothia, C. (1992) One thousand protein families for the molecular biologist. *Nature* **357**, 543–4.

Dahl, H.H.M., Brown, R.M., Hutchison, W.M. et al. (1990) A testis-specific form of the human pyruvate dehydrogenase E1 α subunit is coded for by an intronless gene on chromosome 4. *Genomics* **8**, 225–32.

Downing, A.K., Driscoll, P.C., Harvey, T.S. et al. (1992) Solution structure of the fibrin binding finger domain of tissue-type plasminogen activator determined by 1H nuclear magnetic resonance. *Journal of Molecular Biology* **225**, 821–33.

Eddy, S.R. (1996) Hidden Markov models. *Current Opinion in Structural Biology* **6**, 361–5.

Fischer, D. & Eisenberg, D. (1996) Protein-fold recognition using sequence-derived predictions. *Protein Science* **5**, 947–55.

Gribskov, M., Lüthy, R. & Eisenberg, D. (1990) Profile analysis. *Methods in Enzymology* **183**, 146–59.

Henikoff, S. & Henikoff, T.G. (1992) Amino acid substitution matrix from protein blocks. *Proceedings of the National Academy of Sciences of the USA* **89**, 10915–19.

Hill, R.E. & Hastie, N.D. (1987) Accelerated evolution in the reactive centre regions of serine protease inhibitors. *Nature* **326**, 96–9.

Holm, L., Ouzounis, C., Sander, C. et al. (1993) A database of protein structure families with common folding motifs. *Protein Science* **1**, 1691–8.

Holm, L. & Sander, C. (1994) The FSSP database of structurally aligned protein-fold families. *Nucleic Acids Research* **22**, 3600–9.

Inoue, T., Oz, H.S., Wiland, D., Gharib, S., Deshpande, R., Hill, R.J., Katz, W.S. & Sternberg, P.W. (2004) *C. elegans* LIN-18 is an RYK ortholog in parallel to LIN-17/frizzled in Wnt and functions signalling. *Cell* **118**, 795–806.

Irwin, D.M. & Wilson, A.C. (1989) Multiple cDNA sequences and the evolution of bovine stomach lysozyme. *Journal of Biological Chemistry* **264**, 11387–93.

Lawn, R.M., Schwarz, K. & Patthy, L. (1997) Convergent evolution of apolipoprotein(a) in primates and hedgehog. *Proceedings of the National Academy of Sciences of the USA* **94**, 11992–7.

Leahy, D.J., Hendrickson, W.A., Aukhil, I. et al. (1992) Structure of a fibronectin type III domain from tenascin phased by MAD analysis of the selenomethionyl protein. *Science* **258**, 987–91.

Lesk, A.M. & Fordham, W.D. (1996) Conservation and variability in the structures of serine proteinases of the chymotrypsin family. *Journal of Molecular Biology* **258**, 501–37.

Liepinsh, E., Andersson, M., Ruysschaert, J.M. et al. (1997) Saposin fold revealed by the NMR structure of NK-lysin. *Nature Structural Biology* **4**, 793–5.

Liepinsh, E., Banyai, L., Pintacuda, G., Trexler, M., Patthy, L. & Otting, G. (2003) NMR structure of the netrin-like domain (NTR) of human type I procollagen C-proteinase enhancer defines structural consensus of NTR domains and assesses potential proteinase inhibitory activity and ligand binding. *Journal of Biological Chemistry* **278**, 25982–9.

Lu, W., Yamamoto, V., Ortega, B. & Baltimore, D. (2004) Mammalian RYK is a Wnt coreceptor required for stimulation of neurite outgrowth. *Cell* **119**, 97–108.

McCarrey, J.R. (1990) Molecular evolution of the human *PGK*2 retroposon. *Nucleic Acids Research* **18**, 949–55.

McCarrey, J.R. & Thomas, K. (1987) Human testis-specific *PGK* gene lacks introns and possesses characteristics of a processed gene. *Nature* **326**, 501–5.

McLean, J.W., Tomlinson, J.E., Kuang, W.J. et al. (1987) cDNA sequence of human apolipoprotein(a) is homologous to plasminogen. *Nature* **300**, 132–7.

Nakashima, K., Ogawa, T., Oda, N. et al. (1993) Accelerated evolution of *Trimeresurus flavoviridis* venom gland phospholipase A2 isozymes. *Proceedings of the National Academy of Sciences of the USA* **90**, 5964–8.

Nakashima, K., Nobuhisa, I., Deshimaru, M. et al. (1995) Accelerated evolution in the protein-coding regions is universal in Crotalinae snake venom gland phospholipase A2 isozyme genes. *Proceedings of the National Academy of Sciences of the USA* **92**, 5605–9.

Nobuhisa, I., Nakashima, K., Deshimaru, M. et al. (1996) Accelerated evolution of *Trimeresurus okinavensis* venom gland phospholipase A2 isozyme-encoding genes. *Gene* **172**, 267–72.

Nobuhisa, I., Deshimaru, M., Chijiwa, T. et al. (1997) Structures of genes encoding phospholipase A2 inhibitors from the serum of *Trimeresurus flavoviridis* snake. *Gene* **191**, 31–7.

Ouellette, M., Hettema, E., Wüst, D. et al. (1991) Direct and inverted DNA repeats associated with P-glycoprotein gene amplification in drug resistant Leishmania. *EMBO Journal* **10**, 1009–16.

Patthy, L. (1985) Evolution of the proteases of blood coagulation and fibrinolysis by assembly from modules. *Cell* **41**, 657–63.

Patthy, L. (1987) Detecting homology of distantly related proteins with consensus sequences. *Journal of Molecular Biology* **198**, 567–77.

Patthy, L. (1988) Detecting distant homology of mosaic proteins. Analysis of the sequences of thrombomodulin, thrombospondin, complement components C9, C8a, C8b, vitronectin and plasma cell membrane protein PC-1. *Journal of Molecular Biology* **202**, 689–96.

Patthy, L. (1990) Homology of a domain of the growth hormone/prolactin receptor family with type III modules of fibronectin. *Cell* **61**, 13–14.

Patthy, L. (1991) Homology of the precursor of pulmonary surfactant-associated protein SP-B with prosaposin and sulfated glycoprotein 1. *Journal of Biological Chemistry* **266**, 6035–7.

Patthy, L. (1993) Modular design of proteases of coagulation, fibrinolysis, and complement activation: implications for protein engineering and structure–function studies. *Methods in Enzymology* **222**, 10–21.

Patthy, L. (1996) Consensus approaches in detection of distant homologies. *Methods in Enzymology* **266**, 184–98.

Patthy, L. (2000) The WIF module. *Trends in Biochemical Science* **25**, 12–13.

Pearson, W.R. (1991) Identifying distantly related protein sequences. *Current Opinion in Structural Biology* **1**, 321–6.

Rasmussen, B., Wiberg, F.C., Flodgaard, H.J. et al. (1997) Structure of HBP, a multifunctional protein with a serine proteinase fold. *Nature Structural Biology* **4**, 265–8.

Ray, B.K., Gao, X. & Ray, A. (1994) Expression and structural analysis of a novel highly inducible gene encoding alpha 1-antitrypsin in rabbit. *Journal of Biological Chemistry* **269**, 22080–6.

Ritter, J.K., Chen, F., Sheen, Y.Y. et al. (1992) A novel complex locus UGT1 encodes human bilirubin, phenol, and other UDP-glucuronosyltransferase isozymes with identical carboxyl termini. *Journal of Biological Chemistry* **267**, 3257–61.

Rost, B., Schneider, R. & Sander, C. (1997) Protein-fold recognition by prediction-based threading. *Journal of Molecular Biology* **270**, 471–80.

Stetefeld, J., Jenny, M., Schulthess, T. et al. (2001) The laminin-binding domain of agrin is structurally related to N-TIMP-1. *Nature Structural Biology* **8**, 705–9.

Tamechika, I., Itakura, M., Saruta, Y. et al. (1996) Accelerated evolution in inhibitor domains of porcine elafin family members. *Journal of Biological Chemistry* **271**, 7012–18.

Taylor, W.R. (1986) Identification of protein sequence homology by consensus template alignment. *Journal of Molecular Biology* **188**, 233–58.

Taylor, W.R. & Jones, D.T. (1991) Templates, consensus patterns and motifs. *Current Opinion in Structural Biology* **1**, 327–33.

Tordai, H., Bányai, L. & Patthy, L. (1999) The PAN module: the N-terminal domains of plasminogen and hepatocyte growth factor are homologous with the apple domains of the prekallikrein family and with a novel domain found in numerous nematode proteins. *FEBS Letters* **461**, 63–7.

Trexler, M., Vali, Z. & Patthy, L. (1983) Structure of the ω-aminocarboxylic acid-binding sites of human plasminogen. Arginine 71 and Aspartic acid 56 are essential for binding ligand by kringle 4. *Journal of Biological Chemistry* **257**, 7401–6.

Vinckenbosch, N., Dupanloup, I. & Kaessmann, H. (2006) Evolutionary fate of retroposed gene copies in the human genome. *Proceedings of the National Academy of Sciences of the USA* **103**, 3220–5.

Yokoyama, S. & Yokoyama, R. (1990) Molecular evolution of visual pigment genes and other G-protein-coupled receptor genes. In N. Takahata & J.F. Crow (eds), *Population Biology of Genes and Molecules*, pp. 307–22. Tokyo: Baifukan.

Yoshikawa, S., McKinnon, R.D., Kokel, M. & Thomas, J.B. (2003) Wnt-mediated axon guidance via the *Drosophila* derailed receptor. *Nature* **422**, 583–8.

Useful internet resources

Functional site prediction using phylogenetic information

ELM is a resource for predicting functional sites in eukaryotic proteins (http://elm.eu.org/).

The **ConSurf** server identifies functional regions in proteins by surface mapping of phylogenetic information (http://consurf.tau.ac.il).

Evolutionary Trace Server (http://mammoth.bcm.tmc.edu/server.html).

Evolutionary Trace report_maker is based on Evolutionary Trace to find out information about functional residues (http://mammoth.bcm.tmc.edu/report_maker/index.html).

The **Functional Site Prediction Server** predicts protein functional sites using sequence alignment information as well as Rosetta protein design and Rosetta free energy calculation (http://tools.bakerlab.org/~gcheng/).

Searching for 3D functional sites in a protein structure

PdbFun is a web server for the identification of local structural similarities between annotated residues in proteins. It compares query and target selections with a sequence-independent 3D comparison algorithm (http://pdbfun.uniroma2.it/).

Catalytic Site Atlas is a web server for searching a database of 3D templates of catalytic sites (http://www.ebi.ac.uk/thornton-srv/databases/CSA/).

PDBSiteScan is a web server for scanning a protein structure against its PDBSite database. It permits recognition of potential functional sites in tertiary structures of proteins and allows proteins with functional information to be annotated (http://wwwmgs.bionet.nsc.ru/mgs/gnw/pdbsitescan/).

Prediction of the function of proteins – methods based on function transfer after homology searches

Blast2GO uses BLAST to find homologous sequences to fasta-formatted input sequences and extracts GO terms to each obtained hit by mapping to existing annotation associations (http://www.blast2go.de/)

GOblet uses BLASTable databases with appropriate GO terms to assign a function to the query sequence (http://goblet.molgen.mpg.de/).

GOFigure BLASTs a query DNA or protein sequence against the GO annotated sequences to identify homologs and to predict function (http://udgenome.ags.udel.edu/gofigure/).

GOtcha uses sequence similarity searches to associate GO terms with the query sequence (http://www.compbio.dundee.ac.uk/gotcha/gotcha.php).

The **PFP Protein Function Prediction** server subjects the query sequence to an iterative PSI-BLAST search against UniProt and creates a list of primary GO annotations for the sequence (http://dragon.bio.purdue.edu/pfp/pfp.html).

GOPET is a complete automated tool for assigning molecular function terms to cDNA or protein sequences utilising gene ontology for annotation terms, GO-mapped protein databases for performing homology searches (http://genius.embnet.dkfz-heidelberg. de/menu/cgi-bin/w2h-open/w2h.open/w2h.startthis?SIMGO=w2h.welcome).

ProteomeAnalyst server is based on transfer of function based on homology (http://www.cs.ualberta.ca/%7Ebioinfo/PA/).

JAFA is the Joined Assembly of Function Annotations, a protein function prediction meta-server. JAFA accepts a protein sequence as input, queries several function prediction programs, and produces results assembled from these servers (http://jafa.burnham.org/).

Identification of functional signatures

PROSITE is a database of protein families and domains that also contains sequence patterns that are uniquely associated with a given function in a protein (http://www.expasy.ch/prosite/).

Prediction of the function of proteins – non-homology-based methods

The **Phydbac** (phylogenomic display of bacterial genes) server is an interactive resource for the annotation of bacterial genomes. It uses Phylogenetic Profiling, Genomic Proximity of genes and Domain Fusion events to infer functional predictions for the query (http://igs-server.cnrs-mrs.fr/phydbac/indexPS.html).

ProtFun 2.2 server produces ab initio predictions of protein function from post-translational modifications and localization (http://www.cbs.dtu.dk/services/ProtFun/).

The **Rosetta stone method** finds functionally related proteins by analysis of domain fusions (http://nihserver.mbl.ucla.edu/Rosetta/).

STRING (Search Tool for the Retrieval of Interacting Genes/Proteins) is a database of known and predicted protein–protein interactions (http://dag.embl-heidelberg.de).

Prediction of the function of proteins – structure-based methods

ProFunc is a web server for predicting the likely function of proteins whose 3D structure is known but whose function is not. ProFunc analyses the protein's sequence and structure, identifying functional motifs or close relationships to functionally characterized proteins. It uses a series of methods, including fold matching, residue conservation and surface cleft analysis to identify both the protein's likely active site and possible homologs in the Protein Data Bank (http://www.ebi.ac.uk/thornton-srv/databases/profunc/).

ProKnow infers protein function from protein structure (http://www.doe-mbi.ucla.edu/Services/ProKnow/).

Detection of distant sequence homologies

PSI BLAST (Position-Specific Iterated BLAST) employs iterative improvements of an automatically generated profile to detect weak sequence similarity (http://www.ncbi. nlm.nih.gov/blast/).

HHpred provides homology detection and structure prediction by HMM-HMM comparison (http://protevo.eb.tuebingen.mpg.de/toolkit/index.php?view=hhpred).

InterPro Scan provides an integrated search in PROSITE, Pfam, PRINTS and other family and domain databases (http://www.ebi.ac.uk/InterProScan/).

The **ScanProsite** tool scans protein sequences for the occurrence of patterns, profiles and rules (motifs) stored in the PROSITE database (http://www.expasy.org/tools/scanprosite/).

Pfam HMM search scans a sequence against the Pfam database (http://www.sanger.ac.uk/Software/Pfam/search.shtml).

SMART (Simple Modular Architecture Research Tool) (http://smart.embl-heidelberg.de/).

Block Searcher to search a sequence vs. blocks (http://blocks.fhcrc.org/blocks/blocks_search.html).

FUGUE: Sequence-structure homology recognition is a program for recognizing distant homologs by sequence-structure comparison (http://www-cryst.bioc.cam.ac.uk/fugue/).

PHYRE (Protein Homology/analogY Recognition Engine) is a web-based method for protein-fold recognition using 1D and 3D sequence profiles (http://www.sbg.bio.ic.ac.uk/~phyre/).

PRINTS is a compendium of protein fingerprints. A fingerprint is a group of conserved motifs used to characterize a protein family (http://bioinf.man.ac.uk/dbbrowser/PRINTS/).

P-val FPScan scans PRINTS with a protein query sequence (http://bioinf.man.ac.uk/cgi-bin/dbbrowser/fingerPRINTScan/muppet/FPScan.cgi).

ProDom may be used to scan a comprehensive set of protein domain families automatically generated from the SWISS-PROT and TrEMBL sequence databases (http://www.toulouse.inra.fr/prodom.html).

Compare your sequence with **ProDom** (http://prodom.prabi.fr/prodom/current/html/form.php).

eMOTIF is a collection of conserved motifs of sequence families (http://motif.stanford.edu/emotif/).

eMOTIF search (http://motif.stanford.edu/emotif/emotif-search.html).

Sequence logos

A **Gallery of Sequence Logos** (http://www.lecb.ncifcrf.gov/toms/sequencelogo.html).

Creating sequence logos, pictograms

WebLogo is a web-based tool to generate sequence logos (http://weblogo.berkeley.edu).

Pictogram is a tool to help visualize sequence alignments and consensus sequences (http://genes.mit.edu/pictogram.html).

Detection of distant homologies by comparing 3D structures

The **Dali** server is a network service for comparing protein structures in 3D. Comparing 3D structures may reveal biologically interesting similarities that are not detectable by comparing sequences (http://www.ebi.ac.uk/dali/).

LGA (Local-Global Alignment) is a method for finding 3D similarities in protein structures (http://predictioncenter.org/local/lga/lga.html).

The **Protein Fold Similarity** web server performs a search on the PDBselect database (http://hydra.icgeb.trieste.it/pride/cgi-bin/PDBselectScan.cgi).

LPFC (Library of Protein Family Cores) is useful for building models, threading and exploratory analysis (http://smi-web.stanford.edu/projects/helix/LPFC/).

Databases and Tools for 3-D Protein Structure Comparison and Alignment (http://cl.sdsc.edu/ce.html).

VAST Search is NCBI's structure–structure similarity search service. It compares 3D coordinates of a newly determined protein structure to those in the MMDB/PDB database (http://www.ncbi.nlm.nih.gov/Structure/VAST/vastsearch.html).

Detection of distant homologies by comparing exon–intron structures

The **FINEX** (Fingerprinting of INtron EXon boundaries) program allows sequence
homology searching techniques to be applied to exon–intron structures. In FINEX
the sequence data is replaced with a fingerprint abstracted from the intron/exon
boundary phase and the exon length. A single exon fingerprint can be compared
rapidly against all the entries in a library of fingerprints
(http://www.sanger.ac.uk/Software/analysis/finex/).

Chapter 8: Protein evolution by assembly from modules

The average size of a protein domain of known crystal structure is about 175 residues (Gerstein, 1997). As discussed in Chapter 2, the folded structures of proteins that are larger than ≈200–300 residues usually consist of multiple **structural domains**. The individual structural domains of such **multi-domain proteins** are compact, stable units with a unique three-dimensional structure; the interactions within one domain are more significant than with other domains. The presence of distinct structural domains in multidomain proteins indicates that they fold independently, i.e. the structural domains are also **folding domains**. The validity of this term is supported by the general observation that isolated domains of a multidomain protein fold autonomously into the same three-dimensional structure as in the intact protein.

Frequently, the individual structural and folding domains of a multi-domain protein perform a specific, distinct function that may remain intact in the isolated domain. In the case of such functional autonomy, the structural-folding domain also corresponds to a **functional domain**.

Many of the multidomain proteins are **homomultimeric**, i.e. they contain multiple copies of a single type of structural domain, indicating that internal duplication of gene segments encoding domains has given rise to such proteins. An internal tandem duplication of a gene segment that encodes an entire structural-folding-functional domain thus leads to a novel two-domain protein in which the duplicated domains have the same protein fold and the same function as the common ancestor in the single-domain protein. The fate of the domains of the internally duplicated gene is then determined by the same rules as in the case of whole-gene duplications: the internal duplication may be deleterious, neutral or advantageous; and the duplicated domains may retain the same function or they may diverge in functions just as in the case of paralogous genes. As an example we may refer to the ovomucoid family of serine proteinase inhibitors, which are present in the egg white of birds. Egg white ovomucoid consists of three homologous structural/folding and functional domains, each of which is capable of binding one molecule of either trypsin or another serine proteinase. Homology of the sequence of the three domains indicates that the ovomucoid gene arose from a primordial single-domain gene by two internal duplications.

Many multidomain proteins are **heteromultimeric**, i.e. they contain multiple types of domains – different domains of the protein are not homologous to each other. As an example, we may refer to tissue plasminogen activator, in which a trypsin-like serine-proteinase domain is joined to kringle, finger and epidermal growth factor domains. This type of gene elongation may occur by fusion of two or more gene segments that encode different protein domains. Such **chimeric proteins** are frequently referred to as **mosaic proteins** or **modular proteins**, and the constituent domains are usually referred to as **modules**.

Certain domain types (modules) occur in a wide variety of heteromultimeric and homomultimeric multidomain proteins, indicating that they have some special features that facilitate their duplication and dispersal to other genes. Domains that appear to be predisposed to being shuffled among genes and are versatile building blocks of different types of multidomain proteins are usually referred to as **mobile protein modules**. As for any other type of genetic change, the frequency of the transfer of a given domain type to different proteins (its **mobility**) reflects the probability of such genetic changes and the probability of their fixation.

In this chapter we will describe the various genetic mechanisms that may facilitate gene expansion (or contraction) by internal duplication of modules, or by fusion and insertion (or deletion) of modules, and will evaluate the relative significance of intronic and exonic recombination in such gene rearrangements. The special evolutionary significance of multidomain proteins will also be discussed.

8.1 Modular assembly by intronic recombination

Shortly after the discovery of introns in some vertebrate genes, it was realized that the existence of introns could have dramatic consequences on protein evolution. Gilbert (1978) suggested that recombination within introns could assort exons independently, and middle repetitious sequences present in introns may create hotspots for recombination to shuffle the exonic sequences. This proposal had a great impact on theories of genome evolution. The fascination with the idea of rapid construction of novel genes from parts of old ones led to the formulation of **exon-shuffling** hypotheses, which assigned an all-pervading role to this evolutionary mechanism. It was suggested that introns are the relics of the primordial RNA world (Gilbert, 1986) and that the 'genes in pieces' organization of the eukaryotic genome is in fact the primitive original form (Doolittle, 1978; Darnell & Doolittle, 1986): the **cenancestor** – the common ancestor of all extant organisms – had a similar genome organization. (Various theories belonging to this group are usually referred to as **'introns early' theories**.) According to such hypotheses all extant genes were constructed from a limited number of exon types by exon shuffling and all exons encode protein structural–functional units from which the proteins were assembled (Dorit, Schoenbacher & Gilbert, 1990). The fact that introns are usually missing from prokaryote genomes was explained by assuming that selection eliminated introns from bacteria that thus gained increased efficiency.

A different scenario is suggested by various **'introns late' theories**. According to these hypotheses the prokaryotic genes resemble the original ancestral forms and introns were inserted only later in genes of eukaryotes (Orgel & Crick, 1980; Cavalier-Smith, 1985; Sharp, 1985; Hickey & Benkel, 1986). Consequently, intronic recombination and exon shuffling could not play a major role in the assembly of the most ancient genes.

In the 1980s the 'introns early' theories of exon shuffling dominated the field of protein evolution. It has become part of the folklore of molecular biologists that all protein genes (including those that were formed in the cenancestor, before the prokaryote–eukaryote split) were assembled via intron-mediated recombination from exon modules that coded for folding domains or secondary structural elements. In support of this hypothesis it was frequently claimed that there is a general correlation between protein secondary structure and the exon–intron structure of the genes of the most ancient proteins such as the various glycolytic enzymes, globins, etc. The message of such studies was that exons correspond to building blocks (α-helices, β-sheets, transmembrane helices, signal peptides, etc.) from which all the genes were assembled by intronic recombination (Dorit et al., 1990; Gilbert & Glynias, 1993).

From the mid-1980s it became increasingly difficult to reconcile the introns early hypotheses with new information about the origin and evolution of introns and their possible role in the construction of novel genes by exon shuffling (Patthy, 1987, 1991a,b, 1994, 1995, 1996; Palmer & Logsdon, 1991).

1 In the case of many ancient protein genes no obvious correspondence was observed between protein structure and the location of introns, raising serious doubts about a general correlation between intron–exon structure and protein structure.
2 Contrary to the assumptions of 'introns early' hypotheses and in harmony with the 'introns late' hypotheses it became clear that introns may be inserted into (as well as eliminated from) genes. Consequently, it is incorrect to view all introns as primordial assembly points. In fact, there is now compelling evidence that introns are mostly of recent origin. An important consequence of these findings is that since the exon–intron pattern is the result of intron insertion and removal, the genomic organization of a gene does not necessarily reflect the structure that existed at the time of its formation.
3 Introns suitable for exon shuffling by intronic recombination appeared at a relatively late stage of eukaryotic evolution, therefore exon shuffling could not play a major role in the construction of ancient proteins of the cenancestor.
4 The unquestionable cases of exon shuffling are apparently restricted to 'young' proteins unique to higher eukaryotes. Analysis of the evolutionary distribution of proteins assembled from modules by intronic recombination suggests that exon shuffling became significant at the time of the

appearance of the first multicellular animals and could in fact contribute to the explosive nature of metazoan radiation (Patthy, 1994, 1995, 1996).

5 Analysis of established cases of modular protein evolution has shown that only a special group of exons (exon sets), the 'symmetrical' modules of class 1–1, class 2–2 and class 0–0, are really valuable for exon shuffling (Patthy, 1987). Consequently, the gene structures of proteins created by exon shuffling are characterized not only by showing a correlation between the domain structure of the proteins and location of the introns, but also by a characteristic intron phase distribution. When these criteria are applied to the exon–intron structures of the genes of ancient proteins it is clear that their hypothetical 'modules' do not conform to these rules of exon shuffling (Patthy, 1987, 1991a,b, 1995).

As a result of these developments the 'introns late' hypothesis is gaining increased acceptance: W.F. Doolittle, one of the first proposers of the introns early hypothesis (Doolittle, 1978) has also come to the conclusion that extension of the exon-shuffling hypothesis to the most ancient genes is untenable (Stoltzfus et al., 1994).

To provide the background for the above conclusions, we will first survey the evolution of introns, and the suitability of different intron types to intronic recombination. Then we will discuss the principles, mechanism and evolution of exon shuffling as well as its evolutionary significance.

8.1.1 Introns

Self-splicing introns, catalytic RNAs

The most important feature of self-splicing introns is that their splicing *in vitro* is autocatalytic, and intron excision can occur without the assistance of proteins. Nevertheless, it must be added that splicing of such introns is not an entirely autonomous reaction of RNA. *In vivo*, efficient splicing of these introns is usually assisted by protein factors.

In all cases of self-splicing, the intron folds into a specific secondary/tertiary structure in which the relevant groups are brought into juxtaposition so that the bond breakage and reunion reactions (required for the excision of the intron and the joining of the exons) can occur. Self-splicing introns are usually classified according to the nature of the unit that initiates the splicing reaction. In the case of group I self-splicing introns the reaction is mediated by an external guanosine cofactor; in the case of group II introns the attacking moiety is a specific adenylate that is an integral part of the intron itself (Fig. 8.1).

Group I introns. The 26S rRNA genes of some strains of the ciliate *Tetrahymena thermophila* are interrupted by a single intron. This intron is excised from the precursor in a series of reactions that requires only a guanine

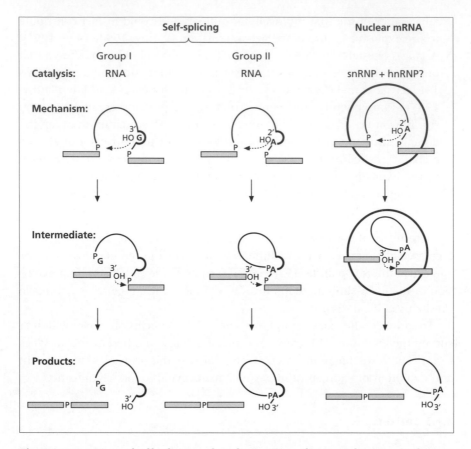

Fig. 8.1 Comparison of self-splicing and nuclear mRNA splicing mechanisms. In the case of group I self-splicing introns the process is catalysed by RNA structures of the intron, utilizing a G cofactor. Self-spicing of group II introns is also catalysed by RNA structures of the intron, and a lariat is formed with an adenosine (A) within the intron. Introns of nuclear pre-mRNA are spliced with the aid of the spliceosome (large circle). Splicing of pre-mRNA introns also proceeds via a lariat intermediate. (Reproduced from Sharp, P.A. (1987) Splicing of messenger RNA precursors, *Science* **235**, 766–71. © 1987 American Association for the Advancement of Science. Reprinted with permission from AAAS.)

nucleotide cofactor (G, GMP, GDP or GTP), which serves as an attacking group in the initial splicing reaction. Introns that use a guanosine cofactor are also found in the genes coding for rRNA of other lower eukaryotes, in fungal mitochondrial pre-mRNAs and in some pre-mRNAs of phage T4.

The binding of the guanosine cofactor to the primary RNA transcript is saturable, indicating that there is a specific binding pocket for guanosine. The guanosine cofactor provides a free 3'-OH group that attacks the 5' splice site to form a phosphodiester bond with the 5' end of the intron. During the first step of splicing, internal base pairings bring together the guanosine cofactor and the 5' splice site so that the 3'-OH of guanosine can attack this splice site. The 5' exon–intron junction is split, generating a 3'-OH at the end of the upstream exon. Another part of the intron holds the downstream exon in position so that it can be attacked by the newly formed 3′-OH of the upstream exon. This second transesterification reaction joins the two exons and leads to the release of the intron (Fig. 8.1).

It must be emphasized that self-splicing crucially depends on the structural integrity of the RNA precursor – much of the intron is needed for self-splicing. The intron has a **well-defined, folded structure** formed by many double-helical loops and stems, maintained primarily by AU and GC pairs (Vicens & Cech, 2006). All group I introns have some short consensus sequence elements whose base pairing generates the characteristic secondary structures in the intron. Each end of the intron pairs with part of an internal guide sequence that brings the exon–intron boundaries near one another. The importance of this folded stem-loop structure is shown by the fact that splicing is blocked when secondary and tertiary structures are disrupted by denaturing agents. Furthermore, there is experimental evidence to show that some of the consensus sequences are essential for splicing. Single base substitutions disrupting AU or GC pairs of stems eliminate splicing activity. Even more striking evidence for the importance of stems is the observation that not only true reversals can restore splicing activity in such mutants. For example if an A_n in an A_nU_m pair is changed to a G_n (resulting in loss of splicing activity) a second mutation changing the original U_m to C_m, could restore activity by restoring base pairing in the stem. In other words, a mutation in one consensus sequence may be reverted by a compensatory mutation in the complementary consensus sequence if it restores pairing.

Group II introns. Group II self-splicing occurs in the pre-mRNAs of mitochondrial genes of yeasts and other fungi and in the chloroplasts of unicellular organisms like *Chlamydomonas*. In this case, splicing does not require an external guanosine cofactor; it is an adenine of the intron itself (close to the 3′ end of the intron) that cuts at the 5′ splice site (Fig. 8.1). The attacking group is the 2′-OH of this adenylate residue. As a result of this reaction a 2′,5′-phosphodiester bond is formed between this A residue and the 5′-terminal phosphate of the intron. Since this adenylate residue is also joined to two other nucleotides by normal 3′,5′ phosphodiester bonds a branch is generated at this site (for this reason the site is called the **branch site**). Thus the 5′ end of the intron loops back to join the RNA of the intron through this branch-site adenine, resulting in the formation of a so-called **lariat** structure. In the final step the two exons are joined and the intron is released. It must be emphasized that group II introns are similar to group I introns inasmuch as they also have well-defined folded structures formed by double-helical stems (Slagter-Jager et al., 2006). Importantly, the integrity of this folded structure is also crucial for splicing activity.

On the other hand, the basic chemistry, the sequence of events and lariat formation are features that make group II introns strikingly similar to the so-called spliceosomal introns characteristic of nuclear pre-mRNAs. Like spliceosomal introns, they also have conserved consensus sequences at the splicing junctions (in this case, GT and APy, at the 5′ and 3′ junctions, respectively) and they also have a branch sequence that resembles the TACTAAC box of spliceosomal introns. The group II introns may be thus regarded as the ancestors of spliceosomal introns.

Spliceosomal introns

Spliceosomal introns are spliced only in the presence of specific proteins and RNAs, which form a complex called a **spliceosome** (Fig. 8.1). Although spliceosomal introns may be very large (>10,000 nucleotides), portions of these introns other than the 5′ and 3′ splice sites and the branch site appear to be nonessential for splicing. As discussed in greater detail in section 1.2, various snRNAs and protein constituents of spliceosomes specifically recognize the branch site and the:

$$5'\frac{C}{A}\textbf{AG}/\text{GTAAGT} \quad \text{and} \quad (\text{Py})_n\text{NCAG}/\textbf{G}\frac{G}{T}3'$$

consensus sequences at the 5′ and 3′ splice sites, respectively (the consensus bases from the flanking exons – **(C,A)AG/G(G,T)** – are in bold) (Shapiro & Senapathy, 1987; Csank, Taylor & Martindale, 1990). It should be noted that conservation at the splice junctions also extends to a few bases of the flanking exons. Segments of exons that conform to the consensus sequences around introns (in this case [C,A]AG/G[G,T]) are usually called **protosplice sites** (Dibb & Newman, 1989).

As long as the crucial splice sites and branch site are unaffected much of an intron can be deleted without altering the site or efficiency of splicing. Similarly, splicing is unaffected by the insertion of long stretches of DNA into the introns of genes. Importantly, chimeric introns created from the 5′ end of one intron and the 3′ end of a different intron may be properly spliced (cf. alternative splicing) provided that the chimeric intron has a complete set of these elements. On the other hand, mutations in these critical regions lead to aberrant splicing (Baralle & Baralle, 2005).

In summary, the spliceosomal intron plays a relatively minor role in its own excision; it contains the short consensus sequences (and intronic splicing enhancer and silencer elements) that may be recognized by the spliceosome, but the splicing reactions are carried out by the spliceosome.

Evolution of spliceosomal introns

Both group I and group II self-splicing mechanisms resemble spliceosome-catalysed splicing in that the initial step in each case is an attack by a ribose hydroxyl group on the 5′ splice site; the newly formed 3′-OH terminus of the upstream exon then attacks the 3' splice site (Fig. 8.1). Furthermore, in each case the reactions are transesterifications in which the phosphate moieties at each splice site are retained in the products. Group II splicing and spliceosome-catalysed splicing of mRNA precursors are even more closely related inasmuch as in both cases the first cleavage is carried out by an adenosine that is part of the intron itself, rather than by an external cofactor. Furthermore, in both cases the intron is released in the form of a lariat. Based on these striking similarities, it is generally accepted that spliceosome-catalysed splicing of mRNA precursors evolved from group II self-splicing introns. A major step in this transition from group II introns was the transfer of the catalytic role of the intron to other molecules. Whereas in the case of self-splicing introns the different exon/intron segments are positioned by intramolecular base pairings, in the case of spliceosomal introns base complementarities with snRNA constituents have become important. In other words, features critical for splicing activity have been transferred to the spliceosome. This has great evolutionary significance, in that the formation of spliceosomes gave spliceosomal introns a greater structural freedom because they were no longer constrained to fulfil the catalytic function themselves.

Although spliceosomal introns of eukaryotes have many features in common, there also seem to be significant differences among spliceosomal introns of fungi, plants and various groups of animals. In general, organisms with more compact genomes tend to have shorter introns (cf. the compact genome of the *Fugu* fish vs. other vertebrate genomes; see section 9.6.3), illustrating the point that introns are tolerant to insertion/deletion as long as splice sites are unaffected. The fact that yeast genes have much fewer and shorter introns than vertebrate genes (see section 9.6.3) may also be a reflection of the compact genome of this organism. Similarly, the fact that in the genes of the nematode *Caenorhabditis elegans* and the fruit fly *Drosophila melanogaster* the average exon/intron ratio is much higher than in the corresponding mammalian genes (see section 9.6.3) may be simply a reflection of the different compactness of their genomes.

Since the compactness of genomes has a direct effect on the repetitiveness of genomes as well as on the exon/intron ratio, it may be expected that the evolution of less compact genomes could be paralleled by an increased significance of intronic recombination. It should also be pointed out that there seem to be major differences in the splicing mechanisms of plant and animal spliceosomal introns.

Phylogenetic distribution of spliceosomal introns

Analysis of the frequency of introns in various groups of organisms has revealed that whereas self-splicing introns may be present in some prokaryotic genomes (e.g. phages, mitochondria, chloroplasts), spliceosomal introns are restricted to nuclear genomes of eukaryotes. Within the group of eukaryotes, spliceosomal introns are practically absent from most genes of the earliest protistan lineages but they are frequent in plant, animal and fungal lineages, suggesting that introns have accumulated in large eukaryote genomes during their evolution from an intron-poor ancestral eukaryote.

A recent systematic analysis of the phylogenetic distribution of spliceosomal introns has confirmed that intron density of annotated eukaryotic genomes varies by more than three orders of magnitude: from only 15 introns in the protist *Encephalitozoon cuniculi* (an intron density of 0.0075 introns per gene) to ~140,000 introns in the human genome (an intron density of 8.4 introns per gene) (Jeffares, Mourier & Penny, 2006). Since the actual intron density of any given genome is the balance of processes leading to intron gain, retention and loss (see below), it is clear that the marked differences in intron density within the group of eukaryotes is the result of differences in the probabilities of these processes.

There is strong evidence that the rate of intron loss is affected by selective pressures that depend on the biology of the organism. For example, Jeffares et al. (2006) have shown that intron density correlates with the logarithm of generation time, i.e. organisms that reproduce rapidly tend to have fewer introns than organisms that have longer life cycles. This correlation is probably due to selection for smaller genomes and selection for genes that can produce proteins quickly. (It is noteworthy in this respect that replication-dependent histones, which have to be synthesized in large amounts in a short period of the cell cycle, are encoded by intronless genes.)

A plausible explanation for the correlation between intron density and generation time is that **r-selected organisms** (i.e. organisms on which natural selection acts on the basis of maximum reproductive and/or growth rate) will experience strong selection pressure to reduce genome size (replication time), gene size (transcription time) and processing times for mRNAs. According to this explanation unicellular eukaryotes are under strong selective pressure to lose introns, whereas in **K-selected** multicellular eukaryotes (i.e. organisms with long life cycles that live in stable environments) there will be much weaker selective pressure to eliminate introns and this results in much higher intron densities.

On the other hand, intron gain and retention may be favoured by selection, especially in complex higher eukaryotes. For example, many introns are required for alternative splicing and some introns may encode a variety of untranslated RNAs (e.g. microRNAs, snoRNAs) that fulfil important biological functions. The fact that up to 60% of alternative splice variants are conserved between mouse and human suggests that many of these splice forms have functional importance. Since loss of such functionally important introns would be deleterious selection may favour the retention of many introns.

In summary, the observed phylogenetic distribution of spliceosomal introns might be explained by their extensive loss from unicellular eukaryotes (by the same reasoning that was invoked originally to explain the virtually complete absence of introns from bacteria) and their accumulation in multicellular K-selected eukaryotes.

Martin and Koonin (2006) have suggested recently that spliceosomal introns (that evolved from group II self-splicing introns) have originally moved into the evolving eukaryotic genome from the α-proteobacterial progenitor of the mitochondria (cf. section 9.6.2.3) and started to invade eukaryotic genes at the outset of eukaryotic evolution. They also argue that this invasion actually triggered the emergence of the nuclear envelope since this allowed mRNA splicing, which is slow, to go to completion so that translation, which is fast, would occur only on mRNA with intact reading frames. In this scenario the rapid spread of introns following the origin of mitochondria provides the selective pressure for nucleus-cytosol compartmentalization.

Some mechanisms that could be responsible for the spread of spliceosomal introns have been clarified recently.

Insertion and spread of spliceosomal introns

Recent studies on *Ac/Ds* and *Spm/dSpm* transposons of maize and the *Drosophila P* and 412 element families have provided evidence for a special mechanism of insertion of spliceosomal introns. In the case of these transposons the major part of the inserted sequence is removed from the transcript utilizing splice sites within the transposable element and in flanking host sequences: insertion of these transposons thus generates novel introns at the site of transposon insertion (Fridell, Pret & Searles, 1990; *Menssen et al.*, 1990; Wessler, 1991a,b; Purugganan & Wessler, 1992).

The presence of splice sites provides a selective advantage for the spread of intron-carrying transposons since splicing mitigates the deleterious impact of transposable element insertion. For example, insertion of an *En/Spm* transposable element into the A2 protein gene does not destroy the activity of the host protein since only 18 bp of the transposon and 3 bp of the target site duplication remain after splicing – the bulk of the transposon is removed. The seven amino acids are added in a region of the recipient A2 protein that tolerates this small insertion. The selective value of splice sites is even clearer in the case of the transposons that carry splice sites in all three reading frames. Alternative use of such multiple splice sites may ensure some read-through of the target protein irrespective of the position of the transposon relative to the reading frame. In other words, insertion of transposons is more likely to be selectively neutral if most of the transposon is removed by splicing; introns would be accepted in positions that can tolerate the insertion of a few amino acids.

The *Drosophila P* and 412 transposons, the *Ac/Ds* and *Spm/dSpm* transposons of maize are not removed perfectly by splicing: part of the transposon persists in the transcript and some host sequences may be deleted. Since intron insertion by this mechanism inevitably will delete or insert a few amino acids at the site of integrations it will be tolerated only in nonessential regions of the target proteins (surface loops connecting structural motifs, interdomain connecting regions, etc.). If this were the major mechanism for intron insertion then introns would be confined to such regions of proteins. In contrast with this expectation, intron insertions have frequently occurred in gene regions that correspond to highly conserved regions of proteins, which do not tolerate insertions or deletions, indicating that a mechanism must exist that can insert 'perfect' spliceosomal introns, i.e. ones that are silent in the sense that they do not affect the sequence of the mRNA and the protein.

There is now experimental evidence that group II introns (the predecessors of spliceosomal introns) can transpose to heterologous sites by reversal of the splicing reaction (Belfort, 1993; Mueller et al., 1993; Lambowitz & Zimmerly, 2004). By reversal of the splicing reaction, an intron can insert to another RNA, this RNA may be reverse transcribed into cDNA, which then can recombine with the gene that gave rise to the original mRNA. In other words, by a process similar to the creation of processed genes or pseudogenes, an intron may be introduced in a new position. Since this pathway is open to all intron types this mechanism can also insert perfect spliceosomal introns.

Such a mechanism implies that intron insertion (by reversal of splicing reactions) will be directed to protosplice sites (i.e. sites of exons that conform to the exon aspect of the consensus sequences of exon–intron junctions). Selection will also favour the acceptance of perfect spliceosomal introns only in such protosplice sites: the inserted intron will be efficiently spliced out only if the exonic environment also conforms to the rules defined by the spliceosome. Introns inserted in a region that does not conform to the protosplice-site consensus will not be spliced. Perfect introns inserted in translated regions of genes are likely to be selectively neutral if spliced properly, but are deleterious if they cannot be spliced.

Two recent studies clearly show that perfect spliceosomal introns are also inserted by reversal of the splicing reaction.

Spliceosomal introns are found not only in protein-coding genes, but also in small nuclear RNAs of some fungi, and in the small- and large-subunit ribosomal DNA genes of ascomycetes, raising questions as to how these introns were inserted. Analyses by Bhattacharya et al. (2000) suggest that these spliceosomal introns are of relatively recent origin, i.e. within the Euascomycetes, and have arisen through aberrant reverse splicing of free pre-mRNA introns into rRNAs. They have also found a nonrandom sequence pattern at sites flanking the rRNA spliceosomal introns (AG-intron-G), closely resembling the protosplice site (see above) postulated for intron insertions in pre-mRNA genes.

In comparative genomic studies on the nematodes *Caenorhabditis elegans* and *Caenorhabditis briggsae* Coghlan and Wolfe (2004) have found that >6000 introns are unique to one or the other species. To study the origins of new introns, they used phylogenetic comparisons to animal orthologs and nematode paralogs to identify cases where an intron-content difference between *C. elegans* and *C. briggsae* was caused by intron insertion rather than deletion. Analyses of the 81 recently gained introns of *C. elegans* and of the 41 introns of *C. briggsae* have revealed that they display a stronger exon splice site consensus sequence than the general population of introns, i.e. novel introns were preferentially inserted into perfect protosplice sites. The fact that genes transcribed in the germline were the primary targets for insertion of novel introns is also consistent with the notion that intron gains were caused by reverse splicing.

It is important to note that this mechanism for the spread of introns has a significant saltatory element: the more introns an organism has, the higher the chance for this process to occur, etc. This phenomenon could thus give rise to an acceleration in the spread of introns, for example in plants, fungi and animals. If intron insertion is silent and creates no or little selective disadvantage, the inserted introns may spread as neutral mutations.

Intron loss

Intron loss also plays a significant role in changing the exon–intron structure of genes. A well-known mechanism whereby all the introns of a gene may be eliminated is the process that gives rise to processed genes and processed pseudogenes (see section 6.2.1). There are various ways by which reverse transcription of RNA may result in elimination of just some of the introns: for example, reverse transcription of perfectly spliced mRNA and recombination with the functional gene. Note that in this case the original gene is mutated by intron deletion. Alternatively, as in the case of the preproinsulin I gene (see section 6.2.1), a partially processed pre-mRNA could give rise to a semi-processed gene (lacking just some of the introns). Note that in this case a new copy (paralog) of the gene is formed.

A key aspect of reverse-transcription-based mechanisms is that only genes actively transcribed in the germline would be susceptible to intron loss, which is clearly not the case. Furthermore, since reverse transcriptases begin from the 3' end of RNA molecules and dissociate in a length-dependent manner, this mechanism is not expected to remove introns from the 5' end of genes as efficiently as from the 3' ends and this is expected to lead to an intron gradient along the genes. In contrast with this expectation, introns in multicellular genomes are evenly distributed throughout the genes (Mourier & Jeffares, 2003; Jeffares et al., 2006). It seems therefore likely that simple

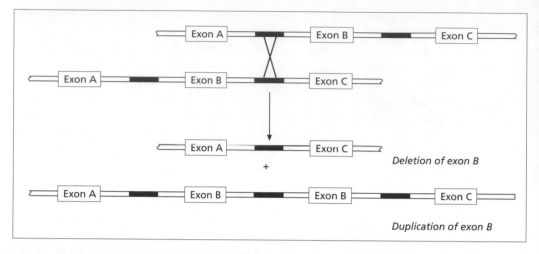

Fig. 8.2 Duplication and deletion of exons by intronic recombination. The black boxes within the introns represent middle repetitious sequences (e.g. Alu repeats) that may facilitate misalignment of nonorthologous introns. (From Patthy, L. (1995) *Protein Evolution by Exon Shuffling.* New York: Springer-Verlag.) With kind permission of Springer Science and Business Media.

genomic deletion – a mechanism that does not have the above limitations – also plays a major role in intron loss.

8.1.2 Internal gene duplications/deletions via recombination in introns

There is compelling evidence for a role of intronic recombination in the duplication, deletion and insertion of exons. Thanks primarily to studies on genes of some medically important proteins the details of exon deletion and exon duplication by intronic recombination is quite well understood (Fig. 8.2). Prominent among these studies were investigations on abnormal low-density lipoprotein (LDL)-receptor genes causing familial hypercholesterolaemia. Comparison of the structures of normal and abnormal LDL-receptor genes revealed that misalignment and recombination involving middle repetitious sequences (frequently Alu repeats) of introns flanking the exons were responsible for the deletion or duplication of domains or entire sets of domains of this multidomain protein. The same type of Alu-mediated intronic recombination affecting a variety of proteins (apolipoprotein B, β-hexosaminidase, C1 inhibitor, antiplasmin, antithrombin, lysyl hydroxylase) was also found to be responsible for some human inherited disorders.

The only difference between the multidomain proteins (such as LDL-receptor) and single-domain proteins (such as the serpins C1 inhibitor, antiplasmin, antithrombin, etc.) is that in the first case deletion/duplication of a domain may yield a structurally 'viable' multidomain protein with an altered number of domains, whereas in the case of single-domain proteins an integral part of a folding domain is deleted/duplicated. Obviously, deletion (or duplication) of an essential segment of a folding domain is more likely to destroy the structural (and functional) integrity of the protein.

Note that the mechanism of Alu-mediated intronic recombination leading to intragenic deletions/duplication is similar to the mechanism of Alu-mediated whole-gene duplications (see section 6.2.1). The only difference is that in the first case recombination occurs in introns, within the boundaries of a gene, whereas in the latter case the recombination points are in intergenic regions.

8.1.3 Fusion of genes via recombination in introns

If the introns involved in recombination belong to different genes then a chimeric protein-coding gene may be produced. Studies on the gene of sterol regulatory element binding protein-2 (SREBP-2) have provided some insights on how this happens. The active component of the SREBP-2 precursor corresponds to its 460–480 residue long N-terminal domain, which must be cleaved off to exert its activity on the expression of the LDL-receptor gene and the genes of enzymes of cholesterol synthesis. This cleavage is blocked by 25-hydroxycholesterol (25-HCS), therefore cells exposed to 25-HCS die of cholesterol deprivation. Yang et al. (1995) have shown that some 25-hydroxycholesterol-resistant mutant hamster cell lines express a constitutively active 460–480 residue N-terminal fragment of the transcription factor from which the downstream regions are missing. Analysis of the genetic basis of this truncation revealed that three different resistant cell lines expressed chimeric mRNAs derived from chimeric genes produced by recombination that involved the intron that follows codon 460 of *SREBP-2* and introns of other genes. Note that in this case there was positive selection (*in vitro*) for the truncation of the *SREBP-2* gene.

Alu-mediated intronic recombination can mediate even the fusion of genes located on different chromosomes. For example, Babcock et al. (2003) were able to trace interchromosomal Alu-mediated fusion between the *IGSF3* gene (immunoglobulin superfamily, member 3) located on chromosome 1p13.1 and the *GGT* gene (gamma-glutamyl transpeptidase) located on chromosome 22q11.2.

8.1.4 Exon shuffling via recombination in introns

It has long been assumed that the same mechanism of intronic recombination that operates during tandem duplications and deletions of exons or fusion of genes may also be responsible for the transfer of exons from one gene or gene segment to another. There are, however, other possibilities. It has been suggested that insertion of exons might occur by the same mechanism as the insertion of introns. According to this hypothesis exon shuffling may be a consequence of the occasional inclusion of exon sequences in the insertion cycle of introns. Alternative splicing by exon skipping (i.e. when splicing joins the 5' donor site of one intron to the 3' site of the downstream intron, skipping the exon that lies in between these introns) may yield exons with flanking introns. If such a composite intron–exon–intron structure is reinserted elsewhere in the genome by the same mechanism that governs mobility of single introns (e.g. by reverse splicing) we have 'exon shuffling'.

Another potential mechanism for exon shuffling is the long interspersed element (LINE)-1 (L1)-mediated 3' transduction. Since during retrotransposition L1 often associates 3' flanking DNA as a read-through transcript it may carry such non-L1 sequences to new genomic locations (Eickbush, 1999; Moran, DeBerardinis & Kazazian, 1999; Ejima & Yang, 2003). For example, if during transcription of an L1 element (located within an intron of a gene) its weak polyadenylation signal is bypassed, the aberrant transcript may include downstream exons of the host gene. The reverse transcriptase and endonuclease encoded by this L1 aberrant transcript may insert a cDNA copy of this complex element into an intron of another gene: we have exon shuffling.

There is one important difference between the 'reverse splicing model' of exon shuffling and the 'intronic recombination model' or 'L1-mediated transduction model' of exon shuffling: whereas in the latter two models exon insertion occurs into a pre-existing intron, the reverse splicing model does not have this requirement. In other words, we can distinguish between these possibilities if we know whether or not an intron was present in the recipient gene at the site of exon insertion and whether there is evidence for the involvement of L1-like elements.

Now there are several examples of exon shuffling where the exon–intron structure of the recipient gene (before exon insertion) can be reconstructed. In the majority of cases it is clear that exon insertion occurred into a pre-existing intron with the appropriate phase (Patthy, 1987, 1991b, 1994). Nevertheless, the fact that these exon-shuffling events occurred hundreds of million years ago prevents the precise reconstruction of the features that mediated the transfer of exons.

The first example came from studies on the complex regulatory proteinases of the trypsin family, where it could be shown that the various class 1–1 domains of the proteinases of the fibrinolytic and blood coagulation cascades were all inserted into a phase 1 intron that was present in the common ancestor of trypsin-like proteinases (Fig. 8.3). Since these complex regulatory proteases are already present in bony fishes (but missing from invertebrates and invertebrate chordates), the events leading to their assembly must have occurred before the appearance of vertebrates, more than 450 million years ago (Hedges, 2002; Jiang & Doolittle, 2003).

A similar conclusion could be drawn from an analysis of the gene structures of members of the zinc metalloproteinase family. Comparison of the genes of interstitial collagenases, stromelysins and type IV collagenases

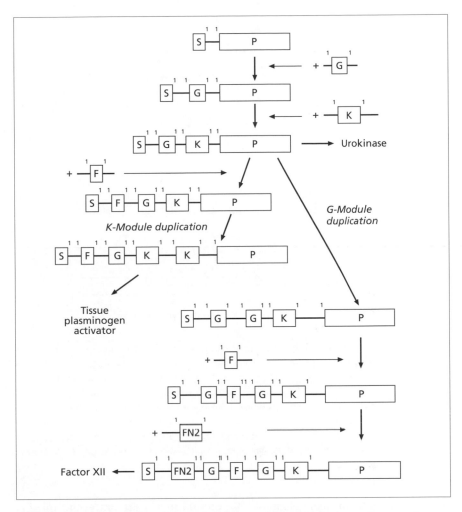

Fig. 8.3 Flowchart of the evolution of urokinase, tissue plasminogen activator and factor XII by acquisition of class 1–1 modules. The numbers indicate the phase of the 5′ and 3′ splice junctions of the introns participating in the assembly process. (For the sake of simplicity introns within domains are not shown.) The phase 1 intron between the signal-peptide domain (S) and trypsin-homolog domain (P) of the ancestral trypsin-type protease was a suitable recipient for class 1–1 modules. Note that insertions and duplications of class 1–1 growth factor (G), kringle (K), finger (F), and fibronectin type II (FN2) modules divide and duplicate phase 1 introns. Modular assembly thus resulted in gene structures in which all intermodule introns are phase 1. (From Patthy, L. (1995) *Protein Evolution by Exon Shuffling*. New York: Springer-Verlag.) With kind permission of Springer Science and Business Media.

(gelatinases) reveals a perfect conservation of the location and phase class of introns. What is important for the present discussion is that a phase 1 intron is present in all genes exactly where the class 1–1 fibronectin type II domain was inserted later to give rise to type IV collagenases (Fig. 8.4). The exon-shuffling events leading to the formation of gelatinases also occurred

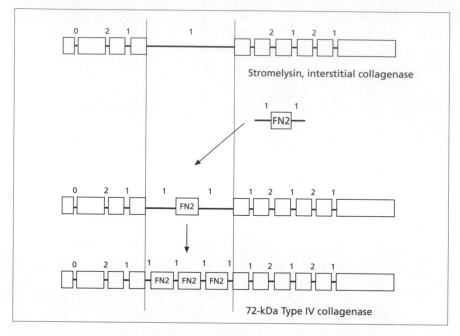

Fig. 8.4 Insertion of class 1–1 fibronectin type II (FN2) module in a phase 1 intron of an ancestral zinc metalloproteinase. Note that the gene structures of stromelysin, interstitial collagenase and type IV collagenases are strikingly similar: their common ancestor must have had a similar exon–intron structure. (The numbers indicate the phase of the introns.) Significantly, a phase 1 intron was present in the ancestral metalloprotease exactly where the class 1–1 fibronectin type II module was inserted to give rise (via internal duplication of the FN2 module) to type IV collagenases. (From Patthy, L. (1995) *Protein Evolution by Exon Shuffling.* New York: Springer-Verlag.) With kind permission of Springer Science and Business Media.

in the vertebrate lineage: gelatinases are present in all vertebrates but missing from invertebrates.

A third example comes from an analysis of the evolution of the α-chains of integrins (Fig. 8.5). The CD11b and p150,95 molecules differ from simpler, more ancestral forms of α-chains (e.g. PS2 of *Drosophila*, glycoprotein IIb) in containing an insertion related to the A-type domains of von Willebrand factor. Comparison of the genes of p150,95, CD11b, PS2 and platelet glyco-protein IIb revealed striking similarities. Significantly, in the PS2 gene of *Drosophila* and the gene of platelet glycoprotein IIb a phase 1 intron is present exactly where the class 1–1 von Willebrand module was inserted into an ancestral integrin α-chain gene to give rise to the CD11b, p150,95 group of integrins in the vertebrate lineage.

In summary, these findings clearly support exon-shuffling models in which exons are inserted into pre-existing introns of the 'target' gene. It seems likely that the same features that facilitate duplication, deletion and fusion of exons by intronic recombination are also responsible for their transfer from the donor to the recipient gene.

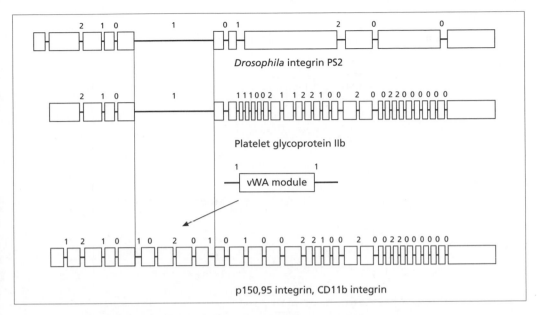

Fig. 8.5 Insertion of a class 1–1 von Willebrand type A (vWA) module in a phase 1 intron of an ancestral integrin. Analysis of the gene structures of *Drosophila* PS2, glycoprotein IIb have shown that a phase 1 intron is present exactly where the class 1–1 von Willebrand type A module has been inserted to give rise to molecules similar to integrins CD11b, LFA-1 and p150,95. Note that the vWA module of CD11b and p150,95 is split by three intramodule introns. (The numbers indicate the phase of the introns.) (From Patthy, L. (1995) *Protein Evolution by Exon Shuffling*. New York: Springer-Verlag.) With kind permission of Springer Science and Business Media.

This assumption is supported by the fact that recombination in Alu repeats is known to predispose Alu-rich regions to translocation and transposition (Hill et al., 2000; Bailey, Liu & Eichler, 2003; Antonell et al., 2005). Transposition of members of the V_K family of immunoglobulin light-chain genes was shown to be facilitated by the Alu repeats surrounding them: the V_K orphans (genes that are located outside the cluster of the V_K gene family) probably arose by alignment of nonadjacent members of the V_K gene family with subsequent excision of the resulting looped-out region, which then could be inserted into other chromosomal locations. In general, sequence homology between duplicated DNA segments (duplicons) increases the chance for misalignment during meiosis and may lead to transposition of genes or genomic segments to other chromosomes (Ji et al., 2000; Chai et al., 2003).

By analogy with transposition of whole genes, recombination of middle repetitious sequences (such as Alu repeats) present in introns that flank a domain may facilitate domain transposition by a similar mechanism. Just as misalignment of different duplicons in a gene cluster can lead to the creation of orphans by their transposition to other chromosomal locations, misalignment of different duplicated domains of a multidomain protein can lead to the transposition of a domain. It is well known that multidomain

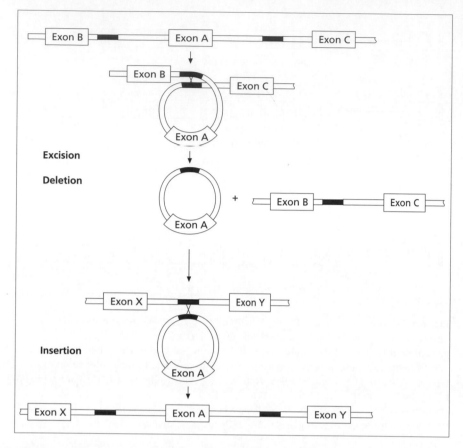

Fig. 8.6 Looping out, excision and reinsertion of modules by intronic recombination. The black boxes within introns represent middle repetitious sequences (e.g. Alu repeats) that may facilitate misalignment of nonhomologous introns. (From Patthy, L. (1995) *Protein Evolution by Exon Shuffling.* New York: Springer-Verlag.) With kind permission of Springer Science and Business Media.

proteins frequently undergo misalignment, leading to the contraction and expansion of such genes by domain duplications. During this process looping out and excision of exons (encoding folding domains) may occur and the excised exons may be inserted in introns of other genes on distant chromosomal locations (Fig. 8.6).

The best examples illustrating the frequency of contraction and expansion of a multidomain protein come from studies on apolipoprotein(a). In the human population this plasminogen-related molecule (Fig. 8.7) shows an appalling variation in structure: the number of kringle domains ranges between 12 and 51 copies, suggesting a high rate of module duplications and deletions. In one apolipoprotein(a) variant 24 of the 37 kringle 4-like units were found to have identical nucleotide sequences, suggesting that they are the products of very recent duplications (McLean et al., 1987; Lawn,

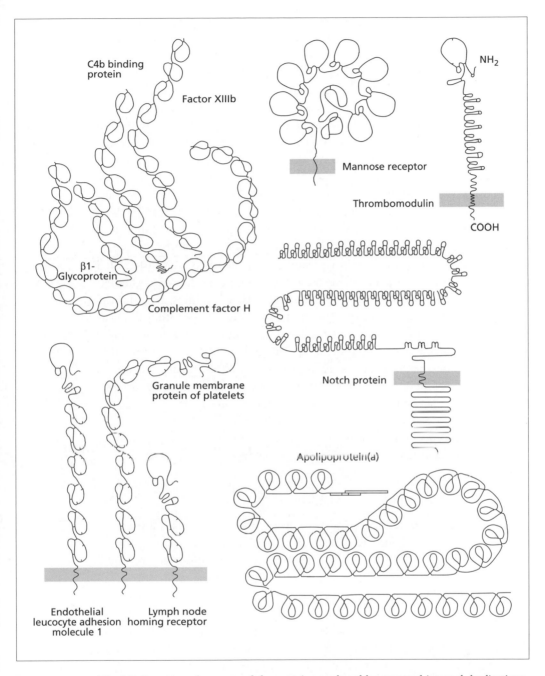

Fig. 8.7 Structure of some modular proteins produced by repeated internal duplications of modules. (Patthy, L. (1991b) Modular exchange principles in proteins. *Current Opinion in Structural Biology* **1**, 351–61.) Reprinted from Current Opinion in Structural Biology, 1, Patthy, L., Modular Exchange principles in proteins, 351–361, Copyright 1991b, with permission from Elsevier.

Schwarz & Patthy, 1997). Duplication and deletion of kringle modules is an ongoing process, as shown by the observation that the isoforms containing different numbers of kringles do not follow a simple Mendelian pattern of inheritance: offspring often have apolipoprotein(a) isoforms that differ from those of their parents. There are many other families of internally repeated multidomain protein genes where variation in repeat number is common. The complement B-type units of various members of the selectin family (lymph node homing receptor, endothelial leucocyte adhesion protein, granule membrane protein of platelets, etc.) also display a remarkable tendency to undergo duplications and deletions (Fig. 8.7). For example, the murine lymphocyte homing receptor has two complement B modules that show complete identity at the nucleotide sequence level including intron regions upstream and downstream of the splice sites, suggesting that duplication of a complement B-type module occurred very recently by intronic recombination. The fact that human, bovine and rat P selectins (GMP-140-related proteins) contain nine, six and eight complement B-type modules, that human E selectin (ELAM-1) contains six complement B-type modules, but that the rabbit homolog has only five of these units, may also suggest that duplication and deletion of these domains occurs in the mammalian lineage. Interestingly, the porcine E-selectin protein has only four complement B modules, but in the gene the 'missing' repeats are found in the intronic region as pseudoexons (Winkler et al., 1996). It must be emphasized that the proteins that show such striking variation in the number of domains (e.g. apolipoprotein(a), selectins) have retained all their interdomain introns. In other words, the presence of interdomain introns may be primarily responsible for the remarkable plasticity of these genes (see below).

Factors favouring intronic recombination

As we have seen above, recombination in spliceosomal introns is a frequent source of the rearrangements of vertebrate genes. An important property of spliceosomal introns that facilitates intronic recombination is that only a tiny portion of these introns is essential for splicing. These short conserved segments (the 5' and 3' splice sites and the branch site) are found in the vicinity of the exon/intron boundaries and are usually separated by very long intron sequences that are dispensable for splicing. The majority of the intron is 'junk' and is thus highly tolerant to deletions and insertions. Middle repetitious sequences (such as Alu repeats, LINE-1) may be inserted in these regions of introns and these may increase the chances of misalignment and recombination involving nonhomologous introns.

These features of vertebrate spliceosomal introns are clearly reflected in their evolutionary behaviour. A comparison of urokinase genes from different species may also be used to illustrate that vertebrate introns are highly tolerant to insertion and deletion as long as their splice sites and branch sites are conserved. The genomic organization of murine, human, porcine and chicken urokinase genes is exactly the same inasmuch as the location and phase class of introns are conserved (Leslie et al., 1990). However, the introns of chicken urokinase gene do not show any discernible sequence similarity with the corresponding introns of mammalian urokinases except for the short segments near the splice junctions and branch sites. Insertions and deletions may drastically change the size of orthologous introns; for example, intron A is 1489 bp in chicken urokinase gene whereas it is only 306 bp in the human homolog. Comparison of mammalian urokinase genes reveals that sequence similarity of orthologous introns is still significant, although independent insertion of Alu family repeats in the human gene and of various insertion elements (B1 and B2 family repeats, etc.) in the murine gene caused major changes in the size of orthologous introns.

Another important feature of pre-mRNA introns is that the splice sites of different introns are more or less equivalent and interchangeable. This is important for exon duplication/exon shuffling since recombination in spliceosomal introns necessarily leads to the formation of a chimeric intron with 5' and 3' splice sites derived from different (not necessarily homologous) introns (Figs 8.2 and 8.6). Such chimeric introns are usually spliced as efficiently as their two progenitors.

The very large size of most vertebrate spliceosomal introns, the presence of middle repetitious sequences in such introns, as well as their tolerance to the structural changes that accompany recombination make them ideal for intronic recombination and exon shuffling. This is not true for non-spliceosomal introns. For example, the self-splicing introns that carry out their own excision have to conserve a unique, catalytically active three-dimensional structure, therefore they are more sensitive to structural changes. They are not tolerant to intronic recombination with nonorthologous introns since such an event is likely to yield chimeras that lack some of the essential elements of a catalytically active RNA and thus will be deficient in splicing. (Deficiency in splicing of a chimeric intron in a protein-coding region is most likely to cause deleterious mutation of the gene.)

What we have said so far in praise of spliceosomal introns holds primarily for vertebrate introns. As mentioned earlier, fungi and plants have fewer and shorter introns than vertebrates, therefore intronic recombination may be less significant in these lineages. The fact that the genes of the best-studied invertebrate genomes (*C. elegans*, *D. melanogaster*) have shorter (and fewer) introns than vertebrates may also have implications for the significance of intronic recombination in these lineages. It is not clear, however, whether the relative compactness of the genes/genomes of these organisms was also characteristic of the common ancestor of metazoa and only the vertebrate lineage acquired a much less compact genome. Indeed, there is now strong evidence that selection has led to a secondary increase in genome compactness in these invertebrate lineages (Bányai & Patthy, 2004). In any case, since the compactness of genomes has a direct effect on the repetitiveness of genomes as well as on the exon/intron ratio, intronic recombination is likely to be less significant for more compact genomes.

It should also be pointed out that there seem to be major differences in the splicing mechanisms of spliceosomal introns of plants, fungi and animals. The spliceosomal introns of plants seem to be less suitable for intronic recombination than those of vertebrates since they are in general significantly shorter (<200 bp) than vertebrate introns. Furthermore, unlike vertebrate introns, the spliceosomal introns of plants are recognized *in toto* rather than as the simple assembly of splice junctions and a branch point and are therefore not tolerant to structural changes (Luehrsen & Walbot, 1992; Waigmann & Barta, 1992). This is also reflected in the fact that splicing of chimeric introns is significantly impaired in yeast and plants (Goguel & Rosbash, 1993). Since formation of chimeric introns is an inevitable consequence of intronic recombination these features of plant and fungal introns may adversely affect their potential role in exon shuffling.

8.1.5 Factors affecting acceptance of mutants created by intronic recombination

There are several levels of selection that influence the probability that a mutant produced by intronic recombination in two nonorthologous introns will be fixed or rejected.

1 The chimeric intron formed as a result of intronic recombination must be spliced correctly, otherwise the intron will remain in the mRNA and will be translated. Since an intron sequence has not been selected for protein-coding function, its translation is likely to run into a premature stop codon and a truncated protein will be produced. Note that nonsense-mediated mRNA decay ensures the rapid degradation of aberrant mRNAs that contain premature termination codons.

2 The two nonorthologous introns involved in intronic recombination must belong to the same phase class, i.e. they must split the reading frame in the same phase. This ensures that the downstream exon will be translated in its original phase. If this phase compatibility rule is violated then there will be a reading-frame shift downstream of the recombinant intron, leading to premature termination and nonsense-mediated mRNA decay. Of course, in the case of deletions, duplications or insertions of entire exons (or exon sets), this requirement holds for introns at both boundaries of the exon. As a consequence, this requirement may be fulfilled only by symmetrical exons, i.e. ones that have introns of identical phase at both boundaries (for details, see below).

3 The mutant protein produced by intronic recombination must be able to fold into a stable protein fold. This is most likely if the exon deleted, duplic-

ated or inserted encodes a protein unit that is able to fold independently, i.e. it corresponds to an autonomous folding domain. Folding may be severely damaged if intronic recombination deletes an integral part of a folding domain, or if a fragment of a folding domain is duplicated or inserted. The significance of this requirement may be illustrated by the fact that the domains used in the construction of the regulatory parts of the blood coagulation, fibrinolytic and complement activation proteinases display remarkable structural independence: the isolated domains can fold efficiently. Obviously, folding autonomy of domains in multidomain proteins is of utmost importance since this minimizes the influence of neighbouring domains. Folding autonomy can ensure that folding of the domain is not deranged when inserted into a novel protein environment. It thus seems that the most widely used modules have been selected for folding autonomy.

Another relevant aspect is whether the folding domain is also a functional domain, i.e. whether its duplication or insertion duplicates or transfers a function to another gene. If a uniquely folded protein with an altered number of folding domains is produced, then natural selection will decide whether having such a mutant protein is advantageous, neutral or deleterious for the organism.

An important aspect of exon shuffling is that the impact of an exon insertion may be mitigated initially by alternative splicing. If an exon is inserted in an intron (e.g. as in Figs 8.3–8.5) the original 5′ and 3′ splice site combination may be more favourable than the two novel combinations, therefore the inserted exon may be skipped in some of the mRNAs. Thus the selective value of the modified protein (encoded by a minor component of the mRNA population) may be tested without the risk of losing the original function.

The intron-phase compatibility rule

The basic steps of module shuffling (gene fusion, insertion and deletion, and duplication of modules by intronic recombination) join the 5′ splice junction of one intron to the 3′ splice junction of another intron (for examples see Figs 8.2 and 8.6). The two intron partners involved in recombination must belong to the same phase class otherwise recombination would shift the reading frame and would lead to loss of protein information downstream of the recombination point (Patthy, 1987).

The symmetrical-exon rule

Insertion, deletion and duplication of a module by intronic recombination could satisfy this phase-compatibility requirement only if the module is

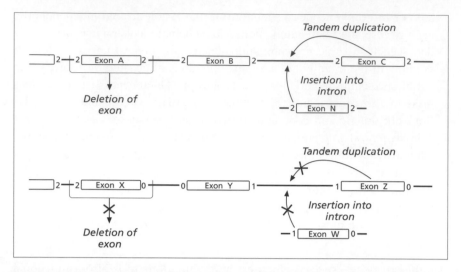

Fig. 8.8 Significance of intron phase in exon shuffling. The boxes represent exons (or exon sets), the connecting lines show the introns. The numbers indicate the phase class of the 5′ and 3′ splice junctions of the intron. Note that deletion, insertion and duplication of modules by intronic recombination joins the 5′ splice junction of one intron to the 3′ splice junction of another intron. If the module is 'symmetrical' (i.e. when the introns at the 5′ and 3′ boundaries of the module are of identical phase) and the recipient also belongs to the same phase class then the reading frame is not affected (upper part). However, deletion, duplication or insertion of 'nonsymmetrical' exons (i.e. the introns at the 5′ and 3′ boundaries of the exon are of different phase) will disrupt the reading frame (lower part). (Reprinted from Patthy, L. (1987) Intron-dependent evolution: preferred types of exons and introns. *FEBS Letters* **214**, 1–7. © 1987, with permission from Elsevier Science.) Reprinted from FEBS Letters, 214, Patthy L., Intron-dependent evolution: preferred types of exons and introns, 1–7, Copyright 1987, with permission from Elsevier.

'symmetrical' in the sense that the two introns flanking the module are of the same phase (Fig. 8.8). Note that **symmetrical modules** have $3n$ bases.

Nonsymmetrical modules (i.e. when the introns at the 5′ and 3′ boundaries of the module are of different phase) when shuffled by intronic recombination will necessarily violate this phase-compatibility rule, resulting in a shift of the reading frame (Fig. 8.8). Note that nonsymmetrical modules have $3n \pm 1$ bases.

In other words, only the three **symmetrical module groups** (class 1–1, class 0–0 and class 2–2, depending on whether phase 1, phase 0 or phase 2 introns are found at both boundaries of the modules) are really useful in exon shuffling; **nonsymmetrical exons/exon sets/modules** (class 0–1, class 0–2, class 1–2, class 1–0, class 2–0, class 2–1) are practically useless in protein evolution by exon shuffling.

It must be emphasized that even symmetrical exons have a phase restriction: a class 0–0 exon can be inserted only into phase 0 introns, a 1–1 exon can be inserted only into phase 1 introns, and a 2–2 exon can be inserted only into phase 2 introns without disrupting the reading frame. (Note that in Figs 8.3–8.5 the class 1–1 modules were always inserted into phase 1 introns.) However, insertion, duplication or deletion of nonsymmetrical exons (like class 0–1, class 0–2, class 1–0, class 1–2, class 2–0, class 2–1) will inevitably lead to a frameshift in all the downstream exons, irrespective of the phase of the 'recipient' intron.

Analysis of the gene structures of modular proteins such as the blood coagulation proteinases (prothrombin, factor IX, factor X, protein C), fibrinolytic proteinases (plasminogen, urokinase, tissue plasminogen activator), fibronectin, EGF-precursor, etc. has revealed that all the introns demarcating the 5' and 3' boundaries of their kringle, epidermal growth factor, fibronectin type I, fibronectin type II, and fibronectin type III modules are phase 1, i.e. the modules are class 1–1 symmetrical modules (Patthy, 1987). In other words, these modules are suitable for exon shuffling since they satisfy the 'symmetrical-exon rule' (see Fig. 8.3).

On the basis of the phase-compatibility rule this observation implied that the class 1–1 modules required a phase 1 target intron for their insertion into the ancestral proteinase gene. Our hypothesis for the evolution of plasma proteinases (Patthy, 1985, 1987) predicted that this target intron was located at the boundary separating the signal peptide domain from the catalytic domain of the common ancestor of these serine proteinases. Analysis of the gene structures of trypsin-type proteinases (Fig. 8.9) has indeed confirmed the prediction that a phase 1 intron must have been present between the signal peptide and catalytic domains of the ancestral proteinase (Patthy, 1987, 1990). During the course of the assembly process insertions and duplications of class 1–1 growth factor, kringle, finger, fibronectin type II, complement B, and von Willebrand factor type A modules, etc. divided, duplicated and thus proliferated phase 1 introns, and this eventually led to gene structures in which all intermodule introns belong to the phase 1 class (Fig. 8.3).

It is thus clear that the intron phase bias (the dominance of a single intron phase class) of these multidomain proteins is a necessary consequence of the intron phase rules of exon shuffling. In other words, evolution by exon shuffling is reflected not only in a correlation between the domain organization of the resulting protein and exon–intron organization of its gene, but also in a characteristic phase distribution of its intermodule introns (Patthy, 1987). The fact that the structure of the gene of a modular protein conforms to these rules may actually be used as evidence that it has evolved by exon shuffling (Patthy, 1988). The general validity of these rules is supported by studies on the genes of numerous modular proteins (Patthy, 1991b, 1994). The genes of selectins provide typical examples of such a correlation: the modules of C-type lectin, growth factor and complement B-type are encoded by discrete class 1–1 exons, i.e. all the intermodule introns are phase 1 (Fig. 8.10).

There are cases where the gene structure does not correlate so well with the modular organization of a protein: the 'expected' introns may be missing from the module boundaries. For example, if we compare the gene of

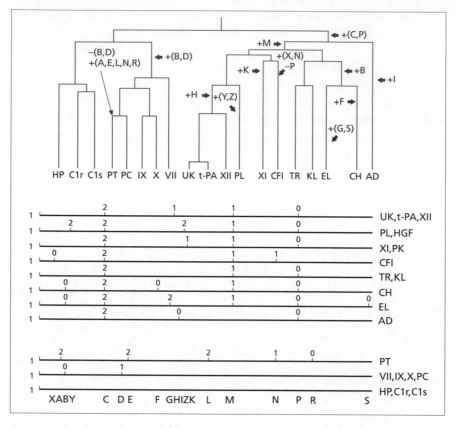

Fig. 8.9 Schematic representation of the exon–intron structure of proteases of
the trypsin family. The upper part of the figure shows the evolutionary tree of the
trypsin family. The abbreviations of the branches are: HP – haptoglobin; C1r and C1s
– complement C1r and C1s; PT – prothrombin; PC – protein C; VII, IX and X – factors
VII, IX and X; UK – urokinase; t-PA – tissue plasminogen activator; XII – factor XII
and hepatocyte growth factor activating protease; PL – plasminogen, apolipoprotein(a),
hepatocyte growth factor (HGF), macrophage stimulating protein; XI – factor XI
and prekallikrein (PK), CFI – complement factor I; TR – trypsin; KL – kallikreins;
EL – elastase; CH – chymotrypsin; AD – adipsin, medullasin, factor D, mast cell
proteasaes, cathepsin G, cytotoxic lymphocyte proteases. The lower part is a schematic
representation of the exon–intron organization of the trypsin-homolog regions only.
The numbers (0,1,2) indicate the location and phase class of the introns. Introns
occupying homologous positions in different genes are identified by the same letter
(A, B . . . Z). The most parsimonious pathway for the insertion (+) and removal (−)
of introns is also shown in the upper part of the figure. Note that all these genes have
a phase 1 intron at the 5′ boundary of the trypsin-homolog region. The ancestral
protease must also have had a phase 1 intron here: between its signal-peptide and
trypsin-homolog domains. (From Patthy, L. (1995) *Protein Evolution by Exon Shuffling.*
New York: Springer-Verlag.) With kind permission of Springer Science and Business
Media.

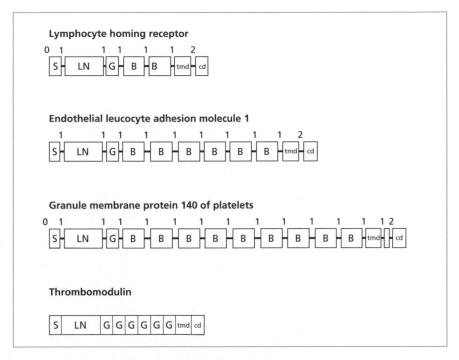

Fig. 8.10 Members of the selectin family (lymphocyte homing receptor, endothelial leukocyte adhesion molecule 1, granule membrane protein of platelets, cf. Fig. 8.7.) were assembled from class 1–1 C-type lectin (LN), epidermal growth factor (G) and complement B-type modules (B). In the genes of selectins all these modules are carried by distinct class 1–1 exons: all introns separating these modules from each other or from the signal-peptide domain (S) and the transmembrane and cytoplasmic domains (tmd, cd) are phase 1. In contrast with this, introns are missing from the boundaries of the class 1–1 C-type lectin and epidermal growth factor modules of the gene of thrombomodulin. (From Patthy, L. (1995) *Protein Evolution by Exon Shuffling*. New York: Springer-Verlag.) With kind permission of Springer Science and Business Media.

factor XIIIb with the gene of the Epstein–Barr virus receptor (Fig. 8.11), we see a striking difference: in the former case, all complement B-type modules are still carried by discrete class 1–1 exons; in the latter case, some exons encoding complement B-type modules have already been fused, whereas other exon modules are split by internal introns. The most plausible explanation for so much weaker a correlation between exon–intron structure and domain structure of the protein is that the original genomic organization of this modular protein has been obscured by subsequent removal and insertion of introns (Patthy, 1994, 1996).

Comparison of the exon–intron structures of human, fly and worm orthologs of mosaic genes assembled from class 1–1 modules by exon shuffling has identified numerous cases where human genes retained the original intermodule introns, but most of these introns were missing from protostome orthologs (Patthy, 1999; Bányai & Patthy, 2004). These studies

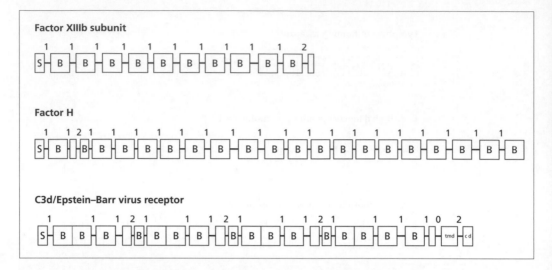

Fig. 8.11 Intron insertion and intron removal may erode the original exon–intron structure of genes. The gene of factor XIII b still reflects its assembly from class 1–1 complement B-type modules: all modules are encoded by distinct exons and all intermodule introns are phase 1. In the related gene of factor H a phase 2 intron has already been inserted in the second complement B-type module, but the other 19 modules are still carried by discrete exons. In the case of the C3d/Epstein–Barr virus receptor/complement receptor 2 gene removal of intermodule introns have fused, insertion of introns (phase 2) have split complement B-type modules resulting in a gene structure in which only five of the 16 modules are carried by distinct class 1–1 exons. Note, however, that the remaining intermodule introns are all phase 1. Abbreviations: S – signal-peptide domain; B – complement B-type module; tmd, cd – transmemembrane and cytoplasmic domain. (From Patthy, L. (1995) *Protein Evolution by Exon Shuffling*. New York: Springer-Verlag.) With kind permission of Springer Science and Business Media.

illustrate how removal of introns erodes the 'original' structure of the genes (i.e. the structure that existed at the time of their formation) and also shows that the rate of intron loss is much higher in the worm and insect lineages than in the chordate lineage.

As a result of continual intron insertion and intron removal, the correlation between modular protein structure and gene structure may get weaker and weaker and may eventually lead to an exon–intron organization that has little or nothing to do with the one that existed at the time of the formation of the gene. There are many cases where a correlation between the modular organization of the protein and the exon–intron structure of its gene can no longer be observed. An extreme example is the gene of thrombomodulin, where the C-type lectin module and the six EGF modules are not flanked by introns (Fig. 8.10). Since thrombomodulin is composed of modules that have been clearly shown in many other cases to be duplicated and inserted into new locations by recombination in phase 1 introns (cf. selectins, Fig. 8.10) it is clear that the ancestor of thrombomodulin also arose by exon shuffling, but its 'original' introns have been removed from its gene.

Since erosion of the original genomic structure progresses with time it is not surprising that old genes usually show less perfect correlation with protein structure than recently assembled ones. This point may be illustrated by the exon–intron structures of laminin genes or *Notch* genes. Laminins are among the oldest modular proteins unique to metazoa, inasmuch as they are already present in Hydrozoa, and their domain organization was essentially unchanged during subsequent evolution. Similarly, Notch-related proteins are present in nematodes, *Drosophila* and vertebrates, indicating that these genes were formed at an early stage of metazoan evolution. Consistent with the great age of these genes, most of the original phase 1 introns are already missing from the boundaries of the modules of the laminin and *Notch* genes even though in other (younger) modular genes the same module types are usually flanked by phase 1 introns.

8.1.6 Classification of modules and mosaic proteins produced by exon shuffling

Module classes and clans of mosaic proteins

A corollary of the phase-compatibility rule is that class 1–1 modules can be joined only to other class 1–1 modules, but not to class 0–0 or class 2–2 modules (and vice versa). As a consequence, there is a cluster of modular proteins constructed from class 1–1 modules. Exchange of modules is possible within this cluster – the '**clan 1**' of class 1–1 module families – but class 2–2 or class 0–0 modules cannot be accepted.

Class 1–1 modules are predominant

In contrast with the large variety of class 1–1 modules (Table 8.1), relatively few class 2–2 or class 0–0 modules are known. Only in the case of the albumin/α-fetoprotein family and the glucagon/vasoactive intestinal peptide family is there evidence that module duplications were mediated by recombination in phase 2 introns. In the genes of prosaposin and surfactant protein precursor the tandemly duplicated saposin modules, and in the case of ovomucoids and ovoinhibitors the duplicated ovomucoid modules, are flanked by phase 0 introns.

The striking predominance of class 1–1 modules is surprising since the three symmetrical module groups (class 1–1, class 0–0 and class 2–2) equally satisfy the phase-compatibility rules of intronic recombination. However, given the small size of the class 2–2 and class 0–0 module pools, it is not surprising that the number of modular proteins assembled from these is much lower than that of clan 1 proteins.

Table 8.1 Class 1–1 modules encountered most frequently as parts of multidomain modular proteins assembled by exon shuffling. The names of the modules usually refer to the name of the protein in which the given module was first identified or to some unique structural or functional feature of the module. Other commonly used names of the modules are given in parentheses. The abbreviations correspond to those used by the SMART database (http://smart.embl-heidelberg.de/).

Module	Abbreviation
A-type module of von Willebrand factor	VWA
Calcium-binding module (gla module)	GLA
Complement B-type module (sushi module)	CCP
Complement C1r module (CUB module)	CUB
C-type lectin module	CLECT
C-type module of von Willebrand factor	VWC
Delta serrate ligand	DSL
D-type module of von Willebrand factor	VWD
EGF-like module of laminin	EGF_Lam
Epidermal growth factor (EGF) module	EGF
Factor V/VIII type C module	FA58C
Fibronectin type-II module	FN2
Fibronectin type-III module	FN3
Finger module (fibronectin type-I module)	FN1
Follistatin module	FIMAC
Follistatin module	FOLN+KAZAL
Frizzled module	FRI
Immunoglobulin module	IG
Kringle module	KR
Kunitz-type trypsin inhibitor module	KU
LCCL domain	LCCL
LDL receptor type-A module	LDLa
Link protein module	LINK
Ly-6 antigen/uPA receptor-like domain	LU
MAM module	MAM
Olfactomedin domain	OLF
P or trefoil domain	PD
PAN (apple) module	PAN_AP
Scavenger receptor module	SR
SEA module	SEA
Somatomedin B-like domains	SO
Thrombospondin type-I module	TSP1
Thyroglobulin module	TY
Whey protein module (WAP module)	WAP
WSC domain	WSC
Zona pellucida (ZP) domain	ZP

The explanation for this imbalance of class 1–1, class 0–0 and class 2–2 modules comes from our present understanding of the evolutionary origin of such mobile modules. When considering the origin of these modules we must remember that the only feature that distinguishes a mobile module from

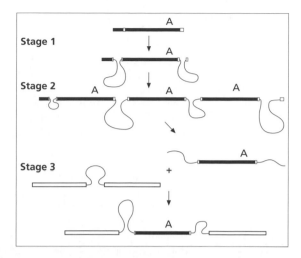

Fig. 8.12 Different stages in the conversion of a domain to a mobile module. The figure shows the modularization of a protein possessing a secretory signal-peptide domain. Stage 1: insertion of introns of identical phase at the amino- and carboxy-terminal boundaries of the domain. Stage 2: tandem duplication stage, leading to homopolymeric multidomain proteins. Stage 3: module is transferred to new locations. (From Patthy, L. (1994) Exons and introns. *Current Opinion in Structural Biology* **4**, 383–92.) Reprinted from Current Opinion in Structural Biology, 4, Patthy, L., Exons and Introns, 383–392, Copyright 1994, with permission from Elsevier.

a protein domain is its mobility: the module is flanked by introns of identical phase that facilitate its dispersal. According to this definition any protein domain may be converted to a **protomodule** if introns of identical phase are inserted at its two boundaries.

There is now evidence that protomodules suitable for exon shuffling may be created by insertion of introns into protein-coding genes at positions corresponding to the N-terminal and C-terminal boundaries of the folding units (Fig. 8.12). Usually, the next step in the evolution of modules is that the symmetrical (class 0–0, class 1–1 or class 2–2) protomodule undergoes internal tandem duplications via recombination in the flanking introns, thereby creating homopolymeric multidomain proteins. Since modules are frequently excised during contraction of genes containing numerous tandem copies, such 'homopolymeric' modular proteins may provide a major source of mobile modules. The excised modules may be inserted into other genes, as illustrated in Fig. 8.6.

A basic element of this modularization pathway is that the modules originally existed as independent proteins. Recent studies have confirmed that this is true for many modules, which for a long time were known only as building blocks of modular proteins. Table 8.2 lists those proteins that consist of only a single globular module. The different steps of the **modularization** process are best illustrated by the Kunitz-type proteinase inhibitor domain

Table 8.2 Protomodules – proteins consisting of a single module.

Module type*	Module class	Protein**
EGF	1–1	TGF-α, amphiregulin, diphteria toxin receptor etc.
CLECT	1–1	C-type lectins
KU	1–1	Bovine pancreatic trypsin inhibitor
FOLN+KAZAL	1–1	Human pancreatic trypsin inhibitor secretory (PST), human acrosin-trypsin inhibitor (HUSI-II) etc.
WAP	1–1	Elafins, WDNM1 gene etc.
IG	1–1	β₂-microglobulin, CD7, Thy-1 etc.
LDLa	1–1	Rous Sarcoma Virus receptor
LINK	1–1	Hermes antigen, surface antigens CD44, pgp-1
PD	1–1	pS2, intestinal trefoil factor
CUB	1–1	Spermadhesins
FA58C	1–1	Discoidin
GLA	1–1	Bone GLA protein
OVO*	0–0	Human pancreatic trypsin inhibitor secretory (PST), human acrosin-trypsin inhibitor (HUSI-II) etc.
SAP*	0–0	Pore-forming peptides

* The abbreviations of class 1–1 modules are identical with those used in Table 8.1. The class 0–0 SAP and OVO designate the saposin and ovomucoid modules, respectively.
** The proteins listed contain only single domain.

(Fig. 8.13(a)). In the gene of bovine pancreatic trypsin inhibitor, phase 1 introns are found at both boundaries of the single inhibitor domain (**protomodule stage**). The lipoprotein-associated coagulation inhibitor consists of three tandem copies of this module (**tandem duplication stage**); each module of this protein is carried by distinct class 1–1 exons. Kunitz-type inhibitor modules have been inserted into the genes of amyloid precursor, collagen VIα3 and collagen VII (**shuffling stage**). In the genes of human amyloid precursor and collagen VII the inhibitor modules are carried by distinct class 1–1 exons.

The Kazal-type proteinase inhibitor domain has been modularized in the class 0–0 route. Acquisition of phase 0 introns on both boundaries of this domain has converted it to the class 0–0 ovomucoid protomodule. The genes of pancreatic secretory trypsin inhibitor and acrosin inhibitor already possess the phase 0 intron at the 5′ boundary of the future ovomucoid module (Fig. 8.13b). The duplication stage is represented by the genes of ovomucoids and ovoinhibitors, where each of the 3–7 ovomucoid repeats are separated by phase 0 introns.

Recent studies have revealed that the globin fold has also made the first major steps to becoming a module. What is even more striking is that in three different animal groups it was modularized independently. In the evolutionary lineage of molluscs the globin fold was converted into a class 2–2 module: the two-domain intracellular globin of the clam *Barbatia reevena* arose by unequal crossing over between two identical or very similar genes for a single-domain globin. In the nematode line, intron insertions have

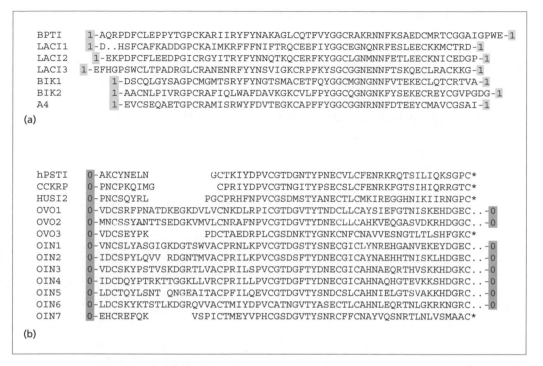

```
BPTI    1-AQRPDFCLEPPYYTGPCKARIIRYFYNAKAGLCQTFVYGGCRAKRNNFKSAEDCMRTCGGAIGPWE-1
LACI1   1-D..HSFCAFKADDGPCKAIMKRFFFNIFTRQCEEFIYGGCEGNQNRFESLEECKKMCTRD-1
LACI2    1-EKPDFCFLEEDPGICRGYITRYFYNNQTKQCERFKYGGCLGNMNNFETLEECKNICEDGP-1
LACI3  1-EFHGPSWCLTPADRGLCRANENRFYYNSVIGKCRPFKYSGCGGNENNFTSKQECLRACKKG-1
BIK1      1-DSCQLGYSAGPCMGMTSRYFYNGTSMACETFQYGGCMGNGNNFVTEKECLQTCRTVA-1
BIK2      1-AACNLPIVRGPCRAFIQLWAFDAVKGKCVLFPYGGCQGNGNNKFYSEKECREYCGVPGDG-1
A4        1-EVCSEQAETGPCRAMISRWYFDVTEGKCAPFFYGGCGGNRNNFDTEEYCMAVCGSAI-1
```
(a)

```
hPSTI   0-AKCYNELN         GCTKIYDPVCGTDGNTYPNECVLCFENRKRQTSILIQKSGPC*
CCKRP   0-PNCPKQIMG        CPRIYDPVCGTNGITYPSECSLCFENRKFGTSIHIQRRGTC*
HUSI2   0-PNCSQYRL      PGCPRHFNPVCGSDMSTYANECTLCMKIREGGHNIKIIRNGPC*
OVO1    0-VDCSRFPNATDKEGKDVLVCNKDLRPICGTDGVTYTNDCLLCAYSIEFGTNISKEHDGEC..-0
OVO2    0-MNCSSYANTTSEDGKVMVLCNRAFNPVCGTDGVTYDNECLLCAHKVEQGASVDKRHDGGC..-0
OVO3    0-VDCSEYPK      PDCTAEDRPLCGSDNKTYGNKCNFCNAVVESNGTLTLSHFGKC*
OIN1    0-VNCSLYASGIGKDGTSWVACPRNLKPVCGTDGSTYSNECGICLYNREHGANVEKEYDGEC..-0
OIN2    0-IDCSPYLQVV RDGNTMVACPRILKPVCGSDSFTYDNECGICAYNAEHHTNISKLHDGEC..-0
OIN3    0-VDCSKYPSTVSKDGRTLVACPRILSPVCGTDGFTYDNECGICAHNAEQRTHVSKKHDGKC..-0
OIN4    0-IDCDQYPTRKTTGGKLLVRCPRILLPVCGTDGFTYDNECGICAHNAQHGTEVKKSHDGRC..-0
OIN5    0-LDCTQYLSNT QNGEAITACPFILQEVCGTDGVTYSNDCSLCAHNIELGTSVAKKHDGRC..-0
OIN6    0-LDCSKYKTSTLKDGRQVVACTMIYDPVCATNGVTYASECTLCAHNLEQRTNLGKRKNGRC..-0
OIN7    0-EHCREFQK      VSPICTMEYVPHCGSDGVTYSNRCFFCNAYVQSNRTLNLVSMAAC*
```
(b)

Fig. 8.13 (a) Different stages in the career of the Kunitz-type inhibitor module. Bovine pancreatic trypsin inhibitor (BPTI) already has phase 1 introns on both boundaries of its single inhibitor domain. The three tandem Kunitz-type inhibitor modules of lipoprotein-associated coagulation inhibitor (LACI), the two modules of bikunin (BIK) and the shuffled module of the precursor of β amyloid (A4) are also flanked by phase 1 introns. (b) Different stages in the career of the Kazal-type inhibitor or ovomucoid module. The single Kazal-type inhibitor domain of human pancreatic secretory trypsin inhibitor (hPSTI), pancreatic cholecystokinin-releasing peptide (CCKRP) and human acrosin-trypsin inhibitor (HUSI2) have a phase 0 intron only at their 5′ boundary. In the ovomucoid (OVO) and ovoinhibitor (OIN) genes the tandem ovomucoid modules are separated by phase 0 introns. The numbers indicate the position and phase class of the introns. The asterisks indicate stop codons. (Modified from Patthy, L. (1994) Exons and introns. *Current Opinion in Structural Biology* **4**, 383–92.) Reprinted from Current Opinion in Structural Biology, 4, Patthy, L., Exons and Introns, 383–392, Copyright 1994, with permission from Elsevier.

converted the globin domain to a class 1–1 module: the nematodes *Ascaris suum* and *Pseudoterranova decipiens* possess extracellular globins with two tandem globin domains; at the boundaries of the globin domains phase 1 introns are found. In the evolutionary line of arthropods modularization and tandem duplication of the globin fold occurred in the class 0–0 route. In the gene of a polymeric globin of the brine shrimp *Artemia salina*, phase 0 introns are found at the boundaries separating its globin domains, indicating that the nine tandem globin units of this protein arose as a result of internal duplications of a class 0–0 globin protomodule.

In terms of the modularization scenario outlined above we may ask why, in general, domains had a greater chance to be modularized as class 1–1 rather than as class 2–2 or class 0–0 modules; i.e. why intron insertion was favoured in phase 1 at the boundaries of would-be modules. As we have discussed before, insertion of spliceosomal introns is not a random process but is targeted to 'protosplice sites', i.e. sites of exons that conform to the consensus sequences that flank exons. In the case of protein-coding genes of animals and plants the consensus sequence for the 5′ splice site is (C,A)AG/GTAAGT, while for the 3′ splice site it is AG/G(G,T)N; the most common sequence of protosplice sites is AG/G.

Since in protein-coding regions protosplice sites are translated, the base preferences of exon–intron junctions are also manifested in some amino acid preferences, according to the reading frame in which they are translated. If the intron in the (C,A)AG/G(G,T)N sequence is a phase 2 intron then it splits an AGG (Arg) codon of the mRNA. By a similar reasoning, if the (C,A)AG/G(G,T)N sequence is split by a phase 1 intron then it splits one of the four codons for glycine (GGN). In the case of phase 0 introns there is a greater freedom of choice of flanking codons for consensus (C,A)AG/G(G,T)N: (Lys,Gln)/(Gly,Val).

Biased amino acid composition of a given segment of a protein may thus be reflected in an absence of protosplice sites or in a biased phase distribution of protosplice sites. In other words, intron insertion may avoid some regions whereas in others insertion in certain phases may be strongly preferred. For example, a highly conserved glycine residue GGN may present a permanent protosplice site as a target for intron insertion (in phase 1). Silent mutations at the third position of such a conserved glycine would have little or no effect on this protosplice site. It may be pointed out that this reasoning suggests that a conserved Arg is a weaker target for intron insertion (in phase 2), since AGG is only one of the six Arg codons, i.e. synonymous substitutions may eliminate the protosplice site.

Consequently, if the boundaries of would-be protomodules are enriched in glycines (GGN) but depleted in arginines (AGG) then these could favour phase 1 insertion of introns over phase 2 insertion. The fact that in general linker regions connecting domains have a biased amino acid composition, where glycines are overrepresented and arginines are underrepresented (Argos, 1990), may thus explain the predominance of phase 1 insertion of introns in would-be linker regions. There is an additional source of bias favouring class 1–1 modules. As will be discussed below, the vast majority of known modular proteins are extracellular and their modules also originate from single-domain extracellular proteins. Since extracellular proteins possess a secretory signal peptide, one of the introns initiating modularization has to be inserted at the boundary separating the secretory signal-peptide domain from the mature protein domain (see Fig. 8.12). The amino acid sequence at this target site is far from random since the signal-sequence cleavage site recognized by the signal peptidase strongly favours small neutral residues such as glycine (see section 2.7.4). In this region glycine is much more abundant than expected by chance, thus intron insertion would be strongly favoured in such glycine residues in phase 1. Analysis of the exon–intron structures of 2208 human genes has indeed revealed that there is a statistically highly significant excess of phase 1 introns in the vicinity of the signal peptide cleavage sites (Tordai & Patthy, 2004).

Since intron insertion at the boundary of the signal-peptide domain is preferred in phase 1 modularization of proteins with secretory signal-peptide domains are most likely to take the class 1–1 route. Thus, the predominance of class 1–1 modules may also be a reflection of the fact that most of them were produced from proteins possessing a secretory signal-peptide domain.

Cellular distribution of modular proteins

A survey of modular proteins produced by exon shuffling reveals that the vast majority is extracellular (Table 8.3). As a reflection of their adaptation

Table 8.3 Major groups of modular proteins produced by exon shuffling.

Group	Proteins
Constituents of the extracellular matrix	Laminin A family, laminin B family, perlecans, agrin, fibronectin, the aggrecan/versican/neurocan family of protcoglycans, tenascins, nidogen/entactin, cartilage link protein, cartilage matrix protein family, cartilage oligomeric matrix protein, osteonectin, mucins, F-spondin, vitronectin, modular collagens (e.g. type I, type V, type VI, type VII collagen, type IX collagen, type XI collagen, type XII and type XIV collagens), surfactant protein A and D, conglutinin, mannose-binding protein, fibulin, fibrillin, etc.
Serine proteinases involved in blood coagulation, fibrinolysis and complement activation	Prothrombin, factor VII, factor IX, factor X, protein C, plasminogen, urokinase, tissue plasminogen activator, factor XII, complement factors C1r, C1s, B, C2, I, plasma prekallikrein, factor XI, etc.
Diverse members of the trypsin family of serine proteinases	Apolipoprotein(a), hepatocyte growth factor, macrophage-stimulating protein, hepatocyte growth factor activator, haptoglobin, enterokinase, etc.
Plasma proteins involved in the regulation of blood coagulation	Factors V and VIII, factor XIIIb chain, protein S, thrombospondins 1–4, von Willebrand factor, lipoprotein-associated coagulation inhibitor, etc.

Continued on p. 206

Table 8.3 (*Continued*)

Group	Proteins
Plasma proteins involved in the regulation of the complement system	Complement factor C6, complement factor C7, complement factors C8α and C8β, complement factor C9, complement factor H-related proteins, C4b-binding proteins, decay-accelerating factor, membrane cofactor protein, properdin, etc.
Metalloproteinases	Type IV collagenases (gelatinases), meprin α and β, bone morphogenetic protein-related proteins, cysteine-rich metalloproteinases, etc.
Diverse receptors, cell adhesion proteins, membrane-associated proteins	The cytokine receptor family, the family of VLDL- and LDL-receptor-related proteins, scavenger receptor-related proteins, complement receptors, the selectin family, netrins, neurexins, integrins Mac-1, LFA-1, p150,95, VLA-2, integrin β4, epidermal growth factor precursor, the Notch family, the mannose receptor/phospholipase A₂ receptor/antigen-processing receptor family, insulin-like growth factor II receptor, interleukin-2-receptor α, TGF-β type III receptor, urokinase receptor, thrombomodulin, the contactin, N-CAM and DCC related proteins, uromodulin, zona pellucida proteins, A4 amyloid protein precursor, etc.
Diverse receptor protein tyrosine phosphatases and receptor protein tyrosine kinases	The LAR family of receptor tyrosine phosphatases, the DDR family of receptor tyrosine kinases, the *ret* proto-oncogene family, the PDGF, FGF, KGF receptors, the *FLT/NYK/FLK-1* receptor protein tyrosine kinases, *trk* neurotrophin receptors, the *Ror* receptor protein tyrosine kinases, the insulin receptor/IGF-I receptor family, the *Axl/Ufo/Sky/Tyro 3* receptor protein tyrosine kinases, the *eph/elk* family of receptor protein tyrosine kinases, the *hyk/tek/tie* family of receptor protein tyrosine kinases etc.
Miscellaneous	Immunoglobulins, thyroglobulin, insulin-like growth factor binding proteins, TGF-β1 binding protein, *cyr61/fisp-12* family of growth factors, leucocyte protease inhibitor, pancreatic apolipoprotein H, spasmolytic polypeptide, follistatin, whey proteins, muscle proteins titin, C-protein, M-protein, skelemin

to the oxidative milieu of the extracytoplasmic space, many of the modules are rich in disulphide bonds (e.g. kringles, epidermal growth factor modules, finger modules, fibronectin type II modules, C-type lectin module, etc.).

A large group of modular extracellular proteins comprises various plasma proteins: the proteinases of the complement, blood coagulation and fibrinolytic cascades, and the nonproteinase factors regulating blood coagulation and complement activation, immunoglobulins, etc.

Another major group includes various constituents of the extracellular matrix (fibronectin, modular collagens, laminins, etc.), and enzymes involved in the remodelling of the extracellular matrix (e.g. type IV collagenases, bone morphogenetic proteins).

A third group of modular proteins contains a variety of membrane-associated or transmembrane proteins with extracellular parts constructed from modules: receptor tyrosine kinases, receptor tyrosine phosphatases, proteins involved in cell–cell or cell–matrix interactions, complement receptors, LDL receptors, neurotrophin receptors, cytokine receptors, etc.

There are just a few exceptions to the generalization that proteins produced from class 1–1 modules occur in the extracytoplasmic space: a few module types (e.g. the immunoglobulin module and the fibronectin type III module) have also been found in the intracellular compartment. For example, the intracellular muscle proteins twitchin, titin, projectin and C-protein are composed of multiple copies of immunoglobulin and fibronectin type III modules.

8.1.7 Genome evolution and the evolution of exon shuffling

The clearcut examples of proteins assembled by exon shuffling are restricted to animals, with practically no counterparts in prokaryotes, plants or fungi. This observation is not too surprising if we consider that the evolution of introns and of modules suggests that exon shuffling is a relatively late development: the exon-shuffling machinery (spliceosomal pre-mRNA introns and protomodules) appeared relatively late during evolution. The importance of the accumulation of a critical mass of module types is illustrated by the fact that most of the modular proteins were assembled from some three dozen class 1–1 modules. The low number of class 0–0 and class 2–2 modules probably precluded a similar burst of exon shuffling in the case of these module types.

To define more precisely the time when exon shuffling became significant in eukaryotes we may analyse the evolutionary distribution of modular proteins that evolved by this mechanism. Thanks to various genome projects on eukaryotic model organisms (*Saccharomyces cerevisiae*, *Plasmodium falciparum*, *Arabidopsis thaliana*, *Caenorhabditis elegans*, *Drosophila melanogaster*, various vertebrates, etc.) sufficient sequence information has already accumulated on protists, plants, fungi and animals to decide whether exon shuffling and modular evolution was significant before their divergence or whether it is a more recent invention of the animal kingdom.

A search for class 1–1 modules and clan 1 modular proteins has revealed many examples from all major groups of metazoa. The fact that proteins composed of class 1–1 modules familiar from vertebrate genes have been found in sponges, hydrozoans, nematodes, molluscs, arthropods and echinoderms indicates that the machinery of exon shuffling was available before the divergence of these metazoan phyla. There can be no doubt that the mechanism of the construction of these modular proteins was the same as that of vertebrate genes since there are several cases of invertebrate genes where the class 1–1 modules are also flanked by the original phase 1 introns. It is noteworthy that modular proteins assembled from class 1–1 modules have already been found in sponges although only a tiny fraction of their genes have been sequenced so far. In contrast with this, there is no evidence yet for related modular proteins in yeast although its complete genome is already known. In the case of plants there are only a few possible examples; e.g. the receptor protein kinase of *Arabidopsis thaliana*, which contains two tandem copies of an EGF-like domain. Interestingly, modular receptor tyrosine kinases (containing class 1–1 EGF and CCP modules typical of animals) are already present in choanoflagellates, the closest unicellular relatives of animals (King & Carroll, 2001).

In addition to metazoa, a few modular proteins containing domains homologous with those of vertebrate proteins have been found in some animal viruses and parasitic protists. It seems likely that these viruses and protists acquired the modular proteins from their multicellular hosts by lateral gene transfer – these proteins assist them in the invasion process. Thus the surface protein of *Plasmodium falciparum* containing four EGF domains facilitates infection by binding to receptors on mosquito epithelial cells, and the thrombospondin-homolog proteins of malarial parasites assist in their entry into hepatocytes by mediating their adherence to these cells (for further details see Chapter 9). Similarly, the vaccinia virus protein containing four complement B-type modules could counteract host immune defences by interfering with the complement cascade.

In summary, the evolutionary distribution of modular proteins that have clearly evolved by exon shuffling is consistent with the suggestion that exon shuffling became significant in the animal kingdom at the time of metazoan radiation. As will be discussed in Chapter 9, studies on the genomes of protists, plants, fungi and various groups of animals have shown that the increase in haploid genome size as we move from single cell eukaryotes to higher eukaryotes is paralleled by a general decrease in genome compactness and an increase in the number and size of introns. The fact that exon shuffling became significant in the metazoan (and especially the vertebrate) lineage is a direct consequence of the evolution of larger, less compact genomes enriched in spliceosomal introns and repetitive elements that facilitate exon shuffling (Patthy, 1999).

8.1.8 Evolutionary significance of exon shuffling

The fact that the overwhelming majority of the proteins regulating blood coagulation, fibrinolysis and complement activation, plus most constituents of the extracellular matrix, cell adhesion proteins and receptor proteins were constructed from modules underlines the extreme value of exon shuffling. Several unique features of this evolutionary mechanism made it so important.

First, in proteins assembled from modules a large collection of binding specificities may coexist making such proteins ideal members of regulatory or structural networks where multiple interactions are critical. For example, different modules of plasma proteinases recognize their substrates or bind inhibitors, cofactors, phospholipid membranes, etc., and through these interactions control the activation and activity of these enzymes. The coexistence of different modules with different binding specificities is also essential for the biological function of multidomain proteins of the extracellular matrix (e.g. fibronectin, proteoglycans, laminins, modular collagens): these interactions are indispensable for the proper organization of the extracellular matrix. The various cell-surface molecules that mediate the interaction of cells with other cells via homophilic and heterophilic interactions, or bind cells to matrix constituents, are also involved in multiple protein–protein interactions.

Second, protein evolution by module acquisition is an extremely powerful evolutionary mechanism. Acquisition of a new domain with a given binding specificity can bring about a sudden change in the specificity of the recipient. One of the most clearcut examples of the advantage of borrowing a module with an established specificity is the case of gelatinases. As discussed in section 8.1.3, these enzymes arose as a result of the insertion of a gelatin-binding fibronectin type II module into an intron of an ancestral metalloproteinase of the collagenase family. The case of gelatinases also illustrates the role of selection in modular protein evolution. Although modules are shuffled at random the chances that an inserted module is accepted depends on whether its insertion is deleterious, neutral or carries a function potentially useful for the host protein. It seems obvious that the insertion of a collagen-binding type II module into an ancestral member of the collagenase family was accepted because its binding specificity for denatured collagens was biologically relevant for a collagenase. Similarly, the observation that the complement B-type modules were inserted several times independently into different components of the complement cascade (complement factors C1r and C1s; factors B and C2; factors C6 and C7; complement factor H, etc.) probably reflects the fact that their binding specificity was meaningful in the context of the complement system. (The role of selection is also obvious in the case of the vaccinia virus protein that contains four complement B-type modules. Lateral transfer of such a modular protein could counteract the immune defence system of the host.)

Shuffling of modules with established specificities is thus an extremely efficient mechanism for the creation of proteins with novel or expanded functions. (Patthy, 2003). It seems logical to assume that the rise of this mechanism had a great impact on evolution.

As discussed above, this powerful evolutionary mechanism has become significant relatively late during evolution, at the time of metazoan radiation, in parallel with a decrease in genome compactness and the evolution of spliceosomal introns. It is interesting in this respect that the explosion of exon shuffling seems to coincide with a spectacular burst of evolutionary creativity: the 'Big Bang' of metazoan radiation. As a result of a mysterious acceleration of evolution, different phyla of metazoa with different body plans appeared almost simultaneously during a very short interval of the Cambrian period. Although numerous hypotheses have been proposed for this Cambrian explosion (changes in atmospheric oxygen, shift of continents and change of climate, etc.) none is generally accepted as a full explanation of this acceleration. It seems probable that invention of modular protein evolution by exon shuffling has contributed significantly to this accelerated evolution of metazoa.

It must be emphasized that most modular proteins produced by exon shuffling are associated with and are absolutely essential for multicellularity. The appearance of the constituents of extracellular matrix is inextricably linked to the appearance of the first multicellular organisms. Clearly, the constituents of the extracellular matrix, membrane-associated proteins mediating cell–cell and cell–matrix interactions, and receptor proteins regulating cell–cell communications such as receptor tyrosine kinases and receptor tyrosine phosphatases are of absolute importance in permitting cells to function in an integrated fashion. It is well known from studies of the developmental biology of *Caenorhabditis*, *Drosophila* and vertebrates that many modular proteins are involved in controlling morphogenesis, differentiation processes and cell-fate decisions, and thus determine the basic body plans of metazoa. In fact, before the *Drosophila* and nematode genome projects most of the modular genes of these animals were identified on the basis of their involvement in basic developmental processes.

In summary, it seems that the evolution and spread of spliceosomal introns and the creation of mobile modules reached a critical point some time before the Cambrian period and this led to a dramatic increase in the efficiency of modular protein evolution. This novel mechanism permitted the rapid creation of diverse multidomain proteins that are essential for the multicellularity of metazoa.

8.1.9 Genome evolution and the evolution of alternative splicing

The increase in the number of introns as we move from single-cell eukaryotes to higher eukaryotes (see Chapter 9) is paralleled by a general increase in the frequency of alternative splicing (Blencowe, 2006). For example, recent analyses of sequence and microarray data suggest that over two-thirds of human genes and over 40% of *Drosophila* genes contain one or more alternative exons, whereas *S. cerevisiae* has only a few known alternative splicing events. As a consequence of alternative splicing, the genomes of higher eukaryotes encode a significantly higher number of proteins than expected just on the basis of the number of their protein-coding genes.

The evolutionary significance of alternative splicing is that a single gene can encode a large variety of transcripts and proteins that may fulfil different functions (Stamm et al., 2005). Functional modifications through alternative splicing include addition of new parts that modify the binding properties, catalytic properties, subcellular localization, stability and post-translational modification of the proteins or may influence some key properties (e.g. stability, targeting or rate of translation) of the mRNA.

The fibronectin gene has become paradigmatic to illustrate generation of protein diversity by alternative mRNA splicing (Kornblihtt et al., 1996). Alternative splicing, domain skipping of the modular fibronectin gene gives rise to multiple fibronectin polypeptides in a cell type, development and age-regulated manner, the different domain variants playing specific roles in blood clotting, adhesion, skin wound healing etc. As another example we may mention alternative splicing of agrin, a key molecule of the vertebrate neuromuscular junction; the single agrin gene encodes several functionally different forms of agrin (Stetefeld et al., 2004). Interestingly, alternative splicing of the agrin gene can give rise to two major types of isoforms that differ in their subcellular localization. In some of these forms the amino-terminal end contains a cleavable signal sequence that targets these proteins to the secretory pathway, in the other forms it contains a noncleaved signal anchor that immobilizes the protein in the plasma membrane (Neumann et al., 2001).

8.2 Modular assembly by exonic recombination

As already discussed, multidomain proteins have several unique properties that endow them with great evolutionary significance. Since the acquisition of a new domain can bring about a sudden and advantageous change in the specificity of the recipient, positive selection may be so powerful that it compensates for the fact that the genetic events leading to mutations that alter domain organization are significantly less frequent than point mutations. In other words, the value of such multidomain proteins may drive evolution in the direction of their creation.

Although exon shuffling by intronic recombination is by far the most powerful mechanism of modular protein evolution, this does not mean it is the only way to exchange domains among protein-coding genes (Patthy, 1996). Some examples show how modular proteins of bacteria may be constructed without the assistance of introns. For example, a modular protein of *Peptostreptococcus magnus* has been shown to be the product of a recent intergenic recombination of two different types of streptococcal surface protein (de Chateau & Bjorck, 1994). The recent transfer of a fragment of a prokaryotic gene to another illustrates that introns are not absolutely essential for exchange of gene fragments. These studies have also shown that gene rearrangements by exonic recombination may be facilitated by the presence of special recombinogenic DNA sequences in intermodule linker regions, and that antibiotics may provide the selective pressure for the creation of advantageous chimeras (de Chateau & Bjorck, 1996). Similarly, promiscuous exonic recombination (facilitated by direct repeats within exonic DNA) is responsible for the remarkable antigenic variation of *Borrelia*, the agent that causes Lyme disease (Zhang et al., 1997). The selective pressure that favours such promiscuous exonic recombination is that evasion of the immune response permits the long-term survival of *Borrelia* in the mammalian host.

Studies on the evolution of the multidomain bacterial proteins of the phosphoenolpyruvate sugar phosphotransferase system suggest that fusion and dissociation of protein domains occurred frequently during the evolution of these enzymes. The three functional domains of the multiphosphoryl transfer protein fruB(HI) are joined by flexible linker regions, and the unusual DNA sequence of these linker regions probably accounts for the high frequency of rearrangements during the evolution of these proteins (Wu, Tomich & Saier, 1990).

The hevein domain of plants behaves in many ways as a mobile module inasmuch as it occurs in a variety of multidomain proteins. For example, various agglutinins have two or four tandem hevein domains, while different chitinases have one or two hevein domains joined to a catalytically active enzyme domain. The tandem duplication and shuffling of this module is thus strikingly similar to modules of animals; however, in this case the genes lack introns at the boundaries of the hevein modules. Recent studies on DNA sequences of chitinases have shown that the hevein domains are flanked by direct repeats, suggesting that they were dispersed by transposition events (Shinshi et al., 1990).

The majority of intracellular proteins involved in cytoplasmic or nuclear signalling processes of eukaryotes are multidomain proteins assembled from a large repertoire of intracellular signalling domains (Bork, Schultz & Ponting, 1997) but there is little or no evidence for a role of exon shuffling in their formation. The most likely explanation for this is that they were formed without the assistance of this mechanism. Comparison of the multidomain proteins involved in these intracellular signalling pathways revealed some striking similarities in single-celled eukaryotes, plants and metazoa (Copley et al., 1999; Aravind & Subramanian, 1999) suggesting that they have originated prior to the last common ancestor of eukaryotes. Accordingly, that there is no clear evidence for a role of exon shuffling in the case of most intracellular modular proteins may simply be due to the fact that they were formed in the early eukaryotes, i.e. organisms with intron-poor genomes (Patthy, 2003).

References

Antonell, A., de Luis, O., Domingo-Roura, X. & Perez-Jurado, L.A. (2005) Evolutionary mechanisms shaping the genomic structure of the Williams–Beuren syndrome chromosomal region at human 7q11.23. *Genome Research* **15**, 1179–88.

Aravind, L. & Subramanian, G. (1999) Origin of multicellular eukaryotes – insights from proteome comparisons. *Current Opinion in Genetics & Development* **9**, 688–94.

Argos, P. (1990) An investigation of oligopeptides linking domains in protein tertiary structures and possible candidates for general gene fusion. *Journal of Molecular Biology* **211**, 943–58.

Babcock, M., Pavlicek, A., Spiteri, E. et al. (2003) Shuffling of genes within low-copy repeats on 22q11 (LCR22) by Alu-mediated recombination events during evolution. *Genome Research* **13**, 2519–32.

Bányai, L. & Patthy, L. (2004) Evidence that human genes of modular proteins have retained significantly more ancestral introns than their fly or worm orthologues. *FEBS Letters* **565**, 127–32.

Baralle, D. & Baralle, M. (2005) Splicing in action: assessing disease causing sequence changes. *Journal of Medical Genetics* **42**, 737–48.

Bailey, J.A., Liu, G. & Eichler, E.E. (2003) An Alu transposition model for the origin and expansion of human segmental duplications. *American Journal of Human Genetics* **73**, 823–34.

Belfort, M. (1993) An expanding universe of introns. *Science* **262**, 1009–10.

Bhattacharya, D., Lutzoni, F., Reeb, V., Simon, D., Nason, J. & Fernandez, F. (2000) Widespread occurrence of spliceosomal introns in the rDNA genes of ascomycetes. *Molecular Biology and Evolution* **17**, 1971–84.

Blencowe, B.J. (2006) Alternative splicing: new insights from global analyses. *Cell* **126**, 37–47.

Bork, P., Schultz, J. & Ponting, C.P. (1997) Cytoplasmic signalling domains: the next generation. *Trends in Biochemical Sciences* **22**, 296–8.

Cavalier-Smith, T. (1985) Selfish DNA and the origin of introns. *Nature* **315**, 283–4.

Chai, J.H., Locke, D.P., Greally, J.M. et al. (2003) Identification of four highly conserved genes between breakpoint hotspots BP1 and BP2 of the Prader–Willi/Angelman syndromes deletion region that have undergone evolutionary transposition mediated by flanking duplicons. *American Journal of Human Genetics* **73**, 898–925.

Coghlan, A. & Wolfe, K.H. (2004) Origins of recently gained introns in *Caenorhabditis*. *Proceedings of the National Academy of Sciences of the USA* **101**, 11362–7.

Copley, R.R., Schultz, J., Ponting, C.P. & Bork, P. (1999) Protein families in multicellular organisms. *Current Opinion in Structural Biology* **9**, 408–15.

Csank, C., Taylor, F.M. & Martindale, D.W. (1990) Nuclear pre-mRNA introns: analysis and comparison of intron sequences from *Tetrahymena thermophila* and other eukaryotes. *Nucleic Acids Research* **18**, 5133–41.

Darnell, J.E. & Doolittle, W.F. (1986) Speculations on the early course of evolution. *Proceedings of the National Academy of Sciences of the USA* **83**, 1271–5.

de Chateau, M. & Bjorck, L. (1994) Protein PAB, a mosaic albumin-binding bacterial protein representing the first contemporary example of module shuffling. *Journal of Biological Chemistry* **269**, 12147–51.

de Chateau, M. & Bjorck, L. (1996) Identification of interdomain sequences promoting the intronless evolution of a bacterial protein family. *Proceedings of the National Academy of Sciences of the USA* **93**, 8490–5.

Dibb, N.J. & Newman, A.J. (1989) Evidence that introns arose at protosplice sites. *EMBO Journal* **8**, 2015–21.

Doolittle, W.F. (1978) Genes in pieces: were they ever together? *Nature* **272**, 581–2.

Dorit, R.L., Schoenbacher, L. & Gilbert, W. (1990) How big is the universe of exons? *Science* **250**, 1377–82.

Eickbush, T. (1999) Exon shuffling in retrospect. *Science* **283**, 1465–7.

Ejima, Y. & Yang, L. (2003) Transmobilization of genomic DNA as a mechanism for retrotransposon-mediated exon shuffling. *Human Molecular Genetics* **12**, 1321–8.

Fridell, R.A., Pret, A.M. & Searles, L.L. (1990) A retrotransposon 412 insertion within an exon of the *Drosophila melanogaster vermilion* gene is spliced from the precursor RNA. *Genes and Development* **4**, 559–66.

Gerstein, M. (1997) A structural census of genomes: comparing bacterial, eukaryotic and archaeal genomes in terms of protein structure. *Journal of Molecular Biology* **274**, 562–76.

Gilbert, W. (1978) Why genes in pieces? *Nature* **271**, 501.

Gilbert, W. (1986) The RNA world. *Nature* **319**, 618.

Gilbert, W. & Glynias, M. (1993) On the ancient nature of introns. *Gene* **135**, 137–44.

Goguel, V. & Rosbash, M. (1993) Splice site choice and splicing efficiency are positively influenced by pre-mRNA intramolecular base pairing in yeast. *Cell* **72**, 893–901.

Hedges, S.B. (2002) The origin and evolution of model organisms. *Nature Reviews in Genetics* **3**, 838–49.

Hickey, D.A. & Benkel, B. (1986) Introns as relict retrotransposons: implications for the evolutionary origin of eukaryotic mRNA splicing mechanisms. *Journal of Theoretical Biology* **121**, 283–91.

Hill, A.S., Foot, N.J., Chaplin, T.L. & Young, B.D. (2000) The most frequent constitutional translocation in humans, the t(11;22)(q23;q11) is due to a highly specific Alu-mediated recombination. *Human Molecular Genetics* **9**, 1525–32.

Jeffares, D.C., Mourier, T. & Penny, D. (2006) The biology of intron gain and loss. *Trends in Genetics* **22**, 16–22.

Ji, Y., Eichler, E.E., Schwartz, S. & Nicholls, R.D. (2000) Structure of chromosomal duplicons and their role in mediating human genomic disorders. *Genome Research* **10**, 597–610.

Jiang, Y. & Doolittle, R.F. The evolution of vertebrate blood coagulation as viewed from a comparison of puffer fish and sea squirt genomes. *Proceedings of the National Academy of Sciences of the USA* **100**, 7527–32.

King, N. & Carroll, S.B. (2001) A receptor tyrosine kinase from choanoflagellates: molecular insights into early animal evolution. *Proceedings of the National Academy of Sciences of the USA* **98**, 15032–7.

Kornblihtt, A.R., Pesce, C.G., Alonso, C.R. et al. (1996) The fibronectin gene as a model for splicing and transcription studies. *FASEB Journal* **10**, 248–57.

Lambowitz, A.M. & Zimmerly, S. (2004) Mobile group II introns. *Annual Reviews in Genetics* **38**, 1–35.

Lawn, R.M., Schwarz, K. & Patthy, L. (1997) Convergent evolution of apolipoprotein(a) in primates and hedgehog. *Proceedings of the National Academy of Sciences of the USA* **94**, 11992–7.

Leslie, N.D., Kessler, C.A., Bell, S.M. et al. (1990) The chicken urokinase-type plasminogen activator gene. *Journal of Biological Chemistry* **265**, 1339–44.

Luehrsen, K.R. & Walbot, V. (1992) Insertion of non-intron sequence into maize introns interferes with splicing. *Nucleic Acids Research* **20**, 5181–7.

Martin, W. & Koonin, E.V. (2006) Introns and the origin of nucleus-cytosol compartmentalization. *Nature* **440**, 41–5.

McLean, J.W., Tomlinson, J.E., Kuang, W.J. et al. (1987) cDNA sequence of human apolipoprotein(a) is homologous to plasminogen. *Nature* **300**, 132–7.

Menssen, A., Hohmann, S., Martin, W. et al. (1990) The *En/Spm* transposable element of *Zea mays* contains splice sites at the termini generating a novel intron from a dSpm element in the A2 gene. *EMBO Journal* **9**, 3051–7.

Moran, J.V., DeBerardinis, R.J. & Kazazian, H.H. Jr. (1999) Exon shuffling by L1 retrotransposition. *Science* **283**, 1530–4.

Mourier, T. & Jeffares, D.C. (2003) Eukaryotic intron loss. *Science* **300**, 1393.

Mueller, M.W., Allmaier, M., Eskes, R. & Schweyen, R.J. (1993) Transposition of group II intron aI1 in yeast and invasion of mitochondrial genes at new locations. *Nature* **366**, 174–6.

Neumann, F.R., Bittcher, G., Annies, M., Schumacher, B., Kroger, S. & Ruegg, M.A. (2001) An alternative amino-terminus expressed in the central nervous system converts agrin to a type II transmembrane protein. *Molecular and Cellular Neuroscience* **17**, 208–25.

Orgel, L.E. & Crick, F.H.C. (1980) Selfish DNA: the ultimate parasite. *Nature* **284**, 604–7.

Palmer, J.D. & Logsdon, J.M. Jr (1991) The recent origins of introns. *Current Opinion in Genetics and Development* **1**, 470–7.

Patthy, L. (1985) Evolution of the proteases of blood coagulation and fibrinolysis by assembly from modules. *Cell* **41**, 657–63.

Patthy, L. (1987) Intron-dependent evolution: preferred types of exons and introns. *FEBS Letters* **214**, 1–7.

Patthy, L. (1988) Detecting distant homologies of mosaic proteins. Analysis of the sequences of thrombomodulin, thrombospondin, complement components C9, C8α and C8β, vitronectin and plasma cell membrane glycoprotein PC-1. *Journal of Molecular Biology* **202**, 689–96.

Patthy, L. (1990) Evolutionary assembly of blood coagulation proteins. *Seminars in Thrombosis and Haemostasis* **16**, 245–59.

Patthy, L. (1991a) Exons – original building blocks of proteins? *BioEssays* **13**, 187–92.

Patthy, L. (1991b) Modular exchange principles in proteins. *Current Opinion in Structural Biology* **1**, 351–61.

Patthy, L. (1994) Exons and introns. *Current Opinion in Structural Biology* **4**, 383–92.

Patthy, L. (1995) *Protein Evolution by Exon Shuffling*. Molecular Biology Intelligence Unit. R.G. Landes, New York: Springer-Verlag.

Patthy, L. (1996) Exon shuffling and other ways of module exchange. *Matrix Biology* **15**, 301–10.

Patthy, L. (1999) Genome evolution and the evolution of exon-shuffling – a review. *Gene* **238**, 103–14.

Patthy, L. (2003) Modular assembly of genes and the evolution of new functions. *Genetica* **118**, 217–31.

Purugganan, M. & Wessler, S. (1992) The splicing of transposable elements and its role in intron evolution. *Genetica* **86**, 295–303.

Shapiro, M.B. & Senapathy, P. (1987) RNA splice junctions of different classes of eukaryotes: sequence statistics and functional implications in gene expression. *Nucleic Acids Research* **15**, 7155–74.

Sharp, P.A. (1985) On the origin of RNA splicing and introns. *Cell* **42**, 397–400.

Shinshi, H., Neuhaus, J.M., Ryals, J. et al. (1990) Structure of a tobacco endochitinase gene: evidence that different chitinase genes can arise by transposition of sequences encoding a cysteine-rich domain. *Plant Molecular Biology* **14**, 357–68.

Slagter-Jager, J.G., Allen, G.S., Smith, D., Hahn, I.A., Frank, J. & Belfort, M. (2006) Visualization of a group II intron in the 23S rRNA of a stable ribosome. *Proceedings of the National Academy of Sciences of the USA* **103**, 9838–43.

Stamm, S., Ben-Ari, S., Rafalska, I. et al. (2005) Function of alternative splicing. *Gene* **344**, 1–20.

Stetefeld, J., Alexandrescu, A.T., Maciejewski, M.W. et al. (2004) Modulation of agrin function by alternative splicing and Ca2+ binding. *Structure* **12**, 503–15.

Stoltzfus, A., Spences, D.F., Zuker, M. et al. (1994) Testing the exon theory of genes: the evidence from protein structure. *Science* **265**, 202–7.

Tordai, H. & Patthy, L. (2004) Insertion of spliceosomal introns in protosplice sites: the case of secretory signal peptides. *FEBS Letters* **575**, 109–11.

Vicens, Q. & Cech, T.R. (2006) Atomic level architecture of group I introns revealed. *Trends in Biochemical Sciences* **31**, 41–51.

Waigmann, E. & Barta, A. (1992) Processing of chimaeric introns in dicot plants: evidence for a close cooperation between 5' and 3' splice sites. *Nucleic Acids Research* **20**, 75–81.

Wessler, S.R. (1991a) The maize transposable *Ds1* element is alternatively spliced from exon sequences. *Molecular Cell Biology* **11**, 6192–6.

Wessler, S.R. (1991b) Alternative splicing of a *Ds* element from exon sequences may account for two forms of wx protein encoded by the *wx-m9* allele. *Maydica* **36**, 317–22.

Winkler, H., Brostjan, C., Csizmadia, V. et al. (1996) The intron–exon structure of the porcine E-selectin-encoding gene. *Gene* **176**, 67–72.

Wu, L.F., Tomich, J.M. & Saier, M.H. Jr (1990) Structure and evolution of a multidomain multiphosphoryl transfer protein. Nucleotide sequence of the fruB (HI) gene in *Rhodobacter capsulatus* and comparisons with homologous genes from other organisms. *Journal of Molecular Biology* **213**, 687–703.

Yang, J., Brown, M.S., Ho, Y.K. & Goldstein, J.L. (1995) Three different rearrangements in a single intron truncate sterol regulatory element binding protein-2 and produce sterol-resistant phenotype in three cell lines. Role of introns in protein evolution. *Journal of Biological Chemistry* **270**, 12152–61.

Zhang, J.R., Hardham, J.M., Barbour, A.G. et al. (1997) Antigenic variation in Lyme disease Borrelia by promiscuous recombination of VMP-like sequence cassettes. *Cell* **89**, 275–85.

Useful internet resources

Databases of intron-containing protein-coding genes

The **Exon-Intron Database** is an exhaustive database of protein-coding intron-containing genes (http://hsc.utoledo.edu/bioinfo/eid/index.html).

IEKB (Intron Exon Knowledge Base) is a relational database containing exon/intron gene structure data for eukaryotic genes derived from ExInt. IE-Kb allows ranking of exons and introns based on properties such as length, phase, composition etc. (http://sege.ntu.edu.sg/wester/iekb/).

Intron Server (IS) is a dynamic database of intron sequences (www.icgeb.trieste.it/introns).

SMART (Simple Modular Architecture Research Tool) provides a platform for the comparative study of complex domain architectures in genes and proteins. Visualization tools allow analysis of gene intron–exon structure within the context of protein domain structure (http://smart.embl-heidelberg.de/).

Chapter 9: Genome evolution and protein evolution

9.1 Evolution of genome size

The amount of DNA present in the haploid genome is called the genome size. The genome size is frequently referred to as the C ('constant') value since the size of the haploid genome is constant and characteristic for a species although it varies enormously among organisms (Table 9.1). Bacterial genome sizes vary from about 6×10^5 bp in some obligatory intracellular parasites and symbionts (such as *Mycoplasma, Buchnera*), to more than 10^7 bp in some cyanobacterial species. The smallest prokaryotes, the mycoplasmas, contain about 400–600 protein-coding genes including some 50 ribosomal proteins, two sets of rRNA genes (5S, 16S and 23S), and about 40 tRNA genes. The gene set of mycoplasmas is assumed to be close to the minimum number sufficient to support life.

The genomes of bacteria consist of **chromosomal DNA** and **extra-chromosomal genetic elements**. The nonchromosomal fraction of the genome is usually much smaller than the chromosomal fraction. The majority of the chromosomal fraction (more than 80%) is dedicated to protein-coding genes required for growth and metabolic functions, about 1% encodes RNA-specifying genes, and the rest comprises intergenic spacers containing regulatory signals. The fact that bacteria do not contain large quantities of **nongenic DNA** indicates that in this case there may be strong selective pressure to eliminate nongenic DNA. As will be illustrated by surveys of the complete genomes of prototypes of bacteria (*Escherichia coli, Haemophilus influenzae, Bacillus subtilis, Mycoplasma genitalium, Mycoplasma pneumoniae, Pelagibacter ubique, Borrelia burgdorferi, Buchnera aphidocola, Carsonella ruddii*), the genomes of bacteria have been shaped by gene duplications, small-scale deletions and insertions, horizontal transfer of genes and extensive loss of DNA in many parasitic or endosymbioyic lineages.

Even though some unicellular eukaryotes (e.g. the yeast *Saccharomyces cerevisiae*) has a small genome (1.3×10^7 bp) similar to those of eubacteria and Archaea, the genome size of eukaryotes is usually much larger than in prokaryotes. The increase in genome size in eukaryotes is probably related to the fact that the eukaryotic genome has multiple origins of replication, therefore eukaryotes are able to replicate much larger amounts of DNA in the same amount of time than prokaryotes. For example, whereas the bacterial genome is usually a single large **replicon** (e.g. in *E. coli* 4600 kb), the mouse genome has about 25,000 replicons with an average length of only about 150 kb. Furthermore, eukaryotes have evolved a very efficient way of

Table 9.1 Genome size, gene number and gene density in different evolutionary lineages.

	Genome size (Mb)	Number of genes	Gene density (genes/Mb)
Eubacteria			
Carsonella ruddii	0.16	182	1138
Mycoplasma genitalium	0.58	479	826
Buchnera aphidicola	0.62	507	817
Mycoplasma pneumoniae	0.82	677	826
Borrelia burgdorferi	0.91	853	937
Rickettsia prowazekii	1.10	834	758
Pelagibacter ubique	1.31	1354	1034
Haemophilus influenzae	1.80	1727	959
Bacillus subtilis	4.20	4100	976
Escherichia coli	4.60	4288	932
Archaea			
Nanoarchaeum equitans	0.49	552	1127
Methanococcus jannaschii	1.70	1738	1022
Archaeoglobus fulgidus	2.20	2436	1107
Sulfolobus solfataricus	3.00	2977	992
Eukarya			
Apicomplexa			
Plasmodium falciparum 3D7	23.00	5300	230
Green algae			
Ostreococcus tauri	12.56	8166	650
Red algae			
Cyanidioschyzon merolae 10D	16.50	5331	323
Diatoms			
Thalassiosira pseudonana	34.00	11242	330
Plants			
Arabidopsis thaliana	125.00	25498	204
Amoebozoa			
Dictyostelium discoideum	34.00	12500	368
Entamoeaba histolytica	23.80	9938	418
Fungi			
Saccharomyces cerevisiae	12.50	5800	464
Schizosaccharomyces pombe	13.80	4824	350
Neurospora crassa	38.60	9200	238
Aspergillus oryzae	37.00	14063	380
Invertebrates			
Caenorhabditis elegans	97.00	19099	196
Caenorhabditis briggsae	104.00	19500	188
Drosophila melanogaster	120.00	13600	113
Anopheles gambiae	278.00	14000	50

Continued on p. 220

Table 9.1 (*Continued*)

	Genome size (Mb)	Number of genes	Gene density (genes/Mb)
Echinoderms			
Strongylocentrotus purpuratus	814.00	23300	29
Urochordates			
Ciona intestinatis	162.00	15852	98
Vertebrates			
Fugu rubripes	320.00	31059	97
Gallus gallus	1000.00	−23000	23
Mus musculus	2500.00	−25000	10
Homo sapiens	2900.00	−25000	9

packaging their DNA with histones (see section 4.4.2), therefore they can accommodate massive amounts of genomic DNA. As a corollary, eukaryotic genome size is much larger than that of bacteria (Table 9.1).

The genome size of eukaryotes does not show a direct relationship to their **organismic complexity**: some closely related species that are very similar in complexity (e.g. onion and lily) have vastly different genome sizes (Cavalier-Smith, 1985). This phenomenon is most striking in the case of some 'sibling' species that are indistinguishable phenotypically, yet have different genome sizes. For example, it is quite common among fishes, amphibians and flowering plants that the genome size is drastically different in sibling species, although there is no difference in their complexity. Such a lack of correlation between genome size (C value) and complexity of genomic information is usually referred to as the **C-value paradox**.

In some eukaryotes, the number of protein-coding genes may be just a few thousand (e.g. the yeast *S. cerevisiae* has just 5800 protein-coding genes), but the genomes of higher eukaryotes may contain as many as about 25,000 protein-coding genes (e.g. human and other vertebrate genomes). This fourfold difference in gene number contrasts with the more than 200-fold difference in the size of yeast (13 Mb) and human (3000 Mb) genomes. This suggests that the vast majority of the genomes of higher eukaryotes consists of DNA that does not encode proteins. It should be pointed out, however, that recent studies indicate that the genomes of higher eukaryotes (in addition to rRNA genes, tRNA genes) contain numerous noncoding RNA genes whose products fulfil important biological roles (cf. Chapter 1). These ncRNAs include regulatory RNAs that control various levels of gene expression, including chromatin architecture/epigenetic memory, transcription, RNA splicing, editing, translation and turnover (Morey & Avner, 2004; Shabalina & Spiridonov, 2004; Costa, 2005; Mattick & Makunin, 2006). Nevertheless, a significant proportion of the genomes of higher eukaryotes is nongenic.

9.2 The role and survival of nongenic DNA

There are several alternative evolutionary explanations for the long-term survival of huge quantities of nongenic DNA that apparently has no functional importance. Some of these hypotheses assigned some function to nongenic DNA such as global regulation of gene expression, genome maintenance, etc. (e.g. Zuckerkandl, 1976; Cavalier-Smith, 1978). At the other extreme we find the opposing views that nongenic DNA is useless junk DNA (Ohno, 1972), a functionless 'parasite' or 'selfish DNA' (Doolittle & Sapienza, 1980; Orgel & Crick, 1980). According to these views the huge amounts of nonfunctional DNA may be carried on from generation to generation if it has no influence on the fitness of the organism and thus there is no selective pressure to eliminate it.

It is now clear that most nongenic DNA can be deleted without detectable phenotypic effects. It is also clear that in higher eukaryotes the large amount of nongenic DNA is no major burden on the organism (thanks to the numerous replicons and efficient DNA packaging). Maintenance and replication of excessive amounts of nongenic DNA seems to have little effect on the fitness of mammals, and it is thus possible that most nongenic DNA is indeed nonfunctional in these organisms. Of course, the definition of 'nonfunctional DNA' has some inherent ambiguity. For example, it is not clear at present to what extent the highly repetitive, tandemly arrayed satellite DNA can be considered 'junk' DNA. Alpha satellite DNA constituting the bulk of the centromeric heterochromatin often contains a binding site for a specific centromere protein, CENP-B. There is evidence that satellite DNA may bind specific chromosomal proteins that may influence chromosome condensation during meiosis and mitosis, and may therefore have an important function. Minisatellite DNA, composed of arrays of tandem repeats, is often located at or close to telomeres, the termini of chromosomes. Telomeric DNA consists of 10–15 kb of tandem hexanucleotide repeat units, TTAGGG, which are added by a specialized enzyme, telomerase. Bys acting as buffers to protect the ends of the chromosomes from degradation, these simple repeats definitely have an important function.

9.3 Repetitiveness of genomic DNA

An important aspect of the structure of genomic DNA is the degree of its repetitiveness. The content of repetitive DNA tends to increase with increasing genome size. It is practically absent from compact genomes of prokaryotes, whereas in the large genomes of animals and plants more than 50% of the genome may be repetitive.

The degree of repetitiveness of genomic DNA of higher eukaryotes may be measured experimentally since the DNA of these organisms may be fractionated according to repetitiveness, based on the kinetics of DNA reassociation following denaturation. The DNA fraction that reassociates almost instantaneously is called **foldback DNA**; its rapid reassociation is due to the fact that it consists of palindromic DNA sequences that form hairpin structures in the renaturation step. The majority of protein-coding genes belong to the **single-copy** or **unique DNA** fraction, which reassociates most slowly. The **middle-repetitive** fraction, consisting of long sequences (thousands of base pairs) present in hundreds of copies in the genome, reassociates more rapidly. The **highly repetitive** fraction, consisting of relatively short sequences (a few hundred base pairs) repeated several hundred thousand times in the genome, reassociate at an even higher rate. The repetitive fractions are of two main types according to the pattern of their distribution in the genome: they may be present in localized tandem arrays or they may be dispersed throughout the genome.

Localized repeated sequences

The localized, tandemly arrayed, highly repetitive DNA sequences may have such a uniform nucleotide composition that during density-gradient separation of fragments of genomic DNA they form bands that are clearly distinct from the main band of genomic DNA. The highly repeated DNA present in such bands is therefore frequently referred to as satellite DNA. The terminology depends on the size of the tandem array: **satellite DNA** usually refers to blocks of more than 100 kb (e.g. centromeric DNA); **minisatellite DNA** refers to blocks of 0.1–20 kb (e.g. telomeric DNA); and **microsatellite DNA** refers to blocks of less than 150 bp. The frequency of changes in the structure of localized repeated sequences is quite high: arrays are continually removed and expanded (e.g. by unequal crossing over, slipped-strand mispairing, telomerase, etc.).

Dispersed repeated sequences

Another class of highly repetitive DNA consists of sequences that are dispersed throughout the genome: they may be present in intergenic regions and in regions flanking genes, as well as in introns of protein-coding genes. Two major kinds of dispersed highly repeated sequences are distinguished according to their size: the **short *in*terspersed repeated *e*lements** (**SINEs**, usually shorter than 500 bp), and the **long *in*terspersed repeated *e*lements (LINEs)**. SINEs usually occur in more than 10^5 copies in the haploid genome. Most SINEs are retrosequences; the Alu family is the best-known representative of SINEs in the human genome (see also section 6.2.1). A typical example of LINEs in the human genome is L1, a 6-kb long repeated element. The L1 sequences have a poly(A) tail, are flanked by short direct repeats and contain two large open reading frames (ORFs), one of which encodes a protein related to reverse transcriptases. The L1 elements are also produced and spread in the genome by reverse transcription and subsequent insertion of the reverse transcripts into the genome. In the human genome there are more than 1,300,000 Alu sequences and about 900,000 L1 sequences, accounting for more than 25% of the genome (Li et al., 2001).

9.4 Mechanisms responsible for increases in genome size

Major increases in genome size are due to global duplications (in which the entire genome or a major part of it, e.g. a chromosome, is duplicated) and regional multiplications, in which a particular sequence is excessively multiplied to generate repetitive DNA. As we have seen above, different mechanisms are responsible for the creation of dispersed repetitive sequences (e.g. retroposition) and for the creation of localized repetitive sequences (unequal crossing over, slipped-strand mispairing, telomerase, etc.).

Genome duplication

Genome duplication (polyploidy) occurs as a consequence of the lack of disjunction of daughter chromosomes following DNA replication. The polymodal distribution of genome sizes in many groups of eukaryotes suggests that polyploidy is a major mechanism in the evolution of genome size in eukaryotes. Recent analyses of complete genome sequences have revealed that many genomes have been duplicated in their evolutionary past (Van de Peer, 2004). As will be discussed later, there is evidence that two rounds of whole-genome duplications occurred in the evolution of the chordate lineage.

Regional multiplications

The majority of the dispersed middle-repetitive DNA fraction of eukaryotes originates from transposable elements like the Alu repeats and L1 repeats. The localized regional increase of genomic DNA may be due to tandem duplications. Localized repeated sequences consisting of very short simple repeated motifs (satellite, minisatellite and microsatellite DNA) may be created by replication slippage, slipped-strand mispairing, hairpin formation, etc. In such processes the DNA polymerase turns back and uses the same template to produce a repeat. Existing tandem repetitive sequences are particularly prone to replication slippage, and the process can therefore produce very long tandem arrays of short repeats. Replication slippage can provide a powerful mechanism for the rapid proliferation of tandemly repeated sequences within the genome. Another mechanism for the formation of simple repeated sequences is hairpin formation (see section 9.3). In the case of relatively long repeated units, the major mechanism for expansion is unequal crossing over or amplification due to multiple replication of the same replicon.

9.5 Compositional organization of eukaryotic genomes

Isochores

Eukaryotic genomes are compartmentalized into distinct regions character-ized by specific nucleotide compositions. For example, although vertebrate genomes show a uniform GC content (40–45%), and there is compositional homogeneity over very long DNA stretches, some stretches of their genomic DNA are GC rich, while others are GC poor. These more or less homo-geneous segments are usually termed **isochores**. The alternation of GC-poor and GC-rich isochores endows the nuclear DNA of vertebrates with a char-acteristic mosaic organization.

Although the significance of isochores is not fully understood at present, there are some interesting correlations that may help decipher their func-tion and evolution (Bernardi, 2004).

1 The genes tend to have a GC content similar to that of the isochore in which they reside. For example, the β-globin gene cluster is embedded in a low-GC stretch and the β- and β-like globin genes are relatively low in GC content, whereas the α-globin gene cluster is embedded in a GC-rich region and α- and α-like globin genes are GC rich.
2 In the genomes of humans and other mammals the location of genes within isochores is nonrandom: most of the genes reside in the most GC-rich isochores, which represent only a small fraction of the genome (Mouchiroud et al., 1991; Lander et al., 2001).
3 Protein-coding genes of GC-poor (and gene-poor) isochores contain more and longer introns than those in GC-rich (and gene-rich) isochores (Duret, Mouchiroud & Gautier, 1995; Lander et al., 2001).
4 There is an important correlation between isochore structure of the genome and its replication: genes localized in GC-rich isochores replicate early, whereas genes localized in GC-poor isochores replicate in the late stage of the cell cycle (Watanabe et al., 2002; Woodfine et al., 2004).
5 There is a positive correlation between GC content and transcriptional activity (Woodfine et al., 2004). The most likely explanation for this correla-tion is that transcriptional activity enables early replication by promoting access of the replication machinery to the DNA.
6 The local rates of recombination positively correlate with GC content (Fullerton, Bernardo, Carvalho & Clark, 2001; Kong et al., 2002).
7 In interphase nuclei the gene-rich/GC-rich chromosomal regions display a much more spread-out conformation compared to the gene-poor/ GC-poor regions in human nuclei (Saccone, Federico & Bernardi, 2002).

As to the molecular processes responsible for the isochore structure of mammalian chromosomes: there is increasing evidence that recombination may be driving the evolution of higher GC content (Meunier & Duret, 2004;

Khelifi et al., 2006). The most widely accepted hypothesis to explain how recombination might influence base composition is the 'biased gene conversion model' that favours fixation of GC alleles (Galtier et al., 2001).

CpG islands

Clusters of the dinucleotide CpG (CpG islands) are present in the majority of the promoter and exonic regions of vertebrate genes, whereas other regions of the mammalian genome contain relatively few CpG dinucleotides (Takai & Jones, 2002). Detection of regions of genomic sequences that are rich in the CpG pattern is important because such regions tend to be associated with genes which are frequently switched on. A large number of experiments have shown that methylation of promoter CpG islands plays an important role in gene silencing, genomic imprinting and X-chromosome inactivation.

The explanation for the decreased occurrence of CpGs in functionally less important regions of vertebrate genomes is that since cytosines in CpG dinucleotides are highly methylated and since methylated cytosines are mutational hotspots this leads to CpG depletion, unless purifying selection conserves these islands (cf. Chapter 3).

In view of their association with promoters and exons, CpG islands are useful markers for genes in organisms containing 5-methylcytosine in their genomes. CpG islands are usually defined as 200-bp regions of DNA with a relativel high G+C content (with base composition >60% G+C) and CpG dinucleotide frequency 0.6 observed/expected as opposed to other regions of the mammalian genome where the average base composition is ~38% G+C and the average CpG frequency is 0.2–0.25 observed/expected.

9.6 Genomes of model organisms

In the past few years the complete sequences of the genomes of hundreds of eubacteria (e.g. *Escherichia coli, Bacillus subtilis, Haemophilus influenzae, Pelagibacter ubique, Borrelia burgdorferi, Mycoplasma pneumoniae, Mycoplasma genitalium, Buchnera aphidicola, Carsonella ruddii*), more than two dozen Archaea (e.g. *Methanococcus jannaschii, Archaeoglobus fulgidus, Nanoarchaeum equitans*), several unicellular eukaryotes (e.g. *Plasmodium falciparum, Ostreococcus tauri, Cyanidioschyzon merolae, Thalassiosira pseudonana, Entamoeaba histolytica, Dictyostelium discoideum*), several fungi (e.g. *Saccharomyces cerevisiae, Schizosaccharomyces pombe, Neurospora crassa, Aspergillus oryzae, Aspergillus nidulans, Aspergillus fumigatus*), plants (e.g. *Arabidopsis thaliana*), protostomes (e.g. *Caenorhabditis elegans, Caenorhabditis briggsae, Drosophila melanogaster, Anopheles gambiae*), echinoderms (e.g. *Strongylocentrotus purpuratus*), urochordates (e.g. *Ciona intestinalis*) and various vertebrates (e.g. *Fugu rubripes, Gallus gallus, Mus musculus, Homo sapiens*) have been determined, permitting a systematic analysis of the evolution of protein-coding genes in all three domains of life.

These studies have also shed light on the relationship between viruses, bacteria, archaea, eukaryotes and the organelles of eukaryotes. The following section will summarize the major conclusions that could be drawn from these comparisons. The generally accepted evolutionary relation of some of the key model organisms discussed here is illustrated in Figs 9.1, 9.2 and 9.3.

9.6.1 Viral genomes

Ever since their discovery, the origin(s) of viruses have been a topic of intense debate. The early theories of the evolutionary origins of viruses can be grouped into two major categories: the first group of theories placed viruses in the earliest phases of evolution and considered them the primitive precursors of cellular systems; the second group of theories proposed that viruses are secondary derivatives of cellular systems that suffered drastic degeneration as a consequence of parasitism. It must be emphasized that these two types of theories are not mutually exclusive: some viruses could be primordial, whereas others could be derivatives of cellular systems.

The availability of complete genome sequences of numerous viruses has shed light on the enormous diversity of viruses and begins to clarify the evolutionary relationship between different groups of viruses and their relationship with cellular organisms. The diversity of viruses is underlined by the fact that they share very few common features. The most important of these are their obligate intracellular parasitism and that their genomic nucleic acid is packaged into a protein capsid. On the other hand, the fact that the genomic nucleic acid of viruses may be either RNA or DNA suggests that they may have had multiple evolutionary origins.

Comparative genomics identified several distinct evolutionary groups of viruses on the basis of the conserved proteins of their replication apparatus. Thus it became clear that retroviruses, hepadnaviruses, plant badnaviruses, tungroviruses and diverse retroposons shared a common ancestor that encoded a reverse transcriptase as the principal replication polymerase. Another group consists of positive strand RNA viruses and several double-stranded RNA viruses that use RNA-dependent RNA polymerases to replicate their genomes. Since reverse transcriptases and RNA-dependent RNA polymerases have descended from an ancestral replicase that utilized an RNA template, these two major classes of viruses might have ultimately descended from an ancient replicon with an RNA genome. It has also been suggested that the diversification of these groups of viruses might be linked to one of the fundamental evolutionary transitions from RNA genomes to the DNA genomes (Leipe, Aravind & Koonin, 1999; Forterre, 2002; Zimmer, 2006).

Several major groups sharing common replication systems can also be distinguished among DNA viruses (Iyer et al., 2006). For example, many small DNA viruses and related plasmids and transposons use rolling circle replication endonuclease to initiate replication (Iyer et al., 2005). The

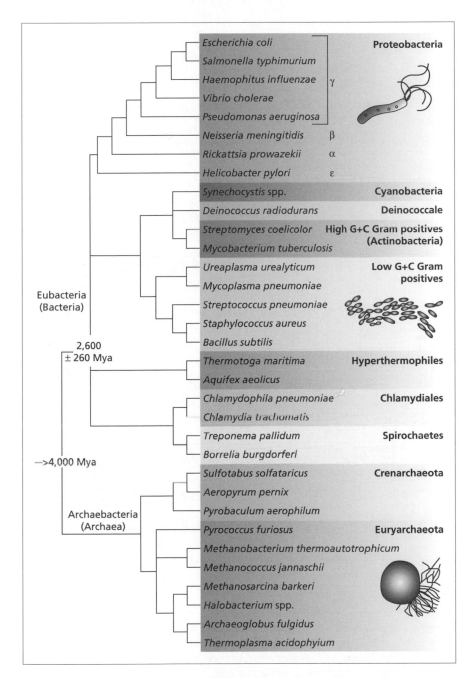

Fig. 9.1 Evolutionary relationship of some prokaryotic model organisms. Times of divergence (million years ago) are indicated at nodes in the tree. (Modified from Hedges, S.B. (2002) The origin and evolution of model organisms. *Nature Reviews in Genetics* **3**, 838–49.)

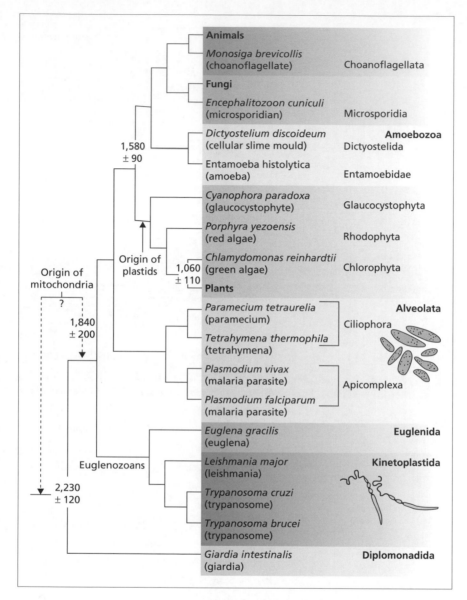

Fig. 9.2 Evolutionary relationship of some eukaryotic model organisms. Times of divergence (million years ago) are indicated at nodes in the tree. The solid arrow indicates the primary endosymbiotic event that has led to the origin of plastids. The dashed arrows reflect the current uncertainty of the origin of mitochondria, presumably a single event that occurred either before the last common ancestor of living eukaryotes or after the divergence of *Giardia intestinalis*. (Modified from Hedges, S.B. (2002) The origin and evolution of model organisms. *Nature Reviews in Genetics* **3**, 838–49.)

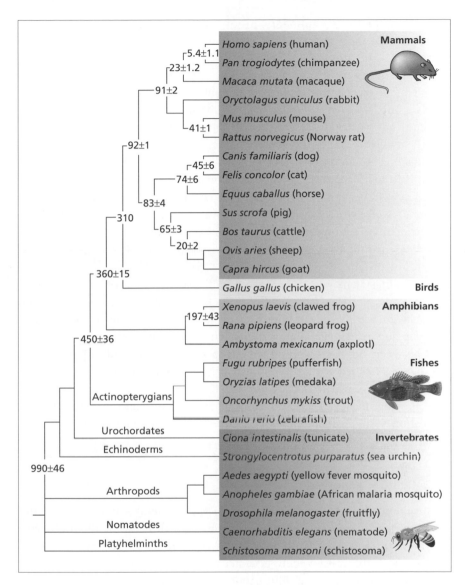

Fig. 9.3 Evolutionary relationship of some metazoan model organisms. Times of divergence (million years ago are indicated at nodes in the tree). (Modified from Hedges, S.B. (2002) The origin and evolution of model organisms. *Nature Reviews in Genetics* **3**, 838–49.)

large eukaryotic DNA viruses are unique among viruses inasmuch as they do not exhibit much dependence on the host replication or transcription systems for completing their replication since all these viruses encode several conserved proteins performing most key lifecycle processes, such as DNA polymerases, helicases and topoisomerases, transcription factors involved in transcription initiation and elongation and chaperones involved in the capsid assembly.

Analysis of the evolutionary history of large eukaryotic DNA viruses suggested that they emerged only after the eukaryotic cell with its full complement of structures was formed. These viruses have subsequently acquired numerous genes from their eukaryotic hosts sometimes resulting in gene complements comparable to those of the eukaryotes. For example, a member of this group, *Acanthamoeba polyphaga* Mimivirus, has a gigantic 1.2-megabase genome that contains 1262 protein-coding genes, i.e. about 2.5 as many as the smallest prokaryotic genomes (e.g. *Mycoplasma genitalium*, *Buchnera aphidicola*). The most unusual aspect of this huge viral genome is the presence of numerous genes encoding proteins involved in translation, such as amino acyl transfer RNA synthetases, translation initiation and elongation factors as well as tRNA genes (Raoult et al., 2004). Evolutionary analyses suggest that numerous multigene families encoded in this genome might have been acquired via extensive lateral gene transfer (Raoult et al., 2004; Desjardins, Eisen & Nene, 2005; Koonin, 2005).

In summary: analysis of the genomes of various groups of viruses suggest that a variety of replicons with distinct replication systems may have coexisted in the precellular phase of evolution, with some of them eventually giving rise to cellular systems. The fact that profound differences exist between the bacterial and the archaeo-eukaryotic replication systems (see below) has led to the hypothesis of two independent origins of DNA replication of cellular organisms (Leipe et al., 1999).

Following the appearance of cellular forms of life, replicons with different replication systems adapted to exploit the highly complex cellular systems as plasmids or viruses.

9.6.2 Cellular genomes

Phylogenetic trees based on ribosomal RNA sequences indicate that the universal tree of cellular life can be divided into three domains: the bacterial, archaeal and eukaryotic domains (Woese, Kandler & Wheelis, 1990). According to the most widely accepted view cellular life began with the prokaryotes, and the first bifurcation in the tree of life separated the bacterial lineage from the archaeal/eukaryotic lineage. In this model eukaryogenesis is the process whereby eukaryotes evolved from prokaryotic (Archaeal) ancestors.

A unique aspect of eukaryotes is that their informational genes seem to be derived from archaea, whereas their metabolic genes are derived from bacteria, i.e. eukaryotes appear to be chimeric (Ribeiro & Golding, 1998; Rivera et al., 1998). The serial endosymbiotic theory explains the origin of nucleated eukaryotic cells by a merging of archaebacterial and eubacterial cells (Margulis & Sagan, 2002). Eukaryotes have subsequently engulfed bacteria that evolved into organelles such as mitochondria and plastids (see below), but the origin of their nucleus is less clear (Martin, 2005).

According to one theory the nucleus of eukaryotes originated by membrane hypertrophy concomitant with the recombination of eu- and archaebacterial DNA that remained attached to eubacterial motility structures that evolved into the microtubular cytoskeleton, including the mitotic apparatus (Margulis et al., 2006). In this model the earliest symbiogenetic fusion integrated archaebacterial thermoplasmas with aerotolerant spirochete bacteria producing the first protist, a chimera that evolved into a stable nucleated protist cell: the last eukaryotic common ancestor. In other words, the nuclear genome was generated by recombination of eu- and archaebacterial DNA that remained attached to the nuclear membrane and to spirochete motility proteins.

On the other hand, the theory of viral eukaryogenesis (Bell, 2001) proposed that the eukaryotic nucleus evolved from a complex DNA virus. According to this theory the virus established a persistent presence in the cytoplasm and evolved into the eukaryotic nucleus by acquiring a set of essential genes from the host genome. Several characteristic features of the eukaryotic nucleus (mRNA capping, linear chromosomes, spatial separation of transcription from translation) were suggested to derive from this viral ancestry. According to Bell's hypothesis phagocytosis and other membrane fusion-based processes characteristic of eukaryotes are derived from viral membrane fusion processes and evolved in concert with the nucleus. This model suggests that coevolution of phagocytosis and the nucleus rendered much of the host archaeal genome redundant since the protoeukaryote could obtain raw materials and energy by engulfing bacteria (both as prey and as endosymbionts).

9.6.2.1 Eubacterial genomes

Bacillus subtilis is an aerobic, endospore-forming **Gram-positive bacterium** commonly found in soil and in association with plants. *B. subtilis* is the best-characterized member of this group of prokaryotes; its complete genome sequence thus permits an insight into the makeup of a self-sufficient Gram-positive bacterium. Its genome of 4,214,810 bp comprises over 4100 protein-coding sequences, with an average size of 890 bp (Kunst et al., 1997). The coding density is thus one gene every 1028 bp, with protein-coding genes covering 87% of the genome sequence.

Of the protein-coding genes, 53% are represented once, 47% constitute various paralogous gene families. Many of the paralogs (about a quarter of the genes) constitute large families of functionally related proteins, mostly involved in the transport of compounds into and out of the cell. The largest family contains 77 putative ATP-binding transport proteins. These ABC (ATP-binding cassette) transporters are extremely important in bacteria since they allow their host to escape the toxic action of many compounds. Other paralogous families comprise 47 transport proteins, 18 amino acid permeases and at least 16 sugar transporters. There are 43 temperature-shock and general stress proteins that display significant similarity to *E. coli* homologs. Regarding the mechanism that gave rise to these paralogs: several duplications are in genes that are located very close to each other, and there are even entire duplicated operons that arose by tandem duplication. Thus, it seems that expansion of these gene families in *B. subtilis* occurred by tandem duplication of individual genes or operons, rather than by genome duplication.

The *B. subtilis* genome contains at least 10 prophages or remnants of prophages, indicating that bacteriophage infection has played an important evolutionary role in lateral gene transfer. It is noteworthy in this respect that although the average GC ratio of the *B. subtilis* genome is 43.5%, there are several AT-rich islands many of which reveal the signatures of bacteriophage lysogens or other inserted elements.

The protein-coding genes of *B. subtilis* constitute three distinct groups that differ in their codon usage. Class 1 comprises the majority of the *B. subtilis* genes (82%), while Class 2 (5%) includes genes that are highly expressed under exponential growth conditions (genes of transcription and translation machineries, core intermediary metabolism, stress proteins, etc.). Class 3 (13%) contains a very high proportion of genes of unknown function (84%). This class is characterized by codons enriched in AT residues; the genes are usually clustered into groups of 15–160 genes and correspond to the AT-rich islands. It is noteworthy that Class 3 genes usually correspond to functions associated with bacteriophages or transposons, and many of these genes are associated with virulence genes identified in pathogenic Gram-positive bacteria. All these observations suggest that these Class 3 genes (phages, virulence factors, etc.) reflect lateral transmission among bacteria.

About 1000 *B. subtilis* genes (24%) have clear orthologous counterparts in *E. coli*. Significantly, about 100 putative operons or parts of operons were conserved between the Gram-negative *E. coli* and the Gram-positive *B. subtilis*. Comparison with *Synechocystis* PCC6803 revealed about 800 orthologs (20% of *B. subtilis* genes).

Mycoplasmas are thought to be derived from Gram-positive bacteria similar to *B. subtilis*. Significantly, 312 of the genes of *Mycoplasma genitalium* (66%) have homologs in *B. subtilis* or other Gram-positive bacteria, supporting the notion that this species has a reduced genome of a Gram-positive bacterium. Mycoplasmas are obligate parasites in a wide range of hosts including humans, other animals and plants. Their parasitic life has led to a drastic reduction in the size and content of their genomes relative to related prokaryotes. *M. pneumoniae* requires the provision of exogenous essential metabolites and completely lacks a cell wall, being surrounded only by a cytoplasmic membrane. *M. genitalium* is a parasite of human genital and respiratory tracts; it contains one of the smallest genomes for a self-replicating organism and thus provided a unique system for defining the minimal functional gene set.

The *M. genitalium* genome is a circular chromosome of 580,070 bp, containing a total of only about 470 predicted coding regions, which include genes for DNA replication, transcription and translation, DNA repair, cellular transport and energy metabolism (Fraser et al., 1995). The 470 predicted coding regions in *M. genitalium* (average size 1040 bp) comprise 88% of the genome – on average one gene every 1235 bp – a value similar to that in *Haemophilus influenzae*, where 1727 predicted coding regions (average size 900 bp) comprise 85% of the genome (one gene every 1042 bp), or that for *E. coli* (one gene every 1081 bp). These data indicate that the reduction in genome size that has occurred in *Haemophilus* and *Mycoplasma* has not resulted in an increase in gene density or a decrease in gene size, but rather in the loss of entire genes.

An intermediate state in genome reduction is represented by the genome of *Mycoplasma pneumoniae* (Himmelreich et al., 1996, 1997). Comparison of the genome of *M. pneumoniae* (816 kb) with the smaller genome of *M. genitalium* (580 kb) is of interest since here the principles of genome reduction are traced most easily. Despite the difference in genome size, the average coding density (88.7%) and the average gene size (1011 bp) are about the same as in *M. genitalium* but *M. pneumoniae* has significantly more ORFs (677) than *M. genitalium* (470). There is one gene every 1205 bp and 1234 bp in *M. pneumoniae* and *M. genitalium*, respectively; hence the reduction of genome size was achieved by loss of entire genes. The reduction of the genome size of *M. pneumoniae* during its evolution from ancestral bacteria was achieved by the loss of complete anabolic (e.g. no amino acid synthesis) and metabolic pathways.

All the 470 proposed ORFs of the smaller *M. genitalium* genome are contained in the larger genome of *M. pneumoniae*. Nevertheless, the *M. genitalium* genome is not simply a truncated *M. pneumoniae* genome. There is evidence for major rearrangements in these genomes. The two genomes can be subdivided into six segments and the order of these segments in the two bacteria is different, suggesting that these segments were rearranged by translocation via recombination. However, within individual segments the order of orthologous genes is well conserved. The 236 kb of *M. pneumoniae* that is missing from the *M. genitalium* genome codes for 209 ORFs, 110 of which are specific to *M. pneumoniae*, while 76 ORFs are amplifications of ORFs existing mainly as single copies in *M. genitalium*.

Escherichia coli is a **Gram-negative bacterium** that colonizes the lower gut of animals. It is a facultative anaerobe that survives when released to the natural environment, allowing its widespread dissemination to new hosts. The complete genome sequence of *E. coli* thus permits an insight into the genetic makeup of a self-sufficient Gram-negative prokaryote (Blattner et al., 1997). On the other hand, *Haemophilus influenzae* is a small, parasitic, nonmotile Gram-negative bacterium. Comparison of the complete genome sequence of *H. influenzae* (Fleischmann et al., 1995) with that of *E. coli* may thus illustrate the effects of a parasitic lifestyle on the genome of a Gram-negative bacterium.

The genome of *E. coli* consists of 4,639,221 bp of circular duplex DNA containing 4288 protein-coding genes. The genome is very compact, with one gene every 1082 bp; the average ORF is 317 amino acids, and the average distance between genes is only 118 bp. Protein-coding genes account for 87.8% of the genome, 0.8% encodes stable RNAs, and 0.7% consists of non-coding repeats. There are regions in the *E. coli* genome with a codon usage, which conflicts with the codon usage characteristic of *E. coli*. The fact that some gene clusters with unusual codon usage correspond to known cryptic prophages supports the hypothesis that such regions have arisen by recent immigration of genetic material by lateral transfer (they still reflect the optimal codon usage of their previous host).

Comparison of all the proteins of *E. coli* with each other has identified numerous paralogous families. Of the 4288 proteins 1345 have at least one paralogous sequence in the *E. coli* genome (earlier estimates suggested that 38–45% of this organism's sequences had arisen from gene duplications). As in *B. subtilis*, the largest family of paralogous proteins in *E. coli* is the family of ABC transporter proteins, which has 80 members. There are 86 pairs of isozymes, or multiple closely related enzymes with identical or nearly identical function in *E. coli*, suggesting that there may be significant functional redundancy in the *E. coli* genome.

Comparison of *E. coli* proteins with proteins of other bacterial genomes has shown that the largest number of matches is found in the 1.83-Mb *H. influenzae* genome (out of the 1703 proteins of *H. influenzae* 1130 (66%) have homologs to *E. coli* proteins). Of the 3168 proteins of the 3.6-Mb *Synechocystis* sp. genome only 675 proteins (21%) have clear *E. coli* homologs. Comparisons with complete genomes of Archaea and eukaryotes have given even lower numbers of matches with *E. coli* genes. The archaeon *Methanococcus jannaschii* (1738 proteins) has 231 homologs (13%), while the eukaryote *S. cerevisiae* (5885 proteins) has 254 homologs (4%).

As mentioned above, about 45% of the genes of *E. coli* have arisen by gene duplication. In the case of the much smaller, reduced genome of *H. influenzae* there is evidence that about 30% of the protein-coding genes were produced by gene duplications. An all-against-all comparison of the *H. influenzae* protein sequences has shown that its 1680 protein sequences belong to 1188 families: 980 (58%) are single representatives of a family, the remaining 700 form 208 families (Brenner et al., 1995). The proportion of duplicated genes is even lower in the extremely reduced genomes of *Mycoplasma* species.

Pelagibacter ubique belongs to a group of very small marine α-proteobacteria that grow by assimilating organic compounds from the ocean's dissolved organic carbon reservoir and can generate metabolic energy either by a light-driven proteorhodopsin proton pump or by respiration. Sequencing the genome of *P. ubique* is of major interest primarily because it is so far the smallest genome (1,308,759 base pairs) and encodes the smallest number of predicted

protein-coding genes (1354 ORFs) of any cell known to replicate independently in nature (Giovannoni et al., 2005).

In contrast with parasitic bacteria and archaea with small genomes,
P. ubique has complete biosynthetic pathways for all 20 amino acids and
appears to encode nearly all of the basic functions of α-proteobacterial cells.
The small genome size is due to the nearly complete absence of nonfunctional or redundant DNA and the absence of all but the most fundamental
metabolic and regulatory functions. For example, intergenic DNA regions are
extremely short (with a median size of only three bases) and no pseudogenes, phage genes or recent gene duplications were found in *P. ubique*.
The genome has the smallest number of paralogous genes observed in any
free-living cell. Furthermore, there is no evidence of DNA originating from
recent horizontal gene-transfer events.

These data are consistent with the hypothesis that the genome of *P. ubique*
is subject to more powerful selection to minimize the cost of cellular replication than in related marine bacteria that have considerably larger genomes.
A possible explanation for this is that the strategy of *P. ubique* is different
from those of other free-living bacteria. Related bacteria with larger genomes
can harbour complex regulatory systems that enable them to implement
various metabolic strategies in response to environmental variation and as
a result of this they can rapidly exploit pulses of nutrients but pay the
price of decreased replication efficiency during the intervening periods. In
contrast with this, *P. ubique* utilizes the ambient dissolved organic carbon
reservoir but keeps replication cost at a minimum by extreme genome
reduction. This explanation is consistent with the observation that *P. ubique*
has a low growth rate that does not vary in response to nutrient addition,
whereas heterotrophic marine bacteria with larger genomes have some of
the highest recorded growth rates and are very responsive to nutrient concentration (Giovannoni et al., 2005).

Borrelia burgdorferi, the aetiological agent of Lyme disease, is an extracellular pathogenic spirochete. Its 911-kb genome is unusual inasmuch as
it is a linear chromosome with telomeres having 26 bp inverted terminal
sequences (Fraser et al., 1997). The 853 genes carried by this chromosome
encode the basic set of proteins for DNA replication, transcription, translation and energy metabolism, but, like *Mycoplasma*, they include no genes
for cellular biosynthetic reactions. Since *Mycoplasma* and *Borrelia* are distantly
related eubacteria (see Fig. 9.1), their similarly reduced genomes are the result
of convergent evolution. Apparently, the parasitic lifestyle has independently
led to gene losses in different evolutionary lineages. The average size of ORFs
(970 bp) is similar to that in other bacteria, with 93% of the chromosome
being occupied by coding sequences. The 11 plasmids of *B. burgdorferi*
contain a total of 430 putative ORFs, including the Vlse lipoprotein recombination cassette that plays an important role in antigenic variation (see
section 8.2).

In summary, comparison of the genomes of parasitic and nonparasitic bacteria indicates that genome reduction can be achieved by elimination of entire genes (primarily duplicated genes), but apparently the coding density is at its upper limit and cannot be further increased. Despite this limiting high coding density in the genome of *M. pneumoniae*, ORFs are not equally distributed over its genome. A lower coding density coincides with regions of lower or higher GC content than the average. Interestingly, the codon usage of the low and high GC content subfractions is clearly influenced by the DNA composition, favouring either codons with G/C or A/T at the third position. The codon usage pattern differs also for the complete genome and for genes that are frequently expressed like the ones coding for ribosomal proteins.

Many bacterial lineages have evolved mutually obligate endosymbiotic associations with animal hosts and bacterial symbioses are especially widespread among insects. In such cases the endosymbiotic bacteria usually produce essential nutrients that are rare in the host diet and the animal host produces specialized cells where bacteria are confined and – similarly to organelles – are maintained through vertical transmission across host generations. As a consequence of such obligate endosymbiosis, the genomes of the bacterial symbionts usually undergo massive reduction.

As an example we may refer to the reductive genome evolution in *Buchnera*, the intracellular symbionts of the aphids. These bacterial symbionts – that evolved from an enterobacterial-like ancestor 200–250 million years ago – support the exploitation of the poor diet of plant phloem sap by aphids through supplementation of deficient nutrients, primarily essential amino acids.

Studies on *Buchnera aphidicola* (van Ham et al., 2003) have revealed that the genome of this γ-proteobecterium is only 617,838 bp (consisting of a 615,980-bp chromosome and a 2399-bp plasmid) and encodes only 507 protein-coding genes. Comparison of this gene repertoire with that of the closely related *E. coli* K-12 genome revealed massive reduction in numbers of genes across all functional categories, most strongly affected are regulatory and transport functions, whereas translation and metabolism of nucleotides and amino acids show more moderate shrinkage.

Studies on the genome of *Carsonella ruddii*, a γ-proteobacterial symbiont of phloem sap-feeding insects, psyllids, provided an even more striking example of the massive reduction in genome size that occurs in obligate endosymbionts (Nakabachi et al., 2006). The complete genome sequence of *C. ruddii* consists of a circular chromosome of just 159,662 base pairs, by far the smallest bacterial genome yet characterized. This small genome has only 182 ORFs. The genome has an extremely high coding density (97%) with many overlapping genes. Genes for translation and amino acid biosynthesis are well represented, but numerous genes considered essential for life are missing, suggesting that *Carsonella* may have achieved organelle-like status. Despite the extensive loss of genes, *Carsonella* retains many genes for biosynthesis of essential amino acids, reflecting the fact that their hosts feed only on plant phloem sap that is poor in essential amino acids.

The minimal genome and redundancy

A survey of the genes of reduced genomes, like that of *M. genitalium*, and comparison of these genes with those of other bacteria permits the description of a minimal set of genes required for survival. Obviously, a minimal cell must contain genes for replication and transcription, translation (rRNA operon, a set of ribosomal proteins, tRNAs and tRNA synthetases), transport proteins to derive nutrients from the environment, biochemical pathways to generate adenosine triphosphate, etc.

Somewhat surprisingly, comparison of the *M. genitalium* and *H. influenzae* genomes showed that they dedicate a similar percentage of their total genomes to genes involved in central intermediary metabolism, energy metabolism, fatty acid and phospholipid metabolism, purine and pyrimidine metabolism, replication, transcription, transport, recombination, repair, etc., even though the number of genes in these categories is considerably fewer in *M. genitalium*. This observation indicates that genome reduction was achieved primarily by elimination of 'redundant' duplicates. The minimal translation machinery seems to require 90 different proteins to proceed, and the complete DNA replication process requires about 30 proteins.

In the case of genes involved in amino acid biosynthesis, the biosynthesis of cofactors, the cell envelope and regulatory functions, entire pathways were eliminated in *M. genitalium* compared with *H. influenzae*: *Mycoplasma* spp. use the metabolic products of their hosts, and *M. genitalium* almost completely lacks enzymes involved in amino acid biosynthesis, *de novo* nucleotide biosynthesis and fatty acid biosynthesis. For example, whereas the *H. influenzae* genome contains 68 genes involved in amino acid biosynthesis, the *M. genitalium* genome contains only one; similarly the *H. influenzae* genome has 228 genes associated with metabolic pathways, whereas the *M. genitalium* genome has just 44.

Although *M. genitalium* is the cellular life form with one of the smallest number of protein-coding genes, this number may still be higher than the actual minimal requirement. The minimal gene set sufficient for supporting a modern-type cell may be identified as the one that is shared by different kinds of parasitic bacteria with reduced genomes. Based on this approach, Koonin and Mushegian (1996) carried out a comparison of *M. genitalium* and *H. influenzae* and concluded that a bacterial version of a minimal gene set consists of about 256 genes. This minimal set contains the complete translation machinery of *M. genitalium*, the basic systems for DNA replication and transcription, a simple repair system, salvage pathways for nucleotide synthesis, anaerobic intermediate metabolism largely limited to glycolysis, some proteins with chaperone-like functions, a protein secretion apparatus and a few metabolite transport systems. With nearly every anabolic pathway eliminated, an organism with such a minimal set of genes can only survive if an external supply of metabolites is available.

Recent studies have attempted to define experimentally the minimal set of genes needed to sustain life of *M. genitalium* (Glass et al., 2006). Using global transposon mutagenesis for gene disruption 382 of the 482 *M. genitalium* protein-coding genes were found to be essential (it is noteworthy that 28% of the essential protein-coding genes encode proteins of unknown function). These studies illustrate that in the case of organisms with reduced genomes the great majority of the genes are essential. In species with larger genomes, however, some of the genes may be dispensable because of functional redundancy.

9.6.2.2 Archaeal genomes

Although cytologically Archaea resemble bacteria, at the molecular level many data indicate that the Archaea are more closely related to eukaryotes than to eubacteria, making them descendants of the 'prokaryotic ancestor' of eukaryotes. Nevertheless, archaeal genomes seem distinctly bacterial in character: they have single circular chromosomes on which genes are tightly packed and often linked in operons. Some archaeal operons are identical in gene order to bacterial operons (Olsen & Woese, 1996; Edgall & Doolittle, 1997).

Methanococcus jannaschii, an archaeon isolated from deep-sea hydrothermal vents, is a methanogen that derives energy by using molecular hydrogen to reduce carbon dioxide to methane. This organism requires no organic nutrients for growth: it has all the biochemical pathways needed to reproduce itself from inorganic constituents. The 1.66-Mb genome of *M. jannaschii* (and its 58-kb and 16-kb extrachromosomal elements) contains a total of 1738 protein-coding genes (Bult et al., 1996), thus it has a coding density very similar to that of eubacteria (955 bp per gene). However, the unique position of Archaea is reflected by the fact that of the open reading frames of the *Methanococcus* genome, only 38% could be assigned a function on the basis of homologies to sequences of known function. This is significantly lower than the values for the *Escherichia*, *Haemophilus* and *Mycoplasma* genome sequences (62%, 58% and 68%, respectively).

Although the majority of genes related to energy production, cell division and metabolism in *M. jannaschii* are most similar to eubacterial versions, the information-processing systems (translation, transcription and DNA replication) of Archaea and eukaryotes resemble one another much more than either genome resembles the bacterial systems. For example, the DNA polymerase of *Methanococcus* resembles that of eukaryotes; the *M. jannaschii* DNA repair enzymes are predominantly of the eukaryotic type; the archaeal and eukaryotic ribosomal proteins tend to be more similar to one another than they are to their eubacterial homologs; all the *M. jannaschii* translation initiation factors are of the eukaryotic type; and the *M. jannaschii* tRNA-charging enzymes are more similar to their eukaryotic counterparts than to their bacterial counterparts. The archaeal transcription initiation system is, again, of the eukaryotic type, and very different from eubacterial transcription initiation.

Key components of a typical eubacterial cell division system, such as the proteins for partitioning chromosomes following replication, are apparently missing from the *Methanococcus* genome. However, no homologs of the eukaryotic cytoskeleton or spindle apparatus have been found in *Methanococcus*, either.

The packaging of nuclear DNA by histones into nucleosomes and chromatin is a definitive feature of eukaryotes, and this chromatin-packaging system seems to have arisen at the origin of the Eukarya. Since histone sequences and the structure of the nucleosome are universally conserved in eukaryotes, it has been long assumed that histones must have predated the emergence of the Eukarya. Indeed, although Archaea lack a nuclear membrane, they do contain histones that help pack DNA into nucleosome-related structures; there are five histone genes in *M. jannaschii*.

Comparison of archaeal histones with eukaryotic histones also illustrates the early phases of the evolution of histones, before their function was fully established. The four nucleosome core histones of eukaryotes, H2A, H2B, H3 and H4, all contain the same histone fold, flanked by different N-terminal and C-terminal domains that extend beyond the nucleosome and provide sites for various post-translational regulatory modifications (see section 4.4.2). These extensions are not, however, absolutely essential for nucleosome assembly or positioning. Significantly, the archaeal histones comprise only the histone fold but lack these regulatory regions (Starich et al., 1996; Reeve, Sandman & Daniels, 1997). Histones of eukaryotes do not exist as folded monomers but form very stable dimers held together by hydrophobic interactions. Whereas eukaryotic histones form only specific (H2A+H2B) and (H3+H4) heterodimers, the more primitive archaeal histones form both homodimers and heterodimers (Grayling, Sandman & Reeve, 1996). Archaeal histone-DNA complexes assembled *in vivo* and *in vitro* visibly resemble nucleosomes. They protect ~60-bp ladders of DNA from micrococcal nuclease digestion, and exposure to formaldehyde cross-links the archaeal histones within these complexes into tetramers (Grayling et al., 1996; Pereira et al., 1997). The archaeal nucleosome appears therefore to be homologous to the structure formed at the centre of the nucleosome by the (H3+H4)2 tetramer. The existence of homodimers lends support to the hypothesis that the four core histones of eukaryotes and the archaeal histones evolved from a common ancestor, one which could only form homodimers. There are two histone-coding genes in *Methanothermus fervidus*, three in *Methanobacterium thermoautotrophicum* and *Methanobacterium formicicum*, and five in *M. jannaschii*, indicating that the original histone gene has undergone different numbers of duplications during the divergence of different lineages of Archaea (Bult et al., 1996; Grayling et al., 1996). Consistent with this, in most cases the primary sequences of the archaeal histones within one species are more similar to each other than to the sequences of histones in other Archaea. The evolution of archaeal/eukaryotic histones thus illustrates that these proteins, which are under extreme functional constraints (and are extremely conservative) within the eukaryote lineage, were much less constrained and evolved at a much higher rate before their present role was fully established.

Archaeoglobales are sulphur-metabolizing strict anaerobic archaea that grow in subsurface oilfields and at extremely high temperatures in hydrothermal environments. A member of this group, *Archaeoglobus fulgidus*, grows at temperatures of 60–95°C, with optimum growth at 83°C. The genome of *A. fulgidus* consists of a single, circular chromosome of 2,178,400 bp that contains 2436 ORFs: there is on average a gene every 895 bp (Klenk et al., 1997). The average size of the *A. fulgidus* ORFs is 822 bp, and they cover 92.2% of this compact genome. As compared with *M. jannaschii*, the larger *A. fulgidus* genome has the distinct characteristic that a larger proportion (30%) of the genes belong to paralogous gene families: 719 of the ORFs belong to 242 paralogous families. It thus seems that extensive gene duplication contributed to its larger genome size. Most of the multigene families are involved in energy metabolism, transport and fatty acid and phospholipid metabolism. As in the case of bacteria, the largest superfamily is the ABC transporter superfamily, with 40 members.

Similarly to *M. jannaschii*, the replication, repair and cell division systems of this archaeon also display both eukaryal and eubacterial features: for example, *A. fulgidus* possesses DNA polymerases related to eukaryal polymerase; it has a homolog of the proofreading e subunit of *E. coli* Pol III. Its cell-division machinery has orthologs of eubacterial *fts* and eukaryal *cdc* genes. Similarly, the *A. fulgidus* transcriptional system shows both eukaryal and eubacterial features: for example, the RNA polymerase contains the large universal subunits and five smaller subunits found in both Archaea and eukaryotes, while transcription initiation has components of the eukaryotic type.

The hyperthermophile *Nanoarchaeum equitans* is a parasite of the crenarchaeon *Ignicoccus*; recent studies on its genome have provided insight into the impact of parasitism on archaeal genomes (Waters et al., 2003). The genome of *N. equitans* consists of a single, circular chromosome of only 490,885 bp and encodes 552 protein-coding genes with an average length of 827 bp. This is one of the smallest microbial genomes and also one of the most compact, with 95% of the DNA predicted to encode proteins or stable RNAs. The reduced genome of *N. equitans* has few pseudogenes or extensive regions of noncoding DNA.

Unlike its *Ignicoccus* host, which gains energy by using hydrogen to reduce elemental sulphur, *N. equitans* has no genes to support a chemolithoautotrophic physiology and lacks the metabolic capacity to synthesize many cell components. This organism lacks almost all known genes that are required for the *de novo* biosyntheses of amino acids, nucleotides, cofactors and lipids. Also missing are genes for glycolysis/gluconeogenesis, the pentose phosphate pathway, the tricarboxylic acid cycle and other known pathways for carbon assimilation.

In contrast to the paucity of metabolic genes, *N. equitans* possesses a complete set of archaeal-type components for information processing (replication, transcription and translation) and completion of the cell cycle. The gene sets for DNA replication and cell cycle are similar to those found in the Euryarchaeota and contain several components usually absent from the Crenarchaeota (e.g. DNA polymerase II, histones).

Phylogenies based on ribosomal proteins and rRNAs place *N. equitans* unequivocally at the most deeply branching position within the Archaea, suggesting that it diverged early within the archaeal lineage and belong to a distinct archaeal kingdom: the Nanoarchaeota. It is interesting in this respect that the *N. equitans* genome is unusual inasmuch as there are numerous examples of 'split genes': whereas in other Archaea single genes encode different domains of multidomain proteins, in *N. equitans* the orthologous domains are frequently encoded by separate genes. Assuming that multidomain proteins evolved from the fusion of simple domains, the 'split genes' of *N. equitans* could reflect the multisubunit ancestral state of these multidomain proteins.

9.6.2.3 Organelle genomes

According to the generally accepted **endosymbiotic theory**, the key organelles of eukaryotic cells such as mitochondria and plastids (e.g. chloroplasts) originated from prokaryotic organisms engulfed by ancestral eukaryotic cells as endosymbionts (primary endosymbiosis). There are several types of proofs supporting the endosymbiotic theory. First, both mitochondria and plastids contain small DNA genomes that are more similar to those of bacteria than to the nuclear genome inasmuch as they are also circular. The organelle genomes, however, are significantly smaller than those of their prokaryotic ancestors since many genes originally present in the genome of the organelle were transferred to the nucleus. In fact, most genes needed for mitochondrial and plastid function are encoded by the nuclear genome. Second, these organelle genomes are surrounded by two or more membranes and the composition of the inner membrane is more similar to the prokaryotic cell membrane than to the other membranes of the cell. Third, the ribosomes of organelles are similar to those found in bacteria and distinct from those involved in translation of nuclear genes.

Mitochondria

There is strong phylogenetic evidence that **mithochondria** originated from α-proteobacteria such as Rickettsiales (Andersson et al., 1998; Andersson & Kurland, 1999; Gray, Burger & Lang, 2001). The first molecular evidence that mitochondria are remnants of an α-proteobacterial ancestor was provided by phylogenetic analyses of ribosomal RNA genes and subsequent analyses of genes that code for components of the respiratory chain complexes in mitochondria have provided additional support for the origin of mitochondria within α-proteobacteria.

Interestingly, almost all members of this group of bacteria live in close association with eukaryotic cells, either as symbionts or as parasites of plants and animals. An obligate intracellular parasite in this group, *Rickettsia prowazekii*, has one of the smallest genomes among bacteria: it is only 1.1 Mb. As a reflection of the fact that *R. prowazekii* (the causative agent of epidemic typhus) is an intracellular parasite that uses metabolites of the host, it has a strongly reduced repertoire of genes involved in amino acid biosynthesis and there are no genes for *de novo* purine and pyrimidine biosynthesis. The functional profile of *R. prowazekii* genes shows similarities to those of mitochondrial genes: no genes required for anaerobic glycolysis are found in either *R. prowazekii* or mitochondrial genomes. On the other hand, a complete set of genes encoding components of the tricarboxylic acid cycle and the respiratory-chain complex is found in *R. prowazekii*, thus ATP production in *Rickettsia* is the same as that in mitochondria.

The mitochondrial genomes represent an even more striking example of reductive evolution: animal mtDNAs are typically small (approximately 16 kbp) circular molecules that encode 37 or fewer tightly packed genes. The mitochondrial genomes of some early diverging protists seem to have retained both gene sequences and genomic arrangements more reminiscent of their putative bacterial ancestors (Gray et al., 1998). For example, the 69-kb mitochondrial genome of the protozoon *Reclinomonas americana* contains a larger collection of genes (97 genes) and preserved the ancestral arrangement of the ribosomal protein genes found in bacteria (Lang et al., 1997). Almost half of the mitochondrial genes in *R. americana* code for components in the translation machinery, while the other half encode enzymes involved in bioenergetic processes. Phylogenetic reconstructions based on these mitochondrial proteins have also revealed a close affiliation between the mitochondria of *R. americana* and α-proteobacteria.

Investigations on the gene content and architecture of the mitochondrial genomes of the closest unicellular relatives of animals, i.e. choanoflagellate and ichthyosporean protists have also revealed that they are radically different from those of metazoa. The circular-mapping mtDNA of the choanoflagellate *Monosiga brevicollis* with its long intergenic regions is four times as large and contains two times as many protein genes as do animal mtDNAs, whereas the ichthyosporean *Amoebidium parasiticum* mitochondrial genome totals >200 kbp. Considering that the closest unicellular relatives of animals possess large, spacious, gene-rich mtDNAs, it seems likely that the distinct compaction characteristic of metazoan mitochondrial genomes occurred simultaneously with the emergence of a multicellular body plan in the animal lineage (Burger et al., 2003).

Hydrogenosomes

Hydrogenosomes, organelles that produce ATP and hydrogen, are found in various unrelated eukaryotes, such as anaerobic flagellates, chytridiomycete fungi and ciliates. Although all of these organelles generate hydrogen, the hydrogenosomes from these organisms are structurally and metabolically quite different. Since hydrogenosomes usually lack an organelle genome this has hampered clarification of their origin.

Like mitochondria, the hydrogenosomes are bound by distinct double membranes and have an inner membrane with some cristae-like projections. According to one hypothesis, the ancestral mitochondrial endosymbiont first gave rise to aerobically functioning mitochondria, which subsequently evolved into hydrogenosomes by the acquisition of genes encoding enzymes essential for an anaerobic metabolism.

Recent studies on the hydrogenosome of *Nyctotherus ovalis* (an anaerobic ciliate that lives in the hindgut of cockroaches) have provided evidence in support of this hypothesis (Boxma et al., 2005). The hydrogenosomes of *N. ovalis* have typical mitochondrial cristae and retained a rudimentary genome encoding components of a mitochondrial electron transport chain. Phylogenetic analyses revealed that these proteins cluster with their homologs from aerobic ciliates. Furthermore, all these genes exhibit a characteristic mitochondrial codon usage. These features, characteristic of anaerobic mitochondria, thus identify the *N. ovalis* organelle as a missing link between mitochondria and hydrogenosomes.

Plastids

The internal structure and biochemistry of **plastids**, the presence of thylakoids and particular chlorophylls is very similar to that of cyanobacteria. Phylogenetic analyses of the complete genome sequences of cyanobacteria and of the higher plant *Arabidopsis thaliana* leave no doubt that the plant **chloroplast** originated from a cyanobacterium (Raven & Allen, 2003). The plastid genomes, however, are at least an order of magnitude smaller than cyanobacterial genomes and they encode only 60–200 proteins whereas cyanobacterial genomes encode at least 1500 proteins. The fact that as many as 5000 nuclear-coded gene products are targeted to plastids suggests that following the engulfment of cyanobacteria many genes were transferred to the nuclear genome.

In green plants chloroplasts are surrounded by two lipid-bilayer membranes, the inner membrane is thought to correspond to the outer membrane of the ancestral cyanobacterium. There is evidence that many groups of protists acquired plastids independently by secondary endosymbiosis, i.e. by engulfing other plastid-containing eukaryotes (Douglas & Penn, 1999; Glockner, Rosenthal & Valentin, 2000; Okamoto & Inouye, 2005). In fact, there are cases of tertiary endosymbiosis. Whereas primary endosymbiosis (engulfment of a photosynthetic cyanobacterium) gave rise to a plastid bound by two membranes, secondary endosymbiosis (engulfment of a eukaryotic alga) resulted in a plastid bound by three or four membranes. Dinoflagellates have undergone tertiary endosymbiosis through the engulfment of an alga containing a secondary plastid (Yoon et al., 2005).

9.6.2.4 Eukaryotic genomes

The genomes of Plasmodium falciparum and Plasmodium yoelii yoelii

The 23-Mb nuclear genome of *Plasmodium falciparum* 3D7, the apicomplexan parasite responsible for malaria in humans, was found to encode about 5300 protein-coding genes suggesting an average gene density of one gene per 4338 base pairs (Gardner et al., 2002). The genome is extremely AT rich (80.6%) and this value rises to approximately 90% in introns and intergenic regions. Introns were predicted in 54% of *P. falciparum* genes, a proportion much higher than observed in *S. cerevisiae* where only 5% of genes contain introns. Nevertheless, the total length of introns accounts for only 6% of the genome, whereas exons correspond to 53% of the genome sequence.

Plasmodium and other members of the phylum apicomplexa also harbour a plastid homologous to the chloroplasts of plants and algae. This plastid, the apicoplast, is known to function in the anabolic synthesis of fatty acids and isoprenoids. The apicoplast arose through a process of secondary endosymbiosis, in which the ancestor of all apicomplexan parasites engulfed a eukaryotic alga, and retained the algal plastid. The 35-kb apicoplast genome encodes only 30 proteins but a total of 551 nuclear-encoded proteins are targeted to the apicoplast. Nuclear-encoded apicoplast proteins include housekeeping enzymes involved in DNA replication and repair, transcription, translation and post-translational modifications, cofactor synthesis, protein import, protein turnover and specific metabolic and transport activities.

Comparative genome analysis with other eukaryotes revealed that, in terms of overall genome content, *P. falciparum* is slightly more similar to *Arabidopsis thaliana* than to other taxa; this may be due in part to the presence of genes derived from plastids or from the nuclear genome of the secondary endosymbiont. There are 237 *P. falciparum* proteins with strong matches to proteins in all completed eukaryotic genomes but no matches to proteins in any complete prokaryotic proteome. This list of 'eukaryote proteins' includes those with roles in cytoskeleton construction and maintenance, chromatin packaging and modification, cell cycle regulation and intracellular signalling.

The genome of this intracellular parasite encodes a large proportion of genes devoted to immune evasion and host–parasite interactions. Erythrocyte invasion by *P. falciparum* involves multiple ligand–receptor interactions (Cowman & Crabb, 2002). For example, merozoite surface proteins, the integral membrane proteins identified on the surface of developing and free merozoites, are important for invasion as suggested by the fact that antibodies directed against these proteins block invasion. Interestingly, the majority of merozoite surface proteins have one or two epidermal growth factor (EGF)-like domains at their COOH terminus.

The presence of typical EGF-like domains in *Plasmodium* is somewhat unexpected since these domains are abundant only in animals but are very rare or absent in other groups of eukaryotes. The EGF domain is not the only 'animal-type' domain of the *P. falciparum* proteome or the proteomes of other apicomplexan protozoa. Recently it has been shown that different lineages of apicomplexan protozoa (e.g. *Plasmodium*, *Cryptosporidium*) have acquired distinct but overlapping sets of multidomain surface proteins constructed from adhesion domains typical of animal proteins, although in no case do they share multidomain architectures identical to those of animals (Deng et al., 2002; Templeton et al., 2004).

A systematic analysis of the *Cryptosporidium parvum* proteome has identified several domains (namely, thrombospondin type I domain, sushi/CCP domain, Notch /Lin domain, neurexin-collagen domain, fibronectin type 2 domain, pentraxin domain, MAM domain, ephrin-receptor domain, the animal signalling protein hedgehog-type HINT domain and the scavenger domain) that have thus far been found only in the surface proteins of animals other than apicomplexans. Domains such as the EGF domain, the LCCL domain, the kringle domain and the SCP domain are found sporadically in some other eukaryotes but they are widespread only in animals. Phylogenetic analyses have revealed specific affinities between apicomplexan and animal versions of these domains, making lateral gene transfer from animals (followed by selective retention of functionally relevant proteins) the most parsimonious explanation for these observations (Pradel et al., 2004).

The 23-Mb genome of *Plasmodium yoelii yoelii*, the parasite causing malaria in rodents, is strikingly similar to that of *P. falciparum* in both genome size and gene number (Carlton et al., 2002). Comparison of the two genomes has also revealed marked conservation of gene synteny within the body of each chromosome. Despite this close similarity, comparison of syntenic regions revealed that the similarity in intergenic regions is negligible. In contrast to intergenic regions, the similarity between species in coding regions is relatively high. The average number of nonsynonymous substitutions per nonsynonymous site is only 26%, whereas synonymous sites are saturated with substitutions (consistent with the lack of similarity in intergenic regions). Thus the substitution rate values are considerably higher than those reported for human–rodent comparisons (i.e. their hosts), which are approximately 7.5% and 45% for nonsynonymous and synonymous substitutions, respectively. The fact that the two *Plasmodium* species diverged more rapidly than their hosts is most likely a consequence of the much shorter generation time of the parasites than their hosts.

The genome of the unicellular green alga Ostreococcus tauri

The green alga lineage evolved shortly after the primary endosymbiosis event that gave rise to early photosynthetic eukaryotes, about 1500 million years ago. The 12.56-Mb nuclear genome of the **unicellular green alga** *Ostreococcus tauri* has 8166 protein-coding genes and the coding sequences cover 82% of the genome. This extremely high gene density is achieved in part by extensive reduction of intergenic regions and other forms of compaction such as gene fusion (Derelle et al., 2006). The average intergenic size is only 196 bp, which is significantly shorter than that of other eukaryotes having a similar genome size.

Another striking feature of the *O. tauri* genome is its heterogeneity. Two chromosomes (2 and 19) are different from the other 18, inasmuch as they have lower G+C content than the 59% G+C of the other 18 chromosomes. These two aberrant chromosomes contain 77% of the 417 transposable elements. Codon usage for genes in the low G+C region of chromosome 2 is different from that of all other chromosomes and many of the genes in the low G+C region contain multiple small introns. Chromosome 2 small introns differ in many respects from the other introns, such as their small size (40–65 bp), composition (they are AT richer than the neighbouring exons), and splice sites and branch points that are less conserved than for other introns. Interestingly, phylogenetic analysis shows that 43% of the genes on this chromosome, including the small intron-containing genes, have green lineage ancestry. In the case of chromosome 19 phylogenetic analyses have shown that only 18% of the peptide-encoding genes are related to the green lineage. Interestingly, most (84%) of the genes with a known function belong to a few functional categories, primarily encoding surface membrane proteins or proteins involved in the building of glycoconjugates. Based on these features, it was suggested that chromosome 19 is of a different origin than the rest of the genome.

The genome of the unicellular red alga Cyanidioschyzon merolae

Cyanidioschyzon merolae is a small unicellular red alga that lives in the extreme environment of acidic hot springs. The 16.5-Mb genome of *C. merolae* 10D contains 5331 genes. A unique characteristic of its genomic structure is the lack of introns in all but 26 genes (0.5% of the protein genes), and all but one of them had only a single intron (Matsuzaki et al., 2004).

Enzymes of the Calvin cycle in plants are known to be a mosaic of enzymes originating from cyanobacteria-like ancestors of an endosymbiont and its eukaryotic host. The fact that this mosaic origin of Calvin cycle enzymes is also conserved in this red alga supports the hypothesis of the existence of single primary plastid endosymbiosis. In other words, it is highly probable that the complex and mosaic origin of Calvin cycle enzymes derived from common ancestors of green plants and red algae, and that no essential changes occurred after the separation of the two lineages, strongly supporting the concept of a single event of primary plastid endosymbiosis.

The genome of the diatom Thalassiosira pseudonana

Whereas green, red, glaucophyte algae and higher plants are derived from a primary endosymbiotic event in which a nonphotosynthetic eukaryote acquired a chloroplast by engulfing a prokaryotic cyanobacterium, diatoms were derived by secondary endosymbiosis whereby a nonphotosynthetic eukaryote acquired a chloroplast by engulfing a photosynthetic eukaryote, probably a red algal endosymbiont.

The 34-Mb nuclear genome of the marine diatom *Thalassiosira pseudonana* was predicted to contain 11,242 protein-coding genes, the average number of introns per genes was 1.4 (Armbrust et al., 2004). The 129-kb plastid genome and the 44,000-bp mitochondrial genome encode 144 and 40 protein-coding genes, respectively. The secondary endosymbiosis hypothesis predicts different possible origins for nuclear-encoded diatom genes: nuclear or mitochondrial genomes of the secondary nonphotosynthetic host, and nuclear, plastid or mitochondrial genomes of the red algal endosymbiont. To infer gene origins in the modern diatom, Armbrust et al., compared its proteome with those of two extant photosynthetic eukaryotes (the green plant *A. thaliana* and the red alga *C. merolae*) and one heterotrophic eukaryote (*Mus musculus*). These comparisons have shown that almost half the diatom proteins have similar alignment scores to their closest homologs in plant, red algal and animal genomes, underscoring the evolutionarily ancient divergence of Plantae (red algae, green algae and plants), animals/fungi, and the unknown secondary host. Interestingly, 806 diatom proteins align with mouse proteins but not with green plant or red alga proteins, suggesting that these 'animal-like' genes were derived from the heterotrophic secondary host.

The nucleomorphs of Bigelowiella natans and Guillardia theta

As mentioned above, the introduction of plastids into different heterotrophic protists by secondary (or tertiary endosymbiosis) has occurred several times independently. Recent studies on chlorarachniophytes and chromophytes provide further insight into the evolution of secondary endosymbiosis since vestiges of the endosymbiont nucleus (known as the **nucleomorph**) are retained in these algae.

Interestingly, although chlorarachniophytes and chromophytes originated by two independent endosymbioses (they acquired their plastids from a green and red alga, respectively), their nucleomorph genomes have evolved similarly to achieve extremely compact, highly reduced genomes.

The nucleomorph genome of the chlorarachniophyte *Bigelowiella natans* is only 373 kb long with only 331 genes; this is the smallest nuclear genome known (Gilson et al., 2006). The genome is eukaryotic in nature, with three linear chromosomes containing densely packed genes (one gene per 1227 bp) with numerous overlaps. The genome contains numerous (852) introns, but these are the smallest introns known, being only 18, 19, 20 or 21 bp in length. It could be shown that these introns are miniaturized versions of the normal-sized introns that were present in the endosymbiont at the time of capture. It is interesting to note that genome compaction was achieved by extreme reduction in the size of introns rather than by loss of introns, raising the possibility that the usual mechanisms for intron removal (e.g. reverse transcriptase) did not operate on the nucleomorph genome. Only 17 of the nucleomorph genes encode proteins that function in the plastid, the other nucleomorph genes are housekeeping genes responsible for maintenance and expression of these plastid proteins. The proteins involved in plastid function are thus encoded by three different genomes: the plastid, nucleomorph and host nuclear genomes.

Chromophyte algae are evolutionary chimeras of a red alga and a nonphotosynthetic host and cryptomonads are the only chromophytes that still retain the enslaved red algal nucleus as a minute nucleomorph. The nucleomorph genome (on three nucleomorph chromosomes) of the cryptomonad *Guillardia theta* is only 551 kb long and encodes only 531 protein-coding genes. The genome is thus extremely gene dense (one gene per 1035 bp), with only 17 short spliceosomal introns (42–52 bp long) and 44 overlapping genes (Douglas et al., 2001). The nucleomorph genome of *G. theta* has a number of curious architectural features that were originally thought to be characteristic of red algae. The fact, however, that chromosomes of the red alga *C. merolae* did not contain overlapped genes such as those of the nucleomorph genome suggests that these features appeared after secondary symbiosis.

The marked evolutionary compaction eliminated nearly all the nucleomorph genes for metabolic functions, but left 30 for chloroplast-located proteins. As in the case of *B. natans*, to allow expression of these proteins, nucleomorphs retain hundreds of genetic housekeeping genes.

Plants

The genome of Arabidopsis thaliana

Arabidopsis thaliana, a small cruciferous weed, was chosen as the first model to analyse plant genome structure because it has a small genome. With a genome size of only about 125 Mb, *Arabidopsis* has one of the smallest plant genomes, similar in size to those of the animal models *Drosophila melanogaster* or *Caenorhabditis elegans*. It is much less compact than the yeast genome. Initial studies on a 1.9 Mb contiguous sequence of the *A. thaliana* genome have suggested that there may be about one gene every 4800 bp (Bevan et al., 1998), thus the expected number of genes in the whole genome was predicted to be about 21,000.

Analysis of the complete 125 Mb genome of *A. thaliana* has revealed that it contains 25,498 protein-coding genes (Arabidopsis Genome Initiative, 2000). The genes of *Arabidopsis* are on average rather small (2 kb) and the majority of genes contain small (usually <200 bp) and few introns; these introns are usually AT rich (Hebsgaard et al., 1996; Bevan et al., 1998). The total length of exons (33,249,250) exceeds the total length of introns (18,055,421).

These studies have also demonstrated that the evolution of *Arabidopsis* involved a whole-genome duplication, followed by subsequent gene loss and extensive local gene duplications, as well as lateral gene transfer from a cyano-bacterial-like ancestor of the plastid (Arabidopsis Genome Initiative, 2000).

There are some noteworthy similarities and differences between the proteomes of multicellular plants and multicellular animals. For example, although the principal components of the cytoskeleton (microtubules, actin filaments) are present in both groups, the *Arabidopsis* proteome appears to lack homologs of proteins that in animal cells link the actin cytoskeleton across the plasma membrane to the extracellular matrix, such as integrin, talin, spectrin, alpha-actinin, vitronectin or vinculin. This apparent lack of 'anchorage' proteins is consistent with the different composition of the cell wall. The regulation of development in both plants and multicellular animals involves cell–cell communication and networks of transcription factors, but the two groups have evolved independent solutions. Plants and animals have used and expanded different transcription factor families as key regulators. For example, in animals pattern formation involves the spatially specific activation of a series of homeobox gene family members, whereas in plants pattern formation is established by the spatially specific activation of members of a different family of transcription factors: the MADS box family. Since plants also have homeobox genes and animals have MADS box genes, this suggests that each lineage invented separately its mechanism of spatial pattern formation.

Independent solution to similar problems has also been found in cell–cell communication. Plants, unlike animals, do not have receptor tyrosine kinases, but the *Arabidopsis* genome has several hundred genes for receptor Ser/Thr kinases, belonging to many different families, defined by their putative extra-cellular domains. Thus, the plant and animal lineages have expanded different families of receptor kinases for a similar set of developmental processes.

Comparative genome analysis between *Arabidopsis* and multicellular animals supports the idea that plants have evolved their own pathways of signal transduction. None of the components of the widely adopted signalling pathways found in vertebrates, flies or worms, such as Wingless/Wnt, Hedgehog, Notch/lin12, JAK/STAT, TGF-beta/SMADs, receptor tyrosine kinase/Ras or the nuclear steroid hormone receptors, is found in *Arabidopsis*. By contrast, brassinosteroids are ligands of the BRI1 Ser/Thr kinase, a member of the largest recognizable class of transmembrane sensors encoded by receptor-like kinase genes in the *Arabidopsis* genome.

Amoebozoa

The genome of Dictyostelium discoideum

The 34-Mb genome of *Dictyostelium discoideum* encodes approximately 12,500 predicted proteins (Eichinger et al., 2005). About 69% of the protein-coding genes contain introns (on average two introns per gene). Since the introns of *Dictyostelium* are short (146 bp) and intergenic regions are small the genome is quite compact: 62% of the genome encodes protein. The genome is extremely A+T rich (77.57%) with some difference between exonic (73%), intronic (88%) and intergenic regions (85%). The A+T richness influences codon usage, inasmuch as codon usage favours codons of the form NNT or NNA over their NNG or NNC synonyms. The extreme A+T richness is reflected not just in the choice of synonymous codons, but also in the amino acid composition of the proteins. Amino acids encoded solely by codons where the first two bases are A or T (e.g. Asn, Lys, Ile, Tyr and Phe) are much more common in *Dictyostelium* proteins than in human proteins; the reverse is true for amino acids encoded solely by codons in which the first two bases are C or G (e.g. Pro, Arg, Ala and Gly).

The *Dictyostelium* genome is rich in gene duplications: 20% of the predicted proteins have arisen by relatively recent duplication. Interestingly, where members of a family are clustered on one chromosome, the physical distance between family members often correlates strongly with their evolutionary divergence (cf. section 6.2.4, see also Fig. 6.2).

Phylogeny of eukaryotes based on complete proteomes suggests that *Dictyostelium* diverged after the plant/animal split, but before the divergence of the fungi, i.e. amoebozoa are a true sister group of the fungi and metazoa. The distribution of proteins shared by *Dictyostelium* and major organism groups is also consistent with this phylogeny: the fact that there are many protein domains present in *Dictyostelium*, metazoa and fungi, but absent in plants, suggests that they probably arose soon after plants diverged and before *Dictyostelium* diverged from the line leading to animals. The major classes of domains in this group of proteins include those involved in small and large G-protein signalling.

The social amoebae, such as the free-living protozoan *Dictyostelium*, are exceptional in their ability to alternate between unicellular and multicellular forms. A broad survey of proteins required for multicellular development shows that *Dictyostelium* possesses cell adhesion and signalling modules normally associated exclusively with animals. For example, G-protein coupled cell surface receptors allow the detection of a variety of environmental and intraorganismal signals such as light, Ca^{2+}, odorants, nucleotides and peptides. They are subdivided into six families that, despite their conserved secondary domain structure, do not share significant sequence similarity. *Dictyostelium* was found to contain members of the secretin family, the metabotropic glutamate/GABA B family and the frizzled/smoothened family of receptors, i.e. families that had been thought to be animal specific. Their occurrence in *Dictyostelium* suggests that they arose before the divergence of the animals and fungi.

Throughout *Dictyostelium* development, cells must modulate their adhesiveness to the substrate, to the extracellular matrix and to other cells in order to create tissues and carry out morphogenesis. To accomplish this, *Dictyostelium* uses a surprising number of components that have been only associated with animals. For example, disintegrin proteins regulate cell adhesiveness and differentiation in a number of metazoa and at least one *Dictyostelium* disintegrin is needed throughout development for cell-fate specification. In animals, tandem repeats of immunoglobulin, cadherin, fibronectin III, E-set domains or EGF/laminin domains are often present in cell adhesion proteins. *Dictyostelium* has 61 predicted proteins containing repeated E-set or EGF/laminin domains and many of these contain additional domains that suggest they have roles in cell adhesion or cell recognition.

The genome of Entamoeba histolytica

Entamoeba histolytica is an intestinal protist parasite, the causative agent of amoebiasis. The 23.8-Mb genome of *E. histolytica* was predicted to contain 9938 protein-coding genes that cover about 49% of the genome. One-quarter of the genes contain introns, with 6% of genes containing multiple introns (Loftus et al., 2005).

Analysis of the *E. histolytica* genome suggests that it evolved by secondary gene loss and lateral gene transfer primarily from bacterial lineages. These evolutionary changes were driven by the characteristic biology of this parasite. First, as a phagocytic resident of the human gut, *E. histolytica* has access to many bacterial and host-derived preformed organic compounds, therefore most pathways for amino acid biosynthesis, *de novo* purine, pyrimidine and fatty acid synthesis were eliminated. Second, *E. histolytica* residing in the anoxic colon is an obligate fermenter, using bacterial-like fermentation enzymes and lacking proteins of the tricarboxylic acid cycle and mitochondrial electron transport chain. An atrophic, mitochondrion-derived organelle has been identified in *E. histolytica* and the genome data support the absence of a mitochondrial genome.

A phylogenetic screen of the *Entamoeba* genome has provided convincing evidence for lateral gene transfer from bacteria. The majority of the laterally transferred genes (58%) encode a variety of metabolic enzymes, the major impact is in the area of carbohydrate and amino acid metabolism, where they have increased the range of substrates available for energy generation including tryptophanase and aspartase, which contribute to the use of amino acids. Several glycosidases and sugar kinases appear to have been acquired through lateral gene transfer and would probably enable *E. histolytica* to use sugars other than glucose; for example, fructose and galactose. It is clear that among the laterally transferred genes, some result in significant enhancements to *E. histolytica* metabolism.

E. histolytica uses complex transduction systems in order to sense and interact with the different environments it encounters. The 270 putative protein kinases include tyrosine kinases with SH2 domains, tyrosine kinase-like protein kinases and receptor Ser/Thr kinases. These Ser/Thr kinases are uncommon in protists, appear to be absent from *Dictyostelium* and have previously been described only in plants, animals and choanoflagellates. The *E. histolytica* receptor Ser/Thr kinases all contain an N-terminal signal peptide, a predicted extracellular domain and a single transmembrane helix followed by a cytosolic tyrosine kinase-like domain.

Fungi

The genomes of Saccharomyces cerevisiae, Schizosaccharomyces pombe, Neurospora crassa, Aspergillus oryzae, Aspergillus nidulans *and* Aspergillus fumigatus

The first completely sequenced eukaryotic genome was that of the yeast *Saccharomyces cerevisiae*. *S. cerevisiae* is a free-living organism, and this genome was the first to provide an insight into the organization of a simple, self-sufficient eukaryote. The nuclear genome of yeast comprises 12.5 Mb of nonribosomal DNA representing a total of about 5800 protein-coding genes (Bassett et al., 1996; Clayton et al., 1997; Dujon, 1997; Oliver, 1997). The introns of protein-coding genes are few (233) and short – they account for less than 1% of the entire genome.

Compared with higher eukaryotes, the most conspicuous feature of the yeast genome is its compactness. This results from the short size of intergenic regions, the scarcity of introns and the infrequency of repetitive sequences. Open reading frames occupy 68% of the yeast genome, and ribosomal DNA (rDNA) accounts for about 9% of the genome leaving little space for noncoding DNA and for all other structural and functional elements. The yeast genome is poor in repeated sequences, with the rDNA contributing most significantly to repetitiveness (rDNA repeats form a long tandem array occupying half of chromosome XII). Nevertheless, compared with eubacterial or archaeal genomes, where ORFs occupy more than 80% of the genomes, the yeast genome is less compact, i.e. it seems to be under weaker selective pressure to reduce genome size. (Whereas in Archaea and eubacteria there is one gene per ~1000–1100 bp, in the case of yeast there is one gene every 2155 bp.) The 'average' yeast gene is a 1450-bp ORF, preceded by an upstream region of 309 bp, followed by a downstream region of 163 bp, making a total of about 1922 bp.

Gene density is not uniform along yeast chromosome maps. In most chromosomes, there exist segments in which gene density is significantly above average, reaching more than 85% in many places. Such chromosome segments are separated by other regions of comparatively much lower gene density (only 50–55%). The yeast genome is moderately biased in composition with an average GC content of about 39%. Long-range compositional variations also exist along yeast chromosomes; usually the major GC-rich peaks exist in the middle of each arm, with pericentromeric and subtelomeric regions being AT rich. In most cases, there is a positive correlation between high GC content and high gene density, as in genomes of more complex eukaryotes.

Variations in GC content also correlate with variations in recombination frequency. Comparison of genetic and physical maps of the yeast genome has shown that the average ratio between genetic and physical map distances is about 0.34 centimorgans (cM) per kb, but smaller chromosomes, such as I and VI, have a higher ratio. Within chromosomes, recombination frequency varies widely and both recombination hotspots (sites of frequent chiasma formation) and coldspots (sites of low chiasma formation) have been identified. Studies on chromosome III may suggest some molecular basis for these local variations in recombination frequency. The average ratio of genetic map distance to physical distance for chromosome III is 0.51 cM per kb; this is similar to values for chromosomes I and VI (0.62 and 0.55 cM per kb, respectively). The variation in the cM per kb ratio for different intervals along the chromosome is at least 10-fold: it is lowest close to the centromere (AT-rich region, gene-poor region) and greatest midway down each arm (GC-rich regions, gene-rich regions). Thus there is an approximate correlation between the pattern of genetic recombination and that of transcription along the chromosome. The basis for an association between high recombination and high transcription may be that both processes require relatively naked DNA within the chromatin.

There is strong evidence that the evolution of the yeast genome has involved frequent duplication of chromosome segments followed by gene shuffling (Seoighe & Wolfe, 1998). Evidence for intragenomic chromosome duplications of syntenic segments comes from the frequent finding that two or several genes situated closely on one chromosome have their homologous loci (in the same order and same orientation) on another chromosome. For example, the presence of S8 ribosomal protein and HSP70 stress protein genes on both chromosome II and chromosome IV suggests that part of one of these chromosomes may have been derived from the duplication of part of the other. In subtelomeric regions, large segments containing several ORFs are often almost identical between several chromosomes, suggesting recent exchange of genetic information.

Duplications of single genes are also frequent. Some are scattered in the genome, while others are in tandem or in inverted orientations. As a result of duplication of single genes and chromosome segments, the *S. cerevisiae* genome displays signs of extensive duplications: duplicated regions represent more than 30% of the entire genome, with 53 duplicated gene clusters on its 16 chromosomes. There are some large paralogous gene families that have arisen by duplication events. For example, there are some 113 protein kinase genes, 29 ABC transporters and 28 multidrug-resistant transport proteins from the major facilitator superfamily. A typical example of evolution through gene duplication and divergence of function is provided by yeast citrate synthase genes. Chromosomes III and XIV, which apparently evolved from a common ancestor, each have a citrate synthase gene at an equivalent position, but *CIT2* on chromosome III encodes a cytoplasmic enzyme, whereas *CIT1* on chromosome XIV encodes a mitochondrial enzyme.

A striking observation of yeast **gene-knockout** experiments was that only a very small fraction of yeast genes confer a lethal phenotype when disrupted. Experiments in which the yeast genome was disrupted at random, suggested that about 70% is dispensable for growth on a rich, glucose-containing medium. In a more specific study, of the 145 novel ORFs found in the chromosome III sequences, 55 have been subjected to gene disruption and for only three of these (5%) did disruption have lethal consequences. The testing of 42 nonlethal disruptants for other types of phenotypic effects showed that in the majority of cases (62%) no phenotypic change was observed. There has been much discussion of such a **redundancy** of the yeast genome, and several kinds of explanation have been put forward to account for this. It has been suggested that once the constraints on genome size imposed by the single, circular bacterial chromosome were released, organisms could afford to use multiple genes to perform identical (or nearly identical) functions in different contexts. An additional explanation is that many yeast genes may be required to deal with special challenges that are common in nature (e.g. rotting fruits) but which are hard to mimic in the laboratory: most of the 'redundancies' may be apparent rather than real. Obvious stresses such as heat, cold, osmotic shock, pH shock or starvation, plus a vast number of other types of conditions may be relevant to the natural history of yeast. An interesting example may illustrate this point. Complete deletion of *YCR32w*, a gene encoding a putative membrane protein, had no obvious effect, thus this gene could have been considered a 'redundant' gene, with no obvious essential function. However, when the mutant was exposed to various physiological challenges, it was discovered that it died when grown on glucose at low pH in the presence of acetic acid. Since *YCR32w* encodes a membrane protein, its product is suspected to be an acetic acid pump. When discussing 'redundancy' of paralogous proteins, one must also remember (see section 3.3) that although a feature may give an organism a selective advantage that is barely detectable in laboratory tests, that selective advantage may nonetheless be sufficient for the feature to be fixed and maintained. Recent studies do indicate that marginal fitness contributions of seemingly 'non-essential' genes of yeast do provide significant selective advantage for the organism (Thatcher, Shaw & Dickinson, 1998).

The genome of the fission yeast *Schizosaccharomyces pombe* was found to have only 4824 protein-coding genes, significantly fewer than the budding yeast *S. cerevisiae*, although its genome (13.8 Mb) is larger. Interestingly, the lower gene density of *S. pombe* is associated with a significantly larger number of introns: the protein-coding genes contain a total of 4730 introns (Wood et al., 2002).

The genome of the filamentous fungus *Neurospora crassa* is much larger (38.6 Mb) than those of fission yeast and budding yeast and it encodes a significantly larger number (about 9200) protein-coding genes (Galagan et al., 2003). The *N. crassa* genome is also more intron rich than the genomes of budding yeast or fission yeast: there is an average of 1.7 introns per gene.

An interesting aspect of the *Neurospora* genome is that 'repeat-induced point mutation', a genome defence mechanism unique to fungi, had a great impact on its evolution. Repeat-induced point mutation is a process that detects and mutates both copies of a sequence duplication, with a strong preference for C to T mutations occurring at CpA dinucleotides. Repeat-induced point mutation requires a minimal duplicated sequence length of about 400 base pairs and greater than about 80% sequence identity between duplicates. The biological significance of this genome defence mechanism is that it may protect the genome against selfish or mobile DNA but it may also prevent gene innovation through gene duplication. In harmony with this expecta tion, *Neurospora* possesses far fewer duplicated genes (in multigene families) than expected and contains no predicted instances of recent paralogous duplication. It thus appears that the emergence of 'repeat-induced point mutation' (and increased genome security) may have arrested the evolution of new genes by this mechanism.

The 37-Mb genome of *Aspergillus oryzae*, a fungus important for the production of fermented foods and beverages in Japan, was found to be 7–9 Mb larger than the genomes of *Aspergillus nidulans* and *Aspergillus fumigatus* (Machida et al., 2005; Galagan et al., 2005). The number of protein-coding genes in *A. oryzae* (14,063) is also significantly higher than in *A. nidulans* (9541) or *A. fumigatus* (9926). Comparison of the three *Aspergillus* species revealed the presence of **syntenic blocks** and *A. oryzae*-specific blocks (lacking synteny with *A. nidulans* and *A. fumigatus*). Phylogenetic analyses suggest that the increase in genome size was due to an *A. oryza*e lineage-specific acquisi- tion of sequences. Interestingly, the blocks of *A. oryzae*-specific sequence are enriched for genes involved in metabolism, particularly those for the synthesis of secondary metabolites, secretory hydrolytic enzymes, amino acid metabolism and amino acid/sugar uptake transporters supports. *A. oryzae* possesses more secretory proteinase genes that function in acidic pH, including aspartic proteinase, pepstatin-insensitive proteinase, serine type carboxypeptidase, reflecting adaptation to acidic pH during the course of its domestication. The *A. oryzae*-specific regions contained genes with highest sequence similarity to bacterial genes, suggesting that these regions have been laterally transferred.

Animals

Nematodes

The genomes of Caenorhabditis elegans and briggsae

The 97-Mb genome of the nematode *Caenorhabditis elegans* was predicted to contain 19,099 protein-coding genes (The C. elegans Sequencing Consortium, 1998). The genome is significantly less compact than that of budding yeast: whereas in budding yeast there is one gene for every 2155 bp, in the case of *C. elegans* this value is about 5078 bp. One source of this decrease in coding density is that the majority of *C. elegans* protein-coding genes – with the exception of most histone genes – has introns, and most have multiple introns. Protein-coding genes have an average of five introns; about 27% of the genome resides in predicted exons and about 26% is predicted to be intronic.

Nevertheless, the *C. elegans* genome is significantly more compact than the human genome, which is estimated to have about one gene per 150,000 bp. This compactness of the *C. elegans* genome (relative to vertebrate genomes) is due to the fact that introns and intergenic distances are significantly shorter. *C. elegans* introns tend to be much shorter than those of vertebrates – the average length is only 48 bp. *C. elegans* genes are usually much smaller than homologous vertebrate genes, and they usually contain fewer introns. For example, the nematode $\alpha_1(IV)$ and $\alpha_2(IV)$ collagen genes are about 9 kb long with 11 and 19 introns, whereas the homologous human type $\alpha_2(IV)$, $\alpha_5(IV)$ and $\alpha_6(IV)$ collagen genes are ~100–200 kb long with 46, 50 and 44 introns (Sibley et al., 1993; Oohashi et al., 1995); similarly, the *C. elegans* osteonectin gene spans 3.6 kb and has five introns, whereas the mammalian homolog spans 26 kb and has nine introns (McVey et al., 1988; Schwarzbauer & Spencer, 1993). In *C. elegans*, coding regions are significantly more GC rich than noncoding regions (46% vs. 30%).

An additional source of genome compactness is that about 25% of the genes are contained in polycistronic transcription units (operons) in which only about 100 bp separate the individual genes (Blumenthal & Spieth, 1996). The operons are transcribed from a single promoter upstream of the first gene but are processed into monocistronic mRNAs by 3' end formation and *trans* splicing. Each of the genes in the operon has a 3' end formation signal (AAUAAA) that results in cleavage and polyadenylation at the 3' end of the RNA. In addition, the next gene downstream is processed by *trans* splicing to form its 5' end. Whereas in bacterial operons, the cistrons are all contained on a single mRNA, in the case of the *C. elegans* operons the genes are translated from separate mRNAs formed from a single polycistronic pre-mRNA. Despite this difference, in both cases proteins belonging to an operon are coexpressed and coregulated. Just as in bacteria, the nematode operons provide a means for coexpression of functionally related genes.

A significant fraction of the *C. elegans* genome sequence is repetitive: tandem repeats account for 2.7% of the genome, inverted repeats account for 3.6% of the genome. Repeat families are distributed nonuniformly with respect to genes and they are more likely to be found within introns than between genes. Although only 26% of the genome is intronic, introns contain 51% of the tandem repeats and 45% of the inverted repeats.

Comparative analyses of the proteomes of *C. elegans* and *S. cerevisiae* have suggested that most of the core biological functions are carried out by orthologous proteins that occur in comparable numbers. On the other hand, those processes of signal transduction and regulatory control that are unique to the multicellular worm appear to use proteins missing from yeast, many of which are multidomain proteins with novel domain architectures (Chervitz et al., 1998). The most obvious examples are the multidomain proteins involved in extracellular signalling and adhesion. Some of the domains of these multidomain protreins (e.g. immunoglobulin, FN3, LRR and vWA domains) are present both in *C. elegans* and *S. cerevisiae*, but they occur in multidomain proteins (and fulfil roles in worm signal transduction pathways) that are not found in yeast. In yeast, these domains are found in intracellular proteins, whereas in the worm they have become prominent extracellular adhesion and signalling domains.

A systematic search of the *C. elegans* genome for transcription factors and signalling gene families that regulate development in a variety of animal species has permitted the identification of genes that are shared by all major groups of metazoa (and may have evolved before the Cambrian explosion), as well as genes that have been more recently invented in some metazoan lineages (Ruvkun & Hobert, 1998). Among the best-known transcriptional regulators the most dramatic expansion in *C. elegans* occurred in the nuclear hormone receptor gene family. Nuclear hormone receptors are DNA-binding proteins with ligand-binding domains for small molecules and regulate diverse developmental and physiological processes. Consistent with a rapid and recent expansion of this gene family in *C. elegans*, many nuclear hormone receptor genes map to one chromosome, suggesting that these genes multiplied in recent evolutionary history and have not yet drifted to other genetic regions.

Of the major pathways mediating extracellular signals in metazoa components of the TGF-beta signalling pathways are present in the worm. There are four TGF-beta ligand family members, two type I receptors, one type II receptor and six Smad proteins that transduce TGF-beta signals from the receptors to the nucleus. There are 28 *C. elegans* receptor tyrosine kinase genes, many of which correspond to orthologs of other vertebrate and other invertebrate genes (e.g. orthologs of EGF receptors, insulin/IGF-I receptors, FGF receptors, ephrin receptor tyrosine kinases). Components of the Wnt/wingless and lin-12/Notch pathways are also present in the worm.

The genome sequence of the nematode *Caenorhabditis briggsae* has also been determined recently (Stein et al., 2003). *C. briggsae* is very similar and closely related to *C. elegans*. Although these two nematodes diverged from a common ancestor about 80 million years ago, they have the same chromosome number, roughly the same genome size (about 100 Mb) and a similar number of protein-coding genes (about 19,000). The number and size of introns are also similar in the two species: exons account for 23–27%, introns account for 26–30% of their genomes. The slight difference in size between the genomes of *C. briggsae* (104 Mbp) and *C. elegans* (100.3 Mbp) is due to repetitive sequence, which accounts for 22.4% of the *C. briggsae* genome in contrast to 16.5% of the *C. elegans* genome.

Of the 19,500 protein-coding genes of *C. briggsae* 12,200 have clear *C. elegans* orthologs (a further 6500 have clearly detectable *C. elegans* homologs) and approximately 800 *C. briggsae* genes have no detectable matches in *C. elegans*. Comparison of orthologous pairs of genes has identified 4379 *C. elegans*-specific introns and 2200 *C. briggsae*-specific introns indicating that intron gains or losses have occurred at a rate of at least 0.5 per gene in the 80 million years since *C. elegans* and *C. briggsae* diverged. The two genomes also exhibit extensive colinearity. Operons of *C. elegans* are highly conserved in *C. briggsae*, with the arrangement of genes being preserved in 96% of cases.

Comparison of *C. briggsae*/*C. elegans* ortholog pairs has also permitted the calculation of the rates of nonsynonymous (K_A) and synonymous (K_S) amino acid substitutions. Orthologous genes had an average K_S of 1.78 and a K_A of 0.11 and a K_A/K_S ratio of 0.06, indicating strong purifying selection. The extent of this purifying selection is even more marked in genes with essential functions. For example, ortholog pairs which exhibit an embryonic lethal phenotype in systematic **RNA interference (RNAi)** screens of the *C. elegans* ortholog partner show a lower K_A/K_S ratio than do pairs for which a wild-type phenotype was observed (K_A/K_S of 0.0445 versus 0.0627).

It is interesting to note that although *C. briggsae* and *C. elegans* diverged at roughly the same time (about 80 million years ago) as mouse and human (about 75 million years ago) and show similar levels of amino acid identity between orthologs (80% for *C. briggsae*/*C. elegans*, 78.5% for mouse/human) *C. briggsae* and *C. elegans* diverged more rapidly than mouse and human. First of all, *C. briggsae*/*C. elegans* are evolving more rapidly at the nucleotide level, with a rate of synonymous substitution of 1.78 substitutions per synonymous site versus 0.6 substitutions per synonymous site in mouse/human. Second, there is a dramatic difference in chromosomal rearrangement rate, which is roughly an order of magnitude higher in the nematodes (4837 conserved syntenic blocks of mean size 37 kbp) than in mouse/human (342 syntenic blocks of mean size 6.9 Mbp). Third, whereas intron gains or losses have occurred at a rate of at least 0.5 per gene in the 80 million years since *C. elegans* and *C. briggsae* diverged, in mouse and human there have been fewer than 0.01 losses or gains per gene in a similar period. The most plausible explanation for these observations is that evolutionary rate is better measured in generation times than in years, and the generation times of the two nematodes are orders of magnitude shorter than those of the two mammals.

Arthropodes

The genomes of Drosophila melanogaster and Anopheles gambiae

The genome of *Drosophila melanogaster* consists of two compartments, a heterochromatic gene-poor 60 Mb region, and a euchromatic gene-rich 120 Mb region. The 60-Mb heterochromatic region consists largely of simple sequence satellite DNA sequences, ribosomal genes and transposable elements. Analysis of the nucleotide sequence of the approximately 120-Mb euchromatic portion of the *Drosophila* genome has revealed that the genome encodes approximately 13,600 protein-coding genes, fewer than the smaller *C. elegans* genome, but with comparable functional diversity (Adams et al., 2000).

There are 56,673 predicted exons, an average of four per gene, occupying 24.1 Mb (20%) of the 120-Mb euchromatic sequence. There are at least 41,000 introns, occupying 20 Mb (17%) of genome sequence. (Since gene-prediction programs are unable to predict the noncoding regions of transcripts, the number of exons and introns and the average transcription unit size are underestimates.) Intron sizes in *Drosophila* are heterogeneous, ranging from 40 bp to more than 70 kb, with a clear peak between 59 and 63 bp. In general, *D. melanogaster* genes tend to have shorter and fewer introns than vertebrates. For example, the fly laminin A gene is 14 kb long with 14 introns, whereas the human homolog occupies more than 260 kb and has 63 introns (Kusche-Gullberg et al., 1992; Zhang, Vuolteenanho & Tryggvason, 1996).

The average gene density in *Drosophila* is one gene per 9000 bp, significantly lower than in the worm. There is significant variation in gene density, regions of high gene density correlate with G+C-rich sequences. In the ~1 Mb adjacent to the centric heterochromatin, both G+C content and gene density decrease.

Comparison of the proteome of *D. melanogaster* with those of *C. elegans* and *S. cerevisiae* has revealed that the nonredundant protein sets (the number of distinct protein families) of flies and worms are similar in size and are only twice that of yeast, but different gene families are expanded in each genome (Rubin et al., 2000). In yeast, the number of families in the nonredundant set is 4383 proteins, whereas the core proteomes of the fly and worm consist of 8065 and 9453 proteins, respectively. The architecture of multidomain proteins and signalling pathways of the fly and worm are, however, far more complex than those of yeast. Whereas the fly and worm have 2130 and 2261 multidomain proteins, yeast has only 672. This difference is primarily due to the numerous extracellular or membrane-associated multidomain proteins involved in cell–cell communication, cell–cell and cell–substrate contacts, which are abundant in flies and in worms but which are absent from yeast.

Studies on the 278-Mb genome of *Anopheles gambiae*, the principal vector of malaria, have revealed that although this genome is more than twice as large as that of the euchromatin portion of *Drosophila* (120 Mb), the number of protein-coding genes is very similar: both species have about 14,000 protein-coding genes (Holt et al., 2002; Zdobnov et al., 2002). Since the number of exons and total coding lengths differ by less than 20%, the size difference between the two genomes is due largely to the increased size of intergenic DNA and intronic DNA. As a corollary, the intron/exon ratio is much higher for *Anopheles* (2.2) than for *Drosophila* (1.2). Since most other families of the diptera order have species with genomes at least as large as that of *A. gambiae*, the most likely explanation for the size difference of the two genomes is that *D. melanogaster* has lost noncoding sequences during the 250 million years since it diverged from *A. gambiae*. Comparison of the exon–intron structures of orthologous genes also illustrates how *Drosophila* has experienced a reduction of intronic regions; equivalent introns in *Drosophila* have only half the length of *Anopheles*. Furthermore, only 11,007 out of the 20,161 *Anopheles* introns present in 1:1 orthologous pairs have equivalent positions in *Drosophila*; indicating that intron losses (or gains) have occurred at a rate of about one per gene per 125 million years. This is similar to the intron gain/loss rate in nematodes since *C. elegans/briggsae* divergence, and much higher than in the mouse/human divergence. Thus, intron–exon structure has apparently evolved more rapidly in nematodes and arthropods than in chordates, probably because insects and worms have a significantly shorter generation time than mammals.

Analyses of the proteomes of the two diptera *A. gambiae* and *D. melanogaster* revealed considerable similarities (Zdobnov et al., 2002). Almost half of the genes in both genomes are interpreted as orthologs and show an average sequence identity of about 56%, which is slightly lower than that observed between the orthologs of the pufferfish and human. Considering that *A. gambiae* and *D. melanogaster* diverged about 250 million years ago, whereas puffersfish and human diverged about 450 million years ago, this indicates that the two insects evolved considerably faster than vertebrates, possibly because insects have a substantially shorter life cycle. The relative abundance of the majority of proteins containing InterPro domains was similar between the mosquito and fly, with insect-specific cuticle and chitin-binding peritrophin A domains and the insect-specific olfactory receptors being similarly overrepresented. However, there are several classes of proteins that contain domains that are overrepresented in mosquito compared to fly, and analysis of other organisms suggests that this difference is due to expansion in *Anopheles* rather than loss in *Drosophila*. For exxample, the serine proteases are well represented in both insect genomes, but *Anopheles* has nearly 100 additional members, perhaps reflecting differences in feeding behaviour and the mosquito's intimate interactions with both vertebrates and parasites. Another noteworthy difference between *Anopheles* and *Drosophila* is the presence of a 19-member odorant receptor family in *Anopheles* and its absence in *Drosophila*. The molecular basis for the distinct preference of *Anopheles* for human blood and the ability to find human prey is unknown, but it almost certainly involves recognition of human-specific odors. It is tempting to speculate that this odorant-receptor family may be important in mosquito-specific behaviour that includes host seeking.

Echinoderms

The genome of the sea urchin Strongylocentrotus purpuratus

Echinoderms – together with hemichordates and chordates – belong to the group of **deuterostomes**. A key distinguishing feature of deuterostomes is that during embryonic development the first opening (the blastopore) becomes the anus, whereas in **protostomes** (like nematodes, arthropodes) it becomes the mouth. According to protein molecular phylogeny, the last common ancestor of deuterostomes diverged from protostomes more than 540 million years ago.

Sequencing the 814-Mb genome of the sea urchin *Strongylocentrotus purpuratus* is a landmark event since its comparison with chordate and protostome genomes permits the distinction between protostome-specific, deuterostome-specific and chordate-specific innovations (Sea Urchin Genome Sequencing Consortium, 2006).

The sea urchin genome is predicted to encode about 23,300 genes and analysis of these genes provided an insight into the evolutionary origins of complex systems characteristic of vertebrates. For example, these studies have revealed the presence of homologs for several components of the complement system of vertebrates. In vertebrates collectins, C1q and mannose-binding protein (MBP) initiate the lectin cascade through members of the MBP-associated protease (MASP)/C1r/C1s family. Although several genes encoding collectins, C1q and MBP are present in the sea urchin genome, there was no evidence for members of the MASP/C1r/C1s family. On the other hand, there are four genes encoding homologs of thioester proteins that initiate the alternative complement pathway in vertebrates (complement factors C3, C4, C5) and there are three homologs of a second member of the alternative pathway, complement factor B. Analysis of the sea urchin genome failed to identify proteins with the same domain organization as those of the terminal complement factors (complement factors C6, C7, C8, C9) that lyse pathogens or pathogen-infected cells with membrane attack complexes in vertebrates. These observations suggest that if the complement system in the sea urchin functions through multiple lectin and alternative pathways in the absence of the lytic functions of the terminal pathway, then the major activity of this primordial complement system may be opsonization.

The echinoderm nervous system differs markedly from those of vertebrates and appears to rely primarily on proteins shared by both protostomes and deuterostomes. The multidomain proteins of key neural adhesion/signalling systems involved in regulating axonal outgrowth (netrin/Unc5/DCC, Slit/Robo and semaphorins/plexins) and synaptogenesis (Agrin/MUSK and neurexin/neuroligins) in both protostomes and vertebrates are all present in the sea urchin genome. There appear to be no genes encoding gap junction proteins suggesting that communication among neurons depends on chemical synapses without ionic coupling. Interestingly, the genes encoding a vertebrate-type neurotrophin-Trk receptor system are present in the sea urchin, whereas they are absent in the genome of the urochordate, *Ciona*. This observation suggests a deuterostome origin for these genes and their potential loss in urochordates. Surprisingly, there is a clear sea urchin ortholog of vertebrate reelin, a large extracellular matrix protein involved in the layered organization of neurons in the vertebrate cerebral cortex.

All the basic multidomain protein constituents of the metazoan basement membrane extracellular matrix (alpha-IV collagen, perlecan, laminin subunits, nidogen and collagen XV/XVIII) are present in sea urchin. The majority of multidomain extracellular matrix constituents characteristic of vertebrates (fibronectins, tenascins, von Willebrand factor, vitronectin, vertebrate-type matrix proteoglycans, complex VWA/FN3 containing multidomain collagens), however, are missing in sea urchins, in harmony with the close functional association of these proteins with the evolution of cartilage and bone of the vertebrate-type endoskeleton.

Among the deuterostomes, only echinoderms and vertebrates produce skeletons but the echinoderm skeleton consists of magnesium calcite, whereas the endoskeleton of vertebrates consists of calcium phosphate. Analysis of the *S. purpuratus* genome revealed major differences in the proteins that mediate biomineralization in echinoderms and vertebrates: sea urchins do not have counterparts of extracellular proteins that mediate biomineral deposition in vertebrates and almost all of the proteins that have been directly implicated in the control of biomineralization in sea urchins are specific to this group.

It is noteworthy that the number of genes predicted in the sea urchin genome is similar to those of higher vertebrates, although whole-genome duplications are believed to have occurred in the chordate lineage (see below). A possible explanation for this apparent contradiction is that many of the duplicated genes have been lost in the vertebrate lineage, whereas many gene families significantly expanded in the echinoderm lineage. As examples, we may refer to the long-established sea urchin-specific expansion of histone genes and an unprecedented expansion of gene families encoding innate immune receptor proteins in the sea urchin genome.

Another point worth noting is that the – with about 29 protein-coding genes per Mb – the genome of the sea urchin is significantly less compact than those of protostomes such as *C. elegans* or *D. melanogaster* and in this respect it is more similar to the genomes of vertebrates (cf. Table 9.1).

Urochordates

The genome of Ciona intestinalis

The first chordates appear in the fossil record at the time of the Cambrian explosion, nearly 550 million years ago. The modern ascidian tadpole represents a plausible approximation to these ancestral chordates. Recent studies on the genome of the sea squirt *Ciona intestinalis* seem to confirm the hypothesis that simple chordates had a gene number very similar to invertebrates (Simmen et al., 1998). For a genome of about 162 Mb, *C. intestinalis* was predicted to encode only about 15,500 genes, a number very close to the 14,000–19,000 genes of *C. elegans and D. melanogaster*.

The 162-Mb *C. intestinalis* genome does indeed contain approximately 15,852 protein-coding genes, similar to the number in other invertebrates, but only half that found in vertebrates (Dehal et al., 2002). The overall organization of the protein-coding genes in the *Ciona* genome is intermediate between that of protostomes and vertebrates. *Ciona* genes are generally compact and densely packed. The average gene density is similar to that in *Drosophila* (one gene per 7.5 kb versus one per 9 kb in fly) but far greater than the density in human (one gene per about 150 kb). *Ciona* genes contain an average of 6.8 exons per gene (versus 5 in *Drosophila* and 8.8 in human).

About 60% of predicted *Ciona* genes have a detectable protostome homolog; these are presumably ancient **bilaterian** genes, with core physiological and/or developmental roles common to all animals. A few hundred *Ciona* genes have stronger similarity to fly and/or worm genes than to any genes in the current vertebrate set, suggesting that they are ancient bilaterian genes that were lost in the vertebrate lineage. This group includes genes encoding chitin synthase, phytochelatin synthase (involved in metal detoxification) and haemocyanin (the primary oxgen carrier in protostomes). Nearly one-sixth of *Ciona* genes lacks a clear protostome homolog yet possess a recognizable vertebrate counterpart. These genes either arose in the deuterostome branch or were lost in flies and ncmatodes.

Vertebrate gene families are typically found in simplified form in *Ciona*, suggesting that ascidians contain the basic ancestral complement of genes involved in cell signalling and development: usually a paralogous family in vertebrates is represented by just a single gene in *Ciona*. This point may be illustrated by the fibroblast growth factor (FGF) gene family that encodes secreted proteins that control cell proliferation and differentiation. There are at least 22 members of the FGF family in mammals, but there is only one *Fgf* gene in *Drosophila* and just two in *C. elegans*. The *Ciona* cDNA project has identified six *Fgf* genes in *C. intestinalis* and phylogenetic analysis indicates that two of the genes correspond to vertebrate *Fgf*8/17/18 and *Fgf*11/12/13/14, respectively. Three of the *Ciona Fgf* genes represent orthologs of vertebrate *Fgf*3/7/10/22, *Fgf*4/5/6 and *Fgf*9/16/20, respectively. In general, the number of gene family members in *Ciona* falls somewhere between those in invertebrate and vertebrate species, and in most cases each of the *Ciona* genes clearly corresponds to more than one vertebrate gene.

The ascidian genome has also acquired a number of lineage-specific innovations, including a group of genes engaged in cellulose metabolism that are related to those in bacteria and fungi. Urochordates are also called tunicates because their bodies are enclosed by a 'tunic' that contains fibres of a cellulose-like carbohydrate called tunicin. Cellulose is produced only by plants and bacteria, therefore its presence in ascidians seems to be a curious lineage-specific evolutionary innovation. Cellulose synthesis and degradation are controlled by a variety of enzymes, including cellulose synthases and endoglucanases. The *Ciona* genome contains at least one potential cellulose synthase similar to those of nitrogen-fixing bacteria and several endoglucanases related to endoglucanases of bacterial and fungal symbionts of termites. These observations raise the possibility that the cellulose synthase and endoglucanase genes were acquired by lateral gene transfer.

The *Ciona* gene set can also be used as a point of reference to identify genes in vertebrates that appeared in vertebrate lineage: these are genes found in fishes and humans, but absent in *Ciona*. For example, the fact that multidomain proteins of the vertebrate-type blood coagulation cascade are already present in fishes but missing from *Ciona* suggests that they were formed in early vertebrates (Jiang & Doolittle, 2003).

Vertebrates

The genome of Fugu rubripes

The genome of the pufferfish, *Fugu rubripes*, is only 365 Mb, just four times larger than that of *C. elegans* (Elgar et al., 1996). Even more striking is that it is about 7.5 times smaller than the human genome, although the **organismic complexity** of fish and mammals does not seem to be drastically different. It was shown that a randomly cloned sequence from the *F. rubripes* genome is 7.5 times more likely to be coding than a random human sequence, suggesting that these organisms have about the same amount of coding sequence.

The striking similarity between fish and mammalian genomes is under-lined by the fact that numerous cases are already known where homo-logous genes in *Fugu* and mammalian genomes show conserved synteny or physical linkage relations. *Fugu* genes usually have the same exon–intron organization as their mammalian homologs. Despite this conservation, the size of the *Fugu* introns is usually drastically shorter than in the mammalian homologs, with the majority of introns between 60 and 150 bp. An illus-trative example is the Huntington's disease gene, which is 'only' 22 kb in *Fugu* as compared with the 180 kb human gene (i.e. an eightfold difference in size). Importantly, the exon–intron organization of this gene is identical to that of the human homolog: reduction occurred at the expense of non-coding regions and introns. The gene structure of complement component *C9* and its linkage to *DOC-2* in the pufferfish also illustrates these features of the *Fugu* genome (Yeo et al., 1997). The 11 exons of the *Fugu C9* gene span 2.9 kb of genomic DNA, whereas the 11 exons of human *C9* span 90 kb, representing a 30-fold difference in size (at the expense of introns). The linkage of *C9* to *DOC-2* in both human and *Fugu* genomes demonstrates that the *Fugu C9/DOC-2* locus is a region of conserved synteny.

The very small genome of the pufferfish, together with its similarity to the human genome, provides a simple means of identifying essential elements that have been conserved despite the large evolutionary distance between the two species, and of distinguishing such elements from regions that are not critical (not conserved). The fact that intron sizes of homologous *Fugu* and mammalian genes differ so drastically is convincing evidence for the contention that the majority of intron sequences are 'junk'. In other words, the compactness of the *Fugu* genome (relative to the human genome) is due to the fact that it contains significantly less noncoding DNA. There are no abundant classes of dispersed repeats in *Fugu*, and all forms of repetitive DNA (including telomeric repeats and ribosomal RNAs) constitute less than 10% of the *Fugu* genome.

Analysis of the compete sequence of the genome of *F. rubripes* has con-firmed that in this compact genome repetitive DNA accounts for less than 15% of the sequence, far below the 35 to 45% observed in mammals (Aparicio et al., 2002). The 31,059 predicted protein-coding genes occupy ~108 Mb, i.e. about one-third of the euchromatic 320-Mb genome. Significant varia-tions in gene density occur across the *Fugu* genome, with clustering into gene-rich and gene-poor regions; in this respect it is similar to mammalian genomes. Despite such variation in gene density, there was much lower variation in overall *Fugu* G+C content than in human, confirming earlier observations that compositional heterogeneity is less marked in cold-blooded than in warm-blooded animals.

The *Fugu* genome is compact partly because introns are shorter compared with the human genome: 75% of introns <425 bp in length, whereas in human 75% of introns <2609 bp. The total numbers of introns are roughly the same (161,536 introns in *Fugu* compared with 152,490 introns in human). Both gain and loss of introns in the *Fugu* lineage have been observed. Comparison of 9874 orthologous *Fugu*/human gene pairs has identified 456 cases of concordance between intronless *Fugu* and human genes; however, 327 human orthologs of intronless *Fugu* genes contained multiple introns and 317 *Fugu* orthologs of human intronless genes contained multiple introns.

Conserved linkages between *Fugu* and human genes indicate the preservation of chromosomal segments from the common vertebrate ancestor, but with considerable scrambling of gene order. The number of conserved segments was found to vary with human chromosomal length, suggesting that the retention of conserved segments is driven largely by the probability of rearrangement, which is a function of chromosome length. These observations indicate that the dominant mode of segmental conservation fits a random breakage model.

It is widely believed that large regional or genome duplications have contributed to the structure of vertebrate genomes (see below). It is now well established that most teleosts contain an excess of duplicate genes in comparison with **tetrapods**. Analysis of the *Fugu* genome provided evidence that these excess duplicate genes arose as a result of whole-genome duplication early during the evolution of ray-finned fishes, probably before the origin of teleosts (Christoffels et al., 2004).

Although 75% of predicted human proteins have a strong match to *Fugu*, approximately 25% of the human proteins had no obvious pufferfish homologs. A minor fraction of the missing proteins were found in *D. melanogaster*, *C. elegans* and *S. cerevisiae*, suggesting that these genes have been probably lost from *Fugu*. Analysis of the group of proteins that are absent in *Fugu* (as well as in protostomes), but present in mammals has revealed that it contains many cell surface receptor-ligand system proteins of the immune system, haematopoietic system and energy/metabolism of warm-blooded animals.

The chicken genome

The draft genome sequence of the red jungle fowl, *Gallus gallus*, is approximately 1 Gb and was predicted to contain an estimated 20,000–23,000 genes (International Chicken Genome Sequencing Consortium, 2004; Burt, 2005; Bourque et al., 2005). The number of genes is thus similar to those predicted in the human and mouse genomes, although the size of the chicken genome is about threefold smaller than those of human and mouse genomes. This difference in genome size is reflected in a significant reduction in interspersed repeat content, pseudogenes and segmental duplications within the chicken genome.

A unique aspect of the chicken genome is the large variability in chromosome size: in addition to a pair of sex chromosomes, chickens have 38 pairs of autosomes: five macro-, five intermediate and 28 microchromosomes. Many sequence characteristics, such as %GC content, CpG island density and gene density show clear relationships with chromosome size and recombination rate. The density of genes is highest on the microchromosomes. The sizes of introns and intergenic regions and density of repetitive elements correlate negatively with gene density and are reduced on microchromosomes.

Although birds diverged from mammals more than 300 million years ago, there are long blocks of conserved synteny in chicken–human comparisons. The chicken genome differs from all other vertebrate genomes studied so far inasmuch as no short interspersed nucleotide elements (SINEs) have been active in this genome for the last ~50 million years. Another noteworthy feature of the chicken genome is the paucity of retroposed pseudogenes; this is probably explained by the high specificity of the reverse transcriptase of the major interspersed repeat element in the chicken genome: the CR1 long interspersed nucleotide element.

The close similarity of chicken and mammalian proteomes is underlined by the fact that about 60% of chicken protein-coding genes have a single human ortholog; for the remainder, orthology relationships are more complex. Chicken and human 1:1 ortholog pairs exhibit lower sequence conservation (about 75.3%) than rodent and human 1:1 ortholog pairs (about 88%), consistent with the greater evolutionary distance of birds and mammals. Most of the genes that are conserved between human and chicken are also conserved in fish: 72% (7606) of chicken–human 1:1 orthologs also possess a single ortholog in the *F. rubripes* genome: these genes are likely to represent a conserved core that is present in most vertebrates.

There are also examples of gene innovations or gene losses that appear to be unique to either chicken or mammals. For example, scales, claws and feathers of birds are formed using an avian-specific family of keratins, whereas hair fibre formation in mammals involves a distinct keratin family, which has greatly expanded within the mammalian lineage. Enamel proteins seem to be absent from the chicken although enamel-associated genes are present in fish and mammals; it seems likely that these genes were lost comcomitant with the loss of teeth in the avian lineage. On the other hand, avidins (one of the many antimicrobial proteins of egg albumen) are represented only in the chicken genome but not in sequenced mammalian genomes. The fact that avidin-homologs are present in oviparous vertebrates, in zebrafish, sea urchin and bacterial genomes indicate a loss of this domain family in mammals, probably concordant with internalization of the embryo during mammalian development.

The mouse genome

The genome of the mouse *Mus musculus* is 2.5 Gb, about 14% smaller than the human genome (2.9 Gb); the difference in genome size probably reflects a higher rate of deletion in the mouse lineage. The mouse and human genomes were predicted to contain a similar number (~20,000–30,000) of protein-coding genes and the proportion of mouse genes with a single identifiable ortholog in the human genome is also very high, approximately 80% (Mouse Genome Sequencing Consortium, 2002).

Despite the high degree of similarity of genome size and gene number, in the 75 million years since the divergence of the human and mouse lineages evolution has altered their genome sequences and caused them to diverge by substitution, deletion/insertion and chomosomal rearrangements. The neutral substitution rate has been roughly half a nucleotide substitution per site since the divergence of the species, with about twice as many of these substitutions having occurred in the mouse compared with the human lineage. The higher neutral substitution rate in the mouse lineage may be attributed to the significant difference between generation times in the two lineages.

Comparison of the repeat content of the mouse and human genomes has revealed that mouse has fewer repetitive sequences than human. Approximately 46% of the human genome can be recognized currently as interspersed repeats resulting from insertions of transposable elements, whereas only 37.5% of the mouse genome is recognized as transposon-derived. This difference may be due partly to a higher deletion rate of nonfunctional DNA in the mouse lineage. Despite extensive chromosomal rearrangements, more than 90% of the mouse and human genomes can be classified into corresponding regions of conserved synteny, reflecting segments in which the gene order in the most recent common ancestor has been conserved in both species.

Several local gene family expansions have occurred in the mouse lineage. Most of these involve genes related to reproduction, immunity and olfaction, suggesting extensive lineage-specific innovation in rodents in these physiological systems. It is noteworthy in this respect that proteins implicated in reproduction, host defence and immune response could be shown to be subject to positive selection. The mouse/human comparisons have also revealed significant conservation of the exon–intron structure of orthologous genes. Out of 1506 orthologous pairs of human–mouse genes, 1289 (86%) were found to have identical number of coding exons. This set of 1289 genes with an identical number of coding exons contains 10,061 pairs of orthologous exons plus 124 intronless genes.

Exon length between orthologous exons is highly conserved: 91% of these human–mouse exon pairs have identical exon length. In contrast, only 1% of orthologous introns have identical length. Consistent with the smaller size of the mouse genome, orthologous mouse introns tend to be shorter: the average human intron in this data set is 4661 bp, whereas the average mouse intron is 3888 bp.

In the set of 1506 orthologous human–mouse gene pairs, there are only 22 cases in which the overall coding length is identical between the gene pairs, but they differ in the number of exons. Most of these cases can be explained by a single intron gain/loss, indicating the slow evolution of exon–intron structure. It should be noted that this rate of intron gain/loss in the mammalian lineage is significantly lower than that observed in the *Drosophila/Anopheles* or *C. elegans/briggsae* comparisons.

The human genome

Initial analysis of the draft sequence of the euchromatic portion of the 2.9-Gb human genome has predicted about 30,000–40,000 protein-coding genes (International Human Genome Sequencing Consortium, 2001; Venter et al., 2001) – about twice as many as in worm or fly. Only about 1.1% of the human genome was predicted to encode exons, 24% was predicted to be in introns, with 75% of the genome being intergenic DNA. Gene density in the human genome (about one gene per 100 kb) was thus predicted to be at least one order of magnitude lower than in the worm or fly genomes. More recent analyses of the finished sequence of the human genome, using improved bioinformatic tools has indicated that the human genome may encode only 20,000–25,000 protein-coding genes, significantly less than previous estimates (International Human Genome Sequencing Consortium, 2004). Nevertheless, the following basic conclusions reached through analysis of the draft sequence remain valid.

Human genes are usually significantly larger than those of worm and fly: many human genes are over 100 kb long, the largest known example being the dystrophin gene at 2.4 Mb. The titin gene has the longest known coding sequence (80,780 bp), has the largest number of exons (178) and the longest single exon (17,106 bp).

Since the typical length of a coding sequence is quite similar in animals (1311 bp for worm, 1497 bp for fly and 1340 bp for human) the larger size of human genes is due to the fact that human genes have more and longer introns. The overall average length of introns is 267 bp for worm and 487 bp for fly, whereas it is more than 3300 bp for human. Alternative splicing appears to be prevalent in the intron-rich genome of humans, with >60% of human genes being subject to alternative splicing. It has been estimated that there are perhaps five times as many primary protein products in the human as in the worm or fly, thanks to extensive use of alternative transcripts.

Human gene size and intron size were also found to vary with the GC content of the isochore in which they reside: GC-rich regions tend to be gene dense with many compact, intron-poor genes, whereas genes in low-GC regions are intron rich, separated by large tracts of apparently non-coding sequence.

About half of the human genome derives from transposable elements, SINEs, LINEs, LTR retroposons and DNA transposon copies comprise 13%, 20%, 8% and 3% of the sequence, respectively. LINE1 and Alu account for 60% of all interspersed repeat sequences. Interestingly, some genomic regions are nearly devoid of repeats. The four regions with the lowest density of interspersed repeats in the human genome are the four home-obox gene clusters, HOXA, HOXB, HOXC and HOXD. The absence of repeats from these regions may be a sign of the presence of large-scale *cis*-regulatory elements that cannot tolerate being interrupted by insertions.

Analysis of the human proteome has permitted the identification of genes that are shared with fly, worm and yeast, representing the conserved core of eukaryotic proteins. Comparison of human proteins with those of fly defined 1308 groups of proteins, each containing at least one predicted ortholog in each species (and many containing additional paralogs). Proteins in these 1308 orthologous groups represent a conserved core of proteins that are mostly responsible for the basic 'housekeeping' functions of the eukaryotic cell, including metabolism, DNA replication and repair, translation etc.

The human proteome also differs markedly from those of the fly, worm and yeast even though very few, if any, of the protein domains are vertebrate inventions. For example, although domains such as FIMAC, FN1 and FN2 are present only in the vertebrate linaeage, the precursors of these domains (follistatin, vWC and kringle domains, respectively) are present in the fly and worm. For example, Ullman and Perkins (1997) have shown that the Factor I/membrane-attack complex (FIMAC) domain identified in various complement proteins of vertebrate immune defence is a distant member of the larger follistatin (FS) superfamily that is represented in invertebrate genomes. Similarly, typical members of the fibronectin type I (FN1) domain family have been found only in the chordate lineage (in proteins like fibronectin, tissue-plasminogen activator, blood coagulation factor XII) but recent structure and sequence comparisons have revealed an evolutionary relationship between the N-terminal subdomain of the von Willebrand factor type C (VWC) modules and the fibronectin type 1 domain (O'Leary et al., 2004). These observations suggest that the FN1 domain evolved in the chordate lineage from a vWC domain (that is present in numerous invertebrate proteins). Finally, the fibronectin type II (FN2) domains appear to be restricted to chordate proteins (such as fibronectin, gelatinases) but the FN2 domain has been shown to be related to the kringle domain (Patthy et al., 1984) that is also widespread among invertebrate proteins.

Some protein families (also present in worm and fly) have significantly expanded in the vertebrate lineage. These gene duplications played a major role in vertebrate evolution as refelected by the fact that many families that are expanded in human relative to fly and worm are involved in distinctive aspects of vertebrate physiology. For example, in vertebrates the immunoglobulin superfamily repertoire has been expanded to fulfil immune functions such as those of antibodies, major histocompatibility proteins, antibody receptors and lymphocyte cell-surface proteins. The fibroblast growth factor, transforming growth factor-beta, wnt and ephrin families involved in the control of development also illustrate vertebrate-specific expansion of gene families and their role in vertebrate evolution. The human genome contains 30 fibroblast growth factor genes as opposed to only two in both fly and worm. There are 42 transforming growth factor-beta genes in the human genome compared with nine and six in the fly and worm, respectively. There are at least eight human ephrin genes (two in the fly, four in the worm) and 12 ephrin receptors (two in the fly, one in the worm). In the wnt-signalling pathway there are 18 wnt family genes (six in the fly, five in the worm) and 12 frizzled receptors (six in the fly, five in the worm). Thus the more sophisticated developmental control mechanisms of vertebrates have evolved from related but simpler systems in invertebrates.

Although very few (if any) protein domain families are restricted to the vertebrate lineage, there are numerous vertebrate-specific multidomain proteins with complex novel architectures, novel combinations of 'old' domains. As discussed in Chapter 8, the majority of these novel architectures were created by exon shuffling. A systematic comparison of distinct protein architecture types found in yeast, worm, fly and human has shown that the human proteome set contained 1.8 times as many protein architecture types as worm or fly and 5.8 times as many as yeast. This is most obvious in the recent evolution of novel extracellular and transmembrane architectures in the human lineage (cf. Chapter 8). Another measure of proteome complexity can be obtained by counting the number of different domain types with which a given domain type co-occurs in multidomain proteins. This measure of protein complexity also illustrates that the human proteome is more complex than those of worm, fly and yeast. For example, the trypsin-like serine protease domain co-occurs with 18 domain types in human (including proteins involved in the mammalian complement system, blood coagulation and fibrinolytic and related systems), whereas it co-occurs with only eight other domains in fly, five in worm and only one in yeast. As discussed in Chapter 8, the creation of novel types of multidomain proteins by exon shuffling has played a major role in vertebrate evolution since the majority of proteins unique for vertebrate biology (proteins of the adaptive immune system, haemostasis, endoskeleton, neuronal functions) was created by this mechanism.

9.6.2.5 Genome duplications in the evolution of early vertebrates

Comparisons of the most fundamental developmental regulatory proteins in various animal phyla have consistently shown that in protostomes and primitive chordates these are encoded by single genes, whereas their vertebrate homologs come in gene families (Sidow, 1996). For example, there is usually a single developmental regulator gene in *Drosophila* but there are 2–4 homologs in vertebrates (e.g. two homologs of *Wnt 5*, *decapentaplegic*, *Eve*; three homologs of *Msx* and *Hedgehog*; four homologs of the *Hox* clusters,

Cdx, MyoD, 60A, Notch, elav, btd/SP). In principle, this phenomenon could be the result of either repeated individual gene duplications or duplications of entire genomes.

As mentioned in the introductory part of this chapter, the fact that certain closely related 'sibling' species of fish and amphibians differ by a factor of two in their chromosome content proves that genome duplications are permissible. Evidence that genome duplications have indeed occurred early on in a common ancestor of chordates comes from comparisons of the DNA content of various **deuterostomes** (Sidow, 1996). These studies have identified the transition from simple chordates to vertebrates as the period in which many gene families expanded by large-scale duplications, giving rise to a significant increase in the number of vertebrate genes.

Since vertebrates were thought to have approximately four times as many genes as *C. elegans* or *D. melanogaster* this has led to the hypothesis that two rounds of genome duplications occurred in the lineage leading to vertebrates. According to this 'two rounds of whole-genome duplication' hypothesis (2R hypothesis) it is to be expected that there are four paralogs in vertebrates for proteins encoded by a single gene in their common 'invertebrate' ancestor. Gene families with only two or three vertebrate paralogs would be those that lost one or two copies after the first or second genome duplications. Several modifications of this 'two tetraploidies' hypothesis have been proposed and there is wide acceptance of the basic scheme (Furlong & Holland, 2002; Blomme et al., 2006).

The case of the *Notch* gene family may serve to illustrate current evidence for two rounds of genome duplications (Sugaya et al., 1997). There is a single *Notch* gene in *Drosophila* but there are four subfamilies of *Notch* in the human genome: *Notch 1, Notch 2, Notch 3* and *Notch 4*, being present on chromosomes 9, 1, 19 and 6, respectively. The fact that 10 genes mapped on 6p21.3 (including *NOTCH4, tenascin TNX*) were found to have counterparts mapped on 9q33–q34 (including *NOTCH1, tenascin C HXB*), indicates chromosome duplication during the course of evolution (Sugaya et al., 1994; Kasahara et al., 1996). Counterparts for a number of other genes have been found on human chromosomes 1, 6, 9 and 19, arguing for extensive homology of parts of these four chromosomes (Katsanis, Fitzgibbon & Fisher, 1996; Sugaya et al., 1997). In a more systematic analysis Popovici et al. (2001) have identified several paralogous regions (paralogons) in mammals and assembled a tentative 'human genome paralogy map' that contains a total of 14 paralogy groups that may be remnants of ancient *en bloc* duplication events. Molecular clock analysis of all protein families in humans that have orthologs in the fly and nematode indicated that a burst of gene duplication activity took place in the period 350–650 million years ago and that many of the duplicate genes formed at this time are located within paralogons, supporting the contention that many of the gene families in vertebrates were formed or expanded by large-scale DNA duplications in an early chordate (McLysaght, Hokamp & Wolfe, 2002).

Recent studies on the genome of the sea squirt *C. intestinalis* seem to confirm the hypothesis that simple chordates had a gene number very similar to invertebrates (Simmen et al., 1998; Dehal et al., 2002). Thus the first genome duplication in early vertebrates probably occurred only after their divergence from tunicates.

The time of these genome duplications may be narrowed down further by two pieces of evidence. First, bony fishes contain about the same number of genes as birds and mammals, with orthologous genes being identifiable by sequence similarity and conserved synteny. This observation indicates that no mainstream genome duplications occurred since the separation of fish and tetrapods. Second, studies on the genes of several gene families (e.g. *Igf, Msx, Wnt, Hox* families) demonstrated that the first genome duplication occurred in a common ancestor of all jawed and jawless vertebrates after the lineage leading to **Cephalochordates** (*Amphioxus*) diverged. It was suggested that a second genome duplication occurred in the common ancestor of jawed vertebrates after they diverged from jawless vertebrates (lampreys) (Sidow, 1996).

The cephalochordate amphioxus, the closest living invertebrate relative of vertebrates, is thus a crucial organism for resolving the question of genome duplications at the origin of vertebrates. The cloning of a number of amphioxus genes in the past has helped form the idea that amplification of gene numbers in vertebrates has occurred on the vertebrate lineage after its divergence from amphioxus (Holland, 1999). Comparison of the major histocompatibility complex (MHC) paralogous regions of various vertebrates with the corresponding region in amphioxus has revealed that duplications occurred after the divergence of cephalochordates and vertebrates but before the radiation of jawed vertebrates. The distribution of human and amphioxus orthologs in their respective genomes also supports the *en bloc* duplication events (Abi-Rached et al., 2002). Recent systematic analyses using completed genomes and a large gene set from the amphioxus *Branchiostoma floridae* has provided evidence that the gene duplication activity was significantly higher after the separation of amphioxus and the vertebrate lineages, the results favouring at least one large duplication event at the origin of vertebrates, followed by smaller scale duplication closer to the bird–mammalian split (Panopoulou et al., 2003).

In summary, since genome analyses seem to support the hypothesis that vertebrate genomes arose by one or two rounds of genome duplication, the ancestral chordate gene number could be remarkably similar to that in flies, worms and sea squirts, i.e. about 15,000 genes. This would also imply that the number of core biochemical pathways and mechanisms was very similar in flies, nematode worms and early chordates (Ohno, 1996).

An important aspect of this genome duplication hypothesis is that it assumes that the vast majority of the duplicated genes thus created have escaped conversion to pseudogenes, although this is thought to be the most probable fate in the case of duplicons produced by individual tandem duplication or retroposition (see section 7.2). The genome duplication hypothesis thus implies that most of the newly duplicated genes were retained despite their massive redundancy, suggesting that most redundant genes were somehow protected from nonfunctionalization. To explain this unusual behaviour we must assume that the fate of individual duplicated genes differs significantly from the fate of genes duplicated as a result of entire genome duplications.

It should be noted that one major difference is that in the case of individual duplications the products of both duplicons have to interact with exactly the same (unduplicated) partners and this may decrease their chances of acquiring a new function and thus decrease their chances of survival. Conversely, in the case of genome duplications, simultaneous duplication of all partners occurs, and the chances of acquiring a unique function are significantly increased, hence entire duplicated partnerships may diverge (coevolve) together.

Let us consider, for example, a hormone/receptor system where neither protein has a meaningful function without binding to the other. In this case, duplication of only the hormone gene creates a situation in which the products of the two hormone genes would compete for the same receptor. The chances of acquisition of a new function (and thus the chance of survival) would be increased if the two genes no longer competed, for example because they are expressed in a different tissue or at a different developmental stage as a result of a regulatory mutation. (Such regulatory changes may be due to a change in the DNA sequence of a regulatory element or due to the generation of a novel element or due to a change in the genomic environment of the genes.) If the hormone receptor gene is also duplicated, the resulting paralogous pairs of hormone/receptor genes have an increased probability of acquiring a new function by regulatory mutations or other types of mutations in which the interacting partners coevolve (Fryxell, 1996).

In general, the acquisition of a novel function by a duplicated gene could be facilitated by pre-existing heterogeneity in proteins that interact directly with the product of the duplicated gene. This point may also be illustrated by the accelerated coevolution of serine proteinases and serine proteinase inhibitors, and the accelerated coevolution of snake venom phospholipases and phospholipase inhibitors (see Chapter 7). The efficiency of such scenarios is underlined by the numerous ligand/receptor pairs that evolved by this strategy (e.g. pairs of paralogs of insulin/insulin receptors, cytokines/cytokine receptors, neurotrophins/neurotrophin receptors TrkA, TrkB, TrkC, etc.). The interacting genes could be duplicated sequentially (as in individual duplications in proteinases and their inhibitors) or simultaneously (by polyploidy). The benefits of simultaneous duplication of receptors and ligands may be best illustrated by insulin and insulin-like growth factor (IGF) and their cognate receptors. Insulin acts by binding to a specific insulin receptor that consists of an extracellular insulin-binding domain and a cytoplasmic tyrosine kinase domain that mediates the action of the hormone. The closely related insulin-like growth factor I is involved in regulating the overall growth of the mammalian embryo and neonate and acts through the IGF-I receptor, which is closely related to the insulin receptor. The insulin receptor and the IGF-I receptors arose by duplication of a common ancestral receptor gene. Since both the ligand pair and the receptor pair arose by gene duplication, it seems that the entire insulin pathway, which regulates energy metabolism, was duplicated to create a second pathway, the IGF-I pathway, which regulates embryonic growth. The facts that (i) the insulin and IGF-I genes are located within paralogous regions of human chromosomes 11 and 12, and that (ii) the insulin receptor and IGF-I receptor genes are located within paralogous regions of human chromosomes 19 and 15 suggest that these two pairs of genes might have diverged simultaneously, following a single polyploidization event.

A strikingly similar situation exists in the case of neurotrophins/Trk neurotrophin receptors. Nerve growth factor (NGF) belongs to a family of related neurotrophic polypeptide hormones, including brain-derived neurotrophic factor (BDNF), neurotrophin-3 (NT-3) and neurotrophin-4 (NT-4). These neurotrophins have distinct but related functions in the development of the nervous system: different neurotrophins bind to neurotrophin receptor tyrosine kinases TrkA, TrkB and TrkC, which are encoded by a receptor gene family derived from a common ancestor gene. It thus seems that gene duplication events in the neurotrophic factor gene family have been correlated with gene duplication events in the corresponding receptor gene family.

The cytokine and cytokine receptor gene families may indicate that polyploidization is not the sole source of such correlation in the evolution of ligands/receptors. For example, the gene for the neutrophil chemoattractant interleukin-8 (IL-8) is located on human chromosome 4, in a cluster of four or five homologous neutrophil chemoattractant genes apparently formed by tandem gene duplication. The chemoattractants encoded by these genes bind to a single family of G-protein-coupled receptors, which are encoded in another gene cluster on human chromosome 2. Thus, the evolutionary expansion of the IL-8 gene family was correlated with specific and repeated recruitment of duplicated members of the IL-8R gene family, to bind to the new ligands.

The lesson that we learn from such ligand/receptor coevolution is that the evolutionary expansion of a gene family is most useful if it is accompanied by the complementary diversification of its partner genes, permitting the duplication and divergence of entire 'pathways'. The widespread occurrence of correlations between functionally related gene families suggests that the coevolution of gene families is a rather general phenomenon (Fryxell, 1996). In other words, despite the initial redundancy of the two pairs of newly duplicated daughter genes, neither will be eliminated if each acquires a unique function (e.g. by regulatory mutation). It should be pointed out that the chances of acquiring a unique function for a duplicated gene increase with the multiplicity of the interactions of its protein product. Significantly, in the case of whole-genome duplications all partners of gene products are duplicated simultaneously, therefore the chances of the acquisition of new functions are dramatically increased over single-gene duplications.

In summary, the chances of the duplicated gene surviving are dramatically increased if the probability of forming a novel regulatory interaction with other genes is high. This probability is higher when a large number of other newly duplicated genes are available to form such novel interactions, and is highest when all genes of the genome are available.

It must be emphasized that regulatory mutations are the most efficient in bringing about sudden changes in the function of a duplicated gene; for example, a mutation in the promoter or enhancer region can lead to drastic changes in tissue-specific or developmental regulation. The probability of gaining a novel enhancer may increase with the amount of noncoding DNA that surrounds the duplicated gene, therefore the chances of *de novo* creation of a novel enhancer, and thus the potential for an advantageous regulatory mutation, is higher in genomes with large introns and much noncoding intergenic DNA (Sidow, 1996). It is possible that the common ancestor of vertebrates was not a compact genome organism like *C. elegans*, but one with a much higher ratio of intergenic to genic DNA. (It should be noted, however, that the genomes of *C. intestinalis* and other extant species belonging to the subphylum Urochordata are only $1.6–1.8 \times 10^8$ bp, similar to those of flies and worms.) According to this hypothesis, the genome duplications of the ancestral chordate could have favoured regulatory mutations.

9.6.3 Value of comparative genomics for the identification of functional elements

The use of comparative genomics to identify functional elements in the genome is based on the principle that regions that fulfil some essential function (protein-coding genes, ncRNA genes and various kinds of gene regulatory elements) are subject to purifying selection: they are therefore more conserved than nonfunctional regions (Boffelli, Nobrega & Rubin, 2004).

The evolutionary distance between the species that are chosen for genome-sequence comparison, however, determines what kind of functional elements can be identified by this approach. If the species compared are distantly related (e.g. human–fish comparisons) only the most conserved functional elements (shared by fish and human) can be identified. On the other hand, if we compare species that are too closely related (e.g. human–chimp comparisons) the high degree of similarity even between nonfunctional orthologous sequences will obscure the functional elements within them. Accordingly, comparison of species that are separated by moderate evolutionary times (e.g. human–mouse comparison) could be most informative. This point may be illustrated by the fact that comparative analyses of the human and mouse genomes allowed much better annotation of both these genomes than would have been possible if only one of these sequences were available. Similarly, by comparing the genomes of four related *Saccharomyces* species, Kellis et al. were able to identify most of the transcription-factor binding sites in the yeast genome and had to revise the total count of yeast genes (Kellis et al., 2003). It appears that comparison of multiple species that are separated by moderate distances is an even more efficient approach for the identification of putative functional elements than pairwise comparisons alone (Thomas et al., 2003).

Nevertheless, comparison of distantly related species (e.g. fish–mouse, fish–human comparisons) proved to be extremely valuable for the identification of crucial functional elements that are conserved over such great evolutionary distances (Aparicio et al., 1995). This point may be illustrated by the case of the *DACH* gene, a gene involved in embryonic development that has a complex expression pattern. In mammals, two large gene deserts surround the *DACH* gene. Human–mouse (moderate distance) comparisons of this interval revealed, in addition to conserved exons, more than 1000 conserved noncoding sequences with >100 bp and 70% identity; such a large number of candidate elements made it impractical to test individual conserved sequences for biological activity. A comparison between human sequence and that of several distant vertebrates (including a frog and three fish species), however, reduced the number of conserved noncoding sequences to 32. *In vivo* mouse transgenic assays of nine of these elements showed that seven of them were enhancers of the endogenous expression of *DACH* (Nobrega et al., 2003).

9.6.4 Finding protein-coding genes in genome sequences

Finding protein-coding genes in genomic sequences and correct identification of their promoter, transcription initiation and termination sites, correct prediction of their exons and introns are some of the most challenging tasks of computational biology in the era of genomics (Brent & Guigo, 2004).

There are three major types of approaches to solve these problems: (i) extrinsic approaches that use expression data (e.g. sequences from cDNAs, ESTs, proteins); (ii) intrinsic (*ab initio* or *de novo*) approaches that use single genomic sequences; and (iii) comparative genomics approaches that use genome sequences from two or more related species.

In the case of the so-called **extrinsic gene-finding approaches**, the target genome is searched for the presence of sequences that are identical with or similar to the known messenger RNA or protein sequences from the same or related species (i.e. identification is based on extrinsic information). For example, if a full-length mRNA sequence is known from the same or related species, it is rather trivial to identify the unique genomic DNA sequences (from transcription iniation site to polyadenylation site) from which it has been transcribed; the full-length mRNA also defines the exon–intron structure of the gene that is relevant for the given mRNA sequence. It should be pointed out, however, that knowledge of an mRNA sequence does not give information on the regulatory elements upstream of the transcription iniation site, nor does it define alternative exon–intron stuctures (used in alternative splicing) of the gene. If only a protein sequence is known, this provides straightforward information only on the coding regions of the gene (from translation initiation to termination site) and defines the exon–intron structure in this region, but it does not provide information on the 5′ and 3′ untranslated regions of the gene. An additional practical limitation of these approaches is that only a fraction of mRNA/protein sequences deposited in databases are full length. Nevertheless, since the amount of such extrinsic information is growing rapidly, these protocols are increasingly valuable for the prediction of the structure of protein-coding genes (Curwen et al., 2004).

In view of the fact that extrinsic information (mRNA, protein, EST) may be lacking (or fragmented) for many genes, it is still important to use **intrinsic gene-prediction approaches** in which the genome sequence is systematically searched for various characteristics of protein-coding genes. These characteristics may be of two major types. The first group includes specific **sequence signals** that indicate the presence of a gene, the position of its promoter region, transcription start and termination sites, positions of exon/intron boundaries etc. The second group comprises **content-type information**: characteristic statistical properties of protein-coding genes.

In the case of protein-coding genes of prokaryotes the promoter sequence signals (Pribnow box, transcription factor binding sites) are quite well defined and can be easily identified. Furthermore, since protein-coding genes of prokaryotes usually lack introns, the genes encode a long, continuous open reading frame. Since stop codons (three out of the 64 triplets) are expected to occur frequently in noncoding nucleic acid sequences but not in translated regions, the statistically significant depletion of stop codons from a genomic region may indicate the presence of a protein-coding gene. Furthermore, codon-usage bias is characteristic of protein-coding genes and this feature may be used to distinguish them from noncoding (intergenic etc.) regions. All these characteristics (and the fact that gene density is very high in prokaryotes) make prokaryotic gene prediction relatively simple and reliable. This point may be illustrated by the fact that the number of protein-coding genes predicted in *M. genitalium* was 470 (Fraser et al., 1995) and subsequent studies modified this number only slightly, to 482 (Glass et al., 2006).

Ab initio gene prediction in eukaryotes is much more difficult than in prokaryotes and their reliability decreases as we move from unicellular eukaryotes to higher eukarytotes. First, unlike prokaryotes where gene density is very high, in eukaryotes protein-coding genes may account only for a few percent of the genome sequence. Accordingly, finding protein-coding genes in genomes of higher eukaryotic genomes is like finding needles in a haystack. Second, eukaryotic protein-coding genes may contain numerous large introns that separate the exons that encode the protein. Since the protein-coding sequence may be split into very small fragments by the introns, some of the statistical tools useful in the case of prokaryotes (avoidance of stop codons, codon usage bias, base composition bias etc.) are less reliable for intron-rich genes of eukaryotes. *Ab initio* gene-prediction protocols of eukaryotic protein-coding genes usually attempt to identify CpG islands and binding sites for a poly(A)-tail to define the 5′ and 3′ boundaries of candidate genes and use sequence signals and content-type informations to define their exon–intron structure.

With the increase in the number of completely sequenced genomes **comparative genomics approaches** become increasingly useful in gene prediction. These approaches are based on the principle that protein-coding genes evolve at a slower rate than nonfunctional genomic regions since they are subject to purifying selection. Accordingly, genes can be detected on the basis of the conservation of their constituent exons, promoter regions etc. by comparing the genomes of related species. It should be noted, however, that conservation of a genomic region is only a reflection of its functional importance but this does not necessarily mean that the conserved region corresponds to a protein-coding gene (see above). Analysis of codon usage and pattern of mutations (e.g. the ratio of nonsynonymous over synonymous substitutions) can discriminate between protein-coding and noncoding conservation.

A significant improvement in gene finding may be achieved by combining informations from cDNAs, ESTs, proteins with information obtained by *ab initio* and comparative genomics approaches (Wheeler et al., 2004). Despite all these improvements, computational gene finders are likely to predict only the coding fraction of a single spliced form of nonoverlapping, canonical protein-coding genes, and they perform poorly with untranslated regions, alternative spliced forms etc. The limitations of (and improvements in) current eukaryotic gene-finding protocols can be illustrated by the fact that within a few years the estimated number of genes in the human genome has changed rapidly from about 120,000 genes down to about 20,000–25,000 (Liang et al., 2000; International Human Genome Sequencing Consortium, 2004).

A recent study has systematically compared the performance of various computational methods to predict human protein-coding genes (Guigo et al., 2006). The methods compared were classified as those using any type of available information (category 1: including methods such as FGENESH++, JIGSAW); single-genome *ab initio* methods (category 2: including methods like GENESCAN); EST-, mRNA-, and protein-based methods (category 3, including methods such as ENSEMBL); dual- or multiple-genome-based methods (category 4, including methods such NSCAN, TWINSCAN). These studies have revealed that none of these strategies produced perfect predictions, but the prediction methods that relied on mRNA and protein sequences (category 3) and those that used combined informations (including expressed sequence information, category 1) were generally the most accurate. It is noteworthy that – at the transcript level – no prediction method correctly identified greater than 45% of the coding transcripts exactly and relatively few prediction methods were able to predict multiple transcripts per gene locus. In summary, the correct prediction of the structure of the transcripts of human genes is still a challenging task.

9.7 The genome of the cenancestor

The availability of numerous complete genome sequences of cellular life forms from all three domains of life (Bacteria, Archaea and Eukarya) also creates the opportunity for the reconstruction of the genome of the common ancestor (cenancestor) of all extant cellular organisms. As mentioned above, from a comparison of bacterial genomes a minimal gene set of ~256 genes was derived that encode the basic functions needed for a single bacterial cell's survival. Of these 256 genes, only 143 were shown to have yeast (eukaryote) orthologs. The explanation for this is that eukaryotes appear to be chimeric inasmuch as only their metabolic genes are derived from bacteria whereas their informational genes (involved in DNA replication etc.) seem to be derived from archaea (Ribeiro & Golding, 1998; Rivera et al., 1998).

The fact that the translation apparatus is basically universal indicates that the most recent common ancestor of all cellular organisms had a fully developed (conserved) translation system. Conversely, the fact that components of the DNA replication machinery show little similarity in the three domains of life suggests that the DNA replication machinery was far from perfected at the cenancestral stage. Nevertheless, the similarities between domains seem to be sufficient to suggest that the cenancestor used DNA.

Many fundamental metabolic reactions (key pathways of amino acid, nucleotide and coenzyme biosyntheses and carbon fixation) are carried out by remarkably similar proteins in Archaea and eubacteria indicating that they shared an autotrophic ancestor. This is in sharp contrast with earlier notions that the cenancestor may have been a complete heterotroph, lacking all intermediary metabolisms. Nevertheless, not all of the central metabolism seems to be universal; for example, methanogenesis seems to be unique for Archaea, chlorophyl-based photosynthesis for the eubacteria.

9.8 Changes in gene number and gene density in different evolutionary lineages

Comparison of model genomes shows that the number of genes increases with genome size (see Table 9.1). In multicellular eukaryotes the increase in gene number and genome size (relative to the single-celled eukaryotes) seems to be paralleled by a decrease in gene density. It seems likely that the increase in the average amount of DNA occupied by a genetic unit (from ~2 kb in yeast, to ~6 kb in plants, to ~14 kb in *Drosophila*, to ~150 kb in mammals) reflects an increased requirement for regulatory functions of intergenic regions in the case of multicellular organisms.

The increase in gene number is primarily due to gene (genome) duplications. As a result of this, the majority of genes exist as multigene families, some of which are very populous. For example, there are about 2000 protein kinases and about 1000 protein phosphatases in mammals, whereas their estimated number in *C. elegans* is 350 and 80, respectively. In multicellular animals, different members of a gene family are usually not exact functional duplicates: usually neither their function nor their expression patterns overlap. Their products are unable to compensate for each other if either gene is mutated, thus there may be no functional redundancy even if the genes encode identical products (if they are subject to different regulation).

Nevertheless, the degree of redundancy may be illustrated by the fact that in *C. elegans* and *D. melanogaster* only about one-third of the genes are essential for viability. There are ~3600 lethal loci in a *Drosophila* genome of 13,600 genes (26%), and the estimates for *Caenorhabditis* range from 2900 to 3500 lethal loci in a genome of ~19,000 genes (15–18%). For comparison, in *Arabidopsis* there are ~500 lethal genes in a genome that contains about 26,000 genes (2%), whereas in *S. cerevisiae*, ~900 genes out of 5800 are lethal (16%). Gene targeting in mammals has revealed many surprising cases of functional gene 'redundancy': knockout of genes that were assumed to fulfil critical roles resulted in either no phenotype or a mild phenotype only (Melton, 1994). One of the most extreme examples is the gene of the extracellular matrix protein tenascin; tenascin-deficient mice developed normally without detectable abnormalities (Saga et al., 1992). It seems that many genes could be redundant because related genes are capable of taking over.

The functional overlap between duplicated pathways may also be illustrated by the neurotrophins/Trk neurotrophin receptors (Davies, 1994). Different members of the neurotrophin family – nerve growth factor (NGF), brain-derived neurotrophic factor (BDNF), neurotrophin-3 (NT-3) and neurotrophin-4 (NT-4) – promote the survival of a characteristic set of developing neurons, although some neurons can be supported by more than one neurotrophin. These neurotrophins bind to different members of the trk family of receptor tyrosine kinases with well-defined binding preferences: TrkA is the preferred receptor for NGF; TrkB is the preferred receptor for BDNF and NT-4; and TrkC is the receptor for NT-3. Despite these preferences, there is cross talk between different receptors and neurotrophins (e.g. NT-3 interacts with both TrkA and TrkB). Such a functional overlap may explain why *trkB* or *trkC* knockout mice do not show any gross defects in the central nervous system (CNS), although TrkB and TrkC mRNA transcripts are widely expressed in the CNS. The fact that expression of these two genes overlaps in many regions means that they could be involved in functionally redundant pathways.

9.9 Proteome evolution

9.9.1 Proteome evolution – classification of proteins by structural features

Having complete genomes from all three domains of life, with the knowledge of their complete set of proteins (**proteomes**), it is possible to study the structural features of entire proteomes. Exhaustive structural classification of the predicted open reading frames of five complete genomes (*S. cerevisiae*, *M. jannaschii*, *H. influenzae*, *M. genitalium* and *M. pneumoniae*) were carried out using a combination of sequence comparison and prediction techniques (Frishman & Mewes, 1997). Based on prediction of α-helices, β-strands, transmembrane regions and low-complexity segments, all predicted and known secondary structures of globular proteins were subdivided into four structural types: all-α, all-β, α–β and irregular. The predicted contents of each class were found to be strikingly similar throughout these model genomes, with 3–4% all-β, 10–16% all-α, 41–55% α–β and 1–4% irregular structures. (This distribution of proteins over folding types differs significantly from that observed in known protein structures deposited in the PDB database: in PDB the structural composition is 51% α–β, 28% all-β and 21% all-α. This difference is primarily due to the fact that PDB is strongly biased in favour of intensely studied proteins.) The number of proteins with at least one predicted hydrophobic region is between 30 and 37% (some of which may be signal peptides of secreted proteins). Proteins with at least two hydrophobic regions constitute 14–18% of *S. cerevisiae*, *H. influenzae*, *M. jannaschii*, *M. genitalium* and *M. pneumoniae*. In other words these simple organisms from the three domains of life were found to have very similar proportions of membrane proteins, despite their significant differences in physiology and the number of intracellular compartments.

The similarity of the proportion of the basic structural types of proteins in Bacteria, Archaea and Eukarya probably reflects the fact that the minimal set of genes of the cenancestor was expanded by gene duplications, rather than by *de novo* creation of novel types of folds. Some estimates (Brenner, Chothia & Hubbard, 1997) indicate that the majority of proteins belong to less than 1000 types of protein folds (see section 2.7.1). In the case of the proteome of *M. genitalium*, of the 468 proteins 22% could be assigned a known protein fold, 18% belong to membrane proteins, but the remainder could not be assigned (Fischer & Eisenberg, 1997).

A structural census of known protein sequences (Gerstein & Levitt, 1997; Gerstein, 1997) has shown that the known protein folds are not equally represented in different groups of organisms, with Eukarya and Bacteria sharing only 156 of the 275 folds, and Archaea, Bacteria and Eukarya sharing only 148 folds. It is noteworthy that the shared folds (which must have been present in the cenancestor) include the versatile TIM barrel. The most common folds in metazoa (such as the immunoglobulin fold, and the globin, zinc finger and serine proteinase-folds) are rare in bacteria. Conversely, in bacteria and in plants protein folds of metabolic enzymes are the most common. Within the eukaryotic lineage the percentage of small protein folds (such as kringle domains, fibronectin domains, etc.) is much larger in complex multicellular animals than in simple eukaryotes (protists, fungi). Moreover, the frequency of such small protein folds increases as one moves from simple metazoa to more complex organisms. These studies thus provide statistical support for our contention (Patthy, 1994) that the spread and proliferation of small protein folds – modules – has become increasingly significant in the metazoan lineage and may have contributed to increased organismic complexity.

9.9.2 Proteome evolution – classification of proteins by homology

Systematic comparison of 17,967 proteins encoded in seven complete genomes (*E. coli, H. influenzae, M. genitalium, M. pneumoniae, Synechocystis* sp., *M. jannaschii, S. cerevisiae*) has led to the delineation of 720 **clusters of orthologous groups** (COGs). A number of these COGs are related to each other: there are 58 superfamilies containing 280 COGs (Tatusov, Koonin & Lipman, 1997). In this natural classification of genes each member of a superfamily has evolved from a single ancestral gene through a series of speciation and duplication events. The collection of COGs included 10,332 distinct gene products, i.e. 58% of the total number of genes in these genomes. The mean number of proteins per COG increases with increasing number of genes in a genome, from 1.2 for *M. genitalium* to 2.9 for yeast, illustrating the fact that gene duplication is the basic mechanism for increasing gene number. The fraction of proteins assigned to a family was greatest in the 'minimal' genomes of parasitic mycoplasmas (70% for *M. genitalium*) and much lower in the larger genomes of *E. coli* (40%) and yeast (26%). This is due to the fact that the most critical and most conserved (best studied) proteins (with basic housekeeping functions) constitute the largest proportion in minimal genomes. Analysis of the lineage distribution of COGs identified 323 COGs that are represented in all three domains: Bacteria, Archaea and Eukarya. Only 114 of these are present in all seven genomes, and most of these are components of the translation and transcription machinery, indicating that these systems were fully established (and conserved) in the cenancestor.

9.9.3 Proteome evolution – classification of proteins by function

The systematic functional classification of proteins usually follows the system of Riley (1993), which distinguishes several categories of proteins according to their specific physiological roles in the cell. These functional categories include: amino acid biosynthesis, biosynthesis of cofactors, cell envelope,

Table 9.2 Protein functions assigned to the three main functional classes. Structural proteins are not used in this classification. (Reprinted from Tamames, J. et al. (1996) Genomes with distinct function composition. *FEBS Letters* **389**, 96–101. © 1994, with permission from Elsevier Science.)

Energy	Information
Transport through membrane	Replication/repair/DNA
Transporters/symporters	Replication
Phosphorylation	Recombination
Amino acid metabolism	Repair
Pathways	Cell division
Modifications	Other nuclear
Nucleotide metabolism	Gene expression/RNA
Pyrimidine	Transcription
Purine	Translation
Lipid metabolism	Ribosomal proteins
Carbohydrate metabolism/energy conservation	Splicing
Preglycolytic sugar modification/degradation	Others
Pentose phosphate cycle	Proteins
Glycolysis	Protein biosynthesis
Electron transport/respiration	Folding
Gluconeogenesis	Internal transport/translocation
Carbohydrate synthesis	Post-translational modification
Storage and starvation	Protein degradation
ppGpp	
Carbamylphosphate	**Communication**
Secondary pathways, other	Signal transduction/regulation
	Interaction with environment
	Recognition
	Adhesion
	Defence (toxic substances)
	Extracellular degradation
	Carbohydrates
	Proteins

cellular processes, central intermediary metabolism, energy metabolism, fatty acid and phospholipid metabolism, metabolism of purines, pyrimidines, nucleosides and nucleotides, regulatory functions, replication, transcription, translation, transport and binding proteins, etc. Tamames et al. (1996) have grouped these functional categories into just three functional superclasses to gain maximal biological insight. The three superclasses represent the following processes: **energy**, **information** and **communication** (Table 9.2). In this classification the first group ('energy') contains all types of proteins that help an organism to maintain itself against the medium. The 'energy' superclass includes proteins related to anabolism, catabolism, intracellular transport, etc. The 'information' superclass contains all the proteins that are involved in storage and retrieval, replication and proliferation of information. It includes proteins related to DNA structure, replication and repair, transcription,

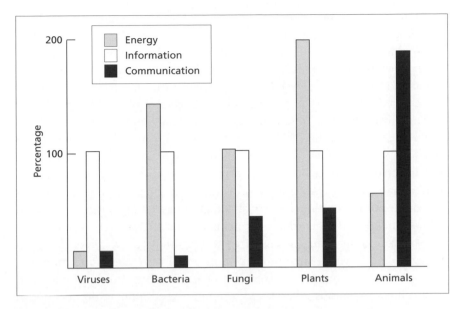

Fig. 9.4 Proteomes of viruses, bacteria, fungi, plants and animals have distinct functional composition. The vertical bars represent the proportion of proteins in the 'energy' and 'communication' categories, relative to those in the 'information' category (for definition of these categories see text). (Modified from Tamames, J. et al. (1996) Genomes with distinct function composition. *FEBS Letters* **389**, 96–101. © 1994, with permission from Elsevier Science.) Modified from FEBS Letters, 389, Tamames J. et al., genomes with distinct functional composition, 96–101, Copyright 1996, with permission from Elsevier.

splicing and translation. The 'communication' superclass contains proteins that mediate the interaction of the organism with the environment (including other cells and tissues of the same organism) and communication between different cellular states. Channels, transporters, cell-cycle proteins, protein–protein and protein–carbohydrate interactions mediating cell–cell communication are included in this superclass.

The justification for the distinction of these three superclasses is supported by the fact that they have distinct evolutionary behaviour in different groups of organisms (Fig. 9.4). For example, viruses have the highest proportion of the 'information' superclass (proteins involved in the control and expression of genetic information, e.g. polymerases), but they are devoid of most proteins in 'energy'. These aspects are explained by the lifestyle of viruses, in that they use their own genetic material, but exploit the host cell's machinery and rely entirely on the energy-supply system of their host.

Self-sufficient bacteria have the highest proportion of their genes dedicated to 'energy' (proteins associated with small molecule transformations and transport) and very few in 'communication'. For example, in a typical self-sufficient prokaryotic cell, *E. coli*, the functional distribution of sequences is: 51% 'energy', 45% 'information' and only 4% 'communication'. As mentioned above, parasitic bacteria (*H. influenzae, M. genitalium, M. pneumoniae*) and Archaea (*N. equitans*) have reduced their genomes by selectively eliminating genes that serve functions that are no longer needed by a parasite (e.g. amino acid biosynthetic genes). In these parasitic organisms the ratio of 'energy' is decreased relative to the 'information' and 'communication' superclasses.

The most significant difference between fungal genomes (e.g. yeast) and bacterial genomes is that fungi have a significantly increased proportion of functions in 'communication'. In plants the proportion of 'energy' is higher than in bacteria or fungi, but the proportion in 'communication' is only slightly higher than in fungi. Animals have a much higher proportion of proteins in 'communication' compared to all other types of organism. In fact, in animals the highest proportion of proteins is associated with intra- and intercellular 'communication'. In this respect, the 'simple' *C. elegans* is similar to higher animals, with a large proportion of proteins in 'communication'.

In general, the trend from simple to complex, and from unicellular to multicellular organisms is paralleled by an increased number of functions in 'communication' (Fig. 9.4). This suggests that the emergence of the eukaryotic cell, and the transition from unicellular to multicellular organisms relied primarily on 'communication' proteins facilitating interaction with the medium or with other cell types or cellular states.

Although the higher proportion of proteins dedicated to communication with the environment may be a natural consequence of the complex organization of multicellular organisms, plant proteomes have a functional composition more similar to bacteria than to multicellular animals. This relatively low number of proteins in 'communication' may reflect the simpler organization of plants, which have fewer cell types than animals. Interestingly, their intercellular communication processes are in many cases carried out by small molecules instead of proteins. It is noteworthy in this respect that receptor tyrosine kinases, so critical for intercellular communication of metazoans are apparently absent from plants (Walker, 1994).

Since all cellular organisms define their genetic identity by having their own set of proteins in the 'information' superclass, this may be considered as a fairly constant set of proteins. If we compare the 'communication'-associated proteins relative to the 'information'-associated proteins we see a continuous increase that parallels an increase in organismal complexity, the ratio being 0.1, 0.4, 0.7 and 1.9 in the case of bacteria, yeast, multicellular plants and animals, respectively (Fig. 9.4). It is worth pointing out here that most modular proteins produced by exon shuffling (see Table 8.3) are associated with and are absolutely essential for multicellularity of metazoa. All these modular proteins (e.g. constituents of the extracellular matrix, membrane-associated proteins mediating cell–cell and cell–matrix interactions, and receptor proteins regulating cell–cell communications, such as receptor tyrosine kinases and receptor tyrosine phosphatases) belong to the 'communication' category and have significantly contributed to the increased organismic complexity of metazoa.

9.9.4 Proteome evolution – evolution of proteome complexity

The complexity of the complete set of proteins – proteomes – of a given entity (cell, organism) is a function of the (i) number and (ii) complexity of proteins it consists of and (iii) the number and dynamics of interactions of these proteins with other proteins and other cellular constituents.

(i) Comparative genomic studies have revealed that the size of eubacterial and archaeal genomes varies in a relatively narrow range, from about 0.6 Mb in the obligatory intracellular parasite *M. genitalium* to about 6 Mb in some free-living bacteria (see Table 9.1). Since gene density in prokaryotes is relatively constant the number of protein-coding genes and proteins change in parallel with genome size. As a corollary, the proteomes of prokaryotes are encoded by about 500–5000 proteins, depending on genome size. In the absence of introns and alternative splicing, the number of protein-coding genes is an appropriate measure of the number of protein sequences present in these organisms.

The genome size of eukaryotes is usually much larger than those of prokaryotes, but the increase in genome size in higher eukaryotes is not accompanied by a comparable increase in gene number: whereas the size of the human genome is a thousand-fold larger than a typical bacterial genome, the number of human protein-coding genes (about 20,000–27,000) is only about fivefold higher than those of bacterial genomes (see Table 9.1). Nevertheless – thanks to the presence of introns in eukaryotic genomes and to alternative splicing – in the case of intron-rich eukaryotic genomes the number of unique protein sequences may be significantly higher than the number of protein-coding genes (Claverie, 2001).

(ii) The complexity of a protein itself is a function of the number (and number of types) of amino acids, secondary structural elements, domains they are constructed from as well as the number, type and dynamics of interactions among these constituents. At one extreme of this scale we find small, low complexity proteins with biased amino acid composition and few secondary structural elements, whereas at the other extreme we find large proteins constructed from multiple types of structural domains. The most important evolutionary mechanisms for increasing complexity of proteins include internal duplication of domains, shuffling of domains and fusion of proteins, thereby creating large multidomain proteins.

In this latter respect there are significant differences between proteins of Archaea, Eubacteria and Eukarya. The proportion of multidomain proteins was found to increase in the order Archaea < Bacteria < Fungi < Plants < Metazoa. At one extreme we find Archaea where only 23% of the proteins contain more than one Pfam-A domain, Metazoa represent the other extreme where 39% of the proteins correspond to multidomain proteins (Tordai et al., 2005). Furthermore, the multidomain proteins of Metazoa tend to be larger than those in Archaea, in harmony with earlier findings that the average protein length is considerably greater in eukaryotes than in prokaryotes.

Mathematical analyses of the distribution of multidomain proteins according to the number of different constituent domains have revealed that their distribution follows a power law, i.e. single-domain proteins are the most abundant, whereas proteins containing larger numbers of domains are increasingly less frequent in all types of organisms, but the likelihood of formation of multidomain proteins increases in the order: Archaea < Bacteria < Eukaryotes. Within the eukaryotic group there were noteworthy differences between animals, plants and fungi: the likelihood of domain joining was found to be significantly greater in Metazoa than in Plants and Fungi (Tordai et al., 2005). This difference is primarily due to the increased propensity to form large extracellular and treansmembrane multidomain proteins in Metazoa from the mobile domains (such as the EGF domain, immunoglobulin domain etc.) through exon shuffling (Patthy, 2003).

The complexity of the multidomain protein repertoire of different organisms may be characterized not only by the proportion and size distribution of multidomain proteins, but also by the number of domain combinations and the number of domain architecture types (i.e. distinct domain organizations). Using graph theory-based tools to compare protein domain organizations of different organisms Ye and Godzik (2004) have shown that the number of domains and the number of domain combinations of each organism increase as we move from prokaryotes to higher eukaryotes. The number of distinct architecture types was also found to increase in parallel with the evolution of higher organisms of greater organismic complexity: Archaea have the lowest values for this parameter, whereas Metazoa, particularly chordates have the highest number of architecture types (Tordai et al., 2005).

The domain-combination analysis of multidomain proteins has also shown that during evolution of eukaryotes significant changes occurred in the structural organization of the domain-combination networks (Tordai et al., 2005). In the case of Protozoa, Fungi and Plants the domain-combination networks consist of two major subclusters containing multidomain proteins constructed from nuclear and cytoplasmic signalling domains, respectively. The domain-combination network of Metazoa is strikingly different from those of other eukaryotes inasmuch as it has a third large, distinct subcluster of multidomain proteins consisting of extracellular (primarily class 1–1) modules. The extracellular subcluster of the domain-combination network is connected to the cytoplasmic signalling subcluster through multidomain transmembrane proteins, such as receptor tyrosine kinases and G-protein coupled receptors. Comparison of domain-combination networks of different eukaryotes thus confirms that the evolution of increased organismic complexity in Metazoa is intimately associated with the generation of novel extracellular and trans-membrane multidomain proteins that mediate the interactions among their cells, tissues and organs (Patthy, 2003).

(iii) One of the most crucial aspects of the complexity of the proteome is the complexity of the **network of interactions** of proteins with other proteins and/or with other cellular constituents, such as DNA, RNA, carbo-hydrates etc. The full range of functional complexity and diversity in biological systems is probably the result of interactions among all these biological entities. The architecture and organization of such interactions are best represented as networks, for example, networks of interacting proteins that reflect biochemical pathways and genetic regulations. The term **inter-actome** is thus defined as the whole set of molecular interactions in cells, but in a narrower sense interactome usually refers to just protein–protein interaction networks.

The interactions among proteins – in multi-subunit proteins and in transient complexes among proteins that also exist independently – are fundamental to cellular functions. Protein interaction networks can be described as graphs where nodes and edges represent proteins and their inter-actions, respectively (Barabasi & Oltvai, 2004). The distribution of protein connectivity in such networks follows a power law, i.e. proteins with few interactions are the most abundant, whereas proteins interacting with larger numbers of other proteins are increasingly less frequent (Giot et al., 2003; Li et al., 2004; Gandhi et al., 2006). For example, the connectivity distribution of the human protein–protein interaction network decreases slowly, closely following a power law (Stelzl et al., 2005). On average, proteins in the human interaction network have 1.87 interaction partners, however, 804 have only one partner, whereas there are 24 highly connected 'hubs' with more than 30 partners.

Park et al. (2005) have recently carried out a comparative interactomics analysis of protein family interaction networks. They have shown that the interactomes of 146 species share a small core network of 47 protein family interaction pairs related to indispensable cellular functions. The functions of protein families constituting this core network are mostly related to pro-tein translation, ribosomal structure and biogenesis, DNA binding and ATP metabolism. The results confirmed previous studies that all species share the same basic protein families and family interactions critical to basic cellular functions. These data also support the notion that the core protein family network shared by all types of life forms was formed in the very early stage of evolution and perform the core biochemical processes of life. A gradual attachment of the interactome seems to have occurred as novel functions, such as cell motility, extracellular structure organization etc. were needed. Park et al. (2005) have also characterized the interaction networks of dif-ferent species by calculating the total number of connections of each inter-acting protein family. These analyses have confirmed that the distribution of protein connectivity follows a power law in all species: i.e. proteins with few interactions are the most abundant, whereas proteins interacting with larger numbers of other proteins are increasingly less frequent.

However, there was a major difference between Eukarya and prokaryotes inasmuch as Eukarya have a much higher number of highly connected 'hub' families than Archaea and Eubacteria. It is also significant that the hub families have more interaction partners in Eukarya (ranging from 24 to 38) than in Eubacteria (ranging from 11 to 17) or in Archaea (ranging from 9 to 11). For example, the number of interacting partners of G proteins is very high in Eukarya and it seems very likely that the growth of this hub is closely associated with the evolution of multicellularity and the involvement of G proteins in signal transduction.

Since Eukarya have more multidomain proteins than Prokarya (see above) it seems very likely that the evolution of more complex protein interaction networks of Eukarya is intimately associated with the evolution of multidomain proteins. In a recent study investigating the relationship between domain-combination networks of multidomain proteins and domain interaction networks Wuchty and Almaas (2006) have found that the combination of domains in the innermost cores strongly coincides with physical interaction. This similarity of the two types of networks is due primarily to the fact that selection has favoured the creation of multidomain proteins from domains (modules) that play crucial roles in protein–protein interactions: these domains are therefore likely to be 'hubs' in both the domain-co-ocurrence and protein–protein interaction networks. For example, in complex cellular signalling pathways there is a great demand for domains that mediate interaction with other constituents of the pathways (e.g. PDZ, PH, ANK, SH3, WD40 domains) thus selection may have favoured the spread of these modules to other multidomain proteins. It is noteworthy that PDZ, PH, ANK, SH3, WD40 domains etc. are hubs in both domain combination networks (e.g. Tordai et al., 2005) and domain interaction networks (e.g. Formstecher et al., 2005).

Recently Rual et al. (2005) used a high-throughput yeast two-hybrid screening system to test pairwise interactions among the products of 8100 human open reading frames and detected approximately 2800 interactions. To gain an insight into the evolution of the interactome, they classified proteins in this interaction network as 'eukaryotic', 'metazoan', 'mammalian' or 'human' and asked whether proteins specific to different evolutionary classes tend to interact with one another. Importantly, the network appeared to be enriched for interactions between proteins of the same evolutionary class but not for interactions between proteins from two different evolutionary classes. This observation indicates that the interactome has evolved through the preferential addition of interactions between lineage-specific proteins. This principle may be illustrated by the fact that the various multidomain proteins that evolved in and are unique to vertebrates (e.g. various constituents of the blood coagulation cascade) interact preferentially with each other.

9.9.5 Proteome evolution and organismic complexity

Organismic complexity is a function of the number and types of cells, tissues and organs they are constructed from, the number of different states of these constituents as well as the number and dynamics of their interactions with other consituents and with the environment. Accordingly, at one extreme of this scale we find simple unicellular prokaryotes, whereas at the other extreme we find complex multicellular organisms with sophisticated dynamic interactions among their various cells, tissues and organs.

The conclusion that the huge human genome might contain only about 20,000–27,000 protein-coding genes came as a surprise since earlier estimates have ranged up to 150,000 genes. This conclusion meant that a rather small increase in gene number could be enough to progress from a 'simple' metazoan organism like *C. elegans* (which has about 19,000 genes) to highly complex organisms like humans. The organismic complexity of humans is thus achieved by a 'surprisingly' small number of protein-coding genes. Claverie (2001) has referred to this lack of a simple linear correlation between gene number and organismic complexity as the *N*-value paradox (in analogy with the *C*-value paradox).

The most plausible explanation for this *N*-value paradox is that organismic complexity depends on the complexity of interactions among genes and all their products – a feature that is not a simple linear function of the actual number of protein-coding genes. First, alternative splicing of protein-coding genes and post-translational modifications of protein products may lead to a much greater diversity at the protein level than expected purely from the number of genes. Second, the complexity of networks of interactions is critically dependent on the number of ways in which the expression of individual protein-coding genes is regulated and this in turn may depend on interactions involving a variety of ncRNAs (see Chapter 1). Third, the complexity of the architecture of interaction networks of an organism would also depend on the number and complexity of protein–DNA, protein–RNA and protein–protein interactions among the constituents of its different cells, tissues and organs (which in turn may depend on the complexity of multi-domain proteins).

Thus, if we wish to assess the contribution of a protein to organismic complexity a relevant aspect is its connectivity (the number of interactions with other molecules): highly connected proteins are more likely to increase complexity than proteins with only a small number of links. Multidomain proteins form a special group among proteins since they are more likely to participate in multiple interactions with multiple partners and thus may be more highly connected than single-domain proteins. In complex multidomain proteins a large number of functions (catalytic activities, different binding activities) may coexist making such proteins indispensable constituents of regulatory or structural networks where multiple interactions (protein–protein, protein–carbohydrate, protein–DNA, protein–RNA etc. interactions) are essential. For example, the domains that constitute multidomain proteins of the intracellular and extracellular signal transducing pathways mediate multiple interactions with other components of the signalling pathways (see Chapter 2).

Similarly, the coexistence of different domains with different binding specificities is also essential for the biological function of multidomain proteins of the extracellular matrix: the multiple, specific interactions among matrix constituents are indispensable for the proper architecture of the extracellular matrix. As a corollary of their involvement in multiple interactions, formation of novel multidomain proteins is likely to contribute significantly to the evolution of increased organismic complexity. According to this view, increased organismic complexity of higher eukaryotes is not only associated with a larger repertoire of complex multidomain proteins but also is a direct result of their complex interactions.

The importance of the creation of novel multidomain proteins is best appreciated if we analyse the correlation between the appearance of novel multidomain proteins and major changes in organismic complexity. There is now overwhelming evidence that the formation of novel extracellular and transmembrane multidomain proteins are intimately associated with novel biological functions and innovations in metazoa (Patthy, 2003). For example, one of the basic aspects of the biology of multicellular animals is that they require mechanisms for intercellular communication and intercellular cohesion. Organization of all multicellular animals depends on molecules that mediate adhesion of cells to other cells, to an extracellular matrix or a basement membrane. Some key multidomain proteins of the basement membrane (e.g. laminins) are present in sponges, hydra, worm, fly and vertebrates, indicating that these fundamental consitutents of the basement membranes were formed very early in the evolution of multicellular animals. Similarly, the linkage of the extracellular matrix to the cytoskeleton via the multidomain transmembrane proteins of the integrin family is a very ancient mode of creating a connection between the intracellular and extracellular space. Consistent with this, integrins are also highly conserved in organisms ranging from sponges, corals, nematodes, fly and echinoderms to mammals. The fact that many types of multidomain transmembrane proteins involved in cell adhesion in vertebrates is already present in the worm and fly (e.g. cadherins, members of the NCAM family, netrin receptors) also indicates that they were formed in an early stage of metazoan evolution.

The multidomain receptor tyrosine kinases and receptor tyrosine phosphatases that are absolutely essential for intercellular communication in metazoa also appear to have emerged very early in metazoan evolution. Phylogenetic analysis of the evolutionary history of metazoan protein tyrosine phosphatases has led to the conclusion that there was a period of explosive gene duplication before the parazoan–eumetazoan split, leading to the formation of most major families of receptor tyrosine phosphatases. Similarly, evolutionary analyses of receptor protein tyrosine kinases have shown that most of the present-day subtypes had been established in a very early stage of the evolution of animals before the parazoan–eumetazoan split.

There is also strong evidence that the formation of novel extracellular multidomain proteins was responsible for the appearance of novel biological functions unique to vertebrates. In contrast with the wide evolutionary distribution and highly conserved structure of basic multidomain protein components of the extracellular matrix, the various types of multidomain collagens typical of vertebrates are missing from invertebrates. Vertebrates use a wide variety of multidomain collagens to construct their endoskeletons (bone, cartilage), to build the tendons that connect bones and to construct interstitial connective tissue that provides structure for vertebrate tissues. It thus appears that the multidomain collagens have evolved in the vertebrate lineage and their formation must have been a *sine qua non* of chordate evolution. Several other multidomain protein constituents of the endoskeleton/extracellular matrix have also been formed during vertebrate evolution: the cartilage aggregating proteoglycan, aggrecan (and other members of this family), cartilage link protein, cartilage matrix protein (and other matrilins), thrombospondins are all missing from invertebrates.

There are many novel multidomain proteins associated with sophisticated vertebrate haemostasis and defence mechanisms. The coagulation, fibrinolytic and complement cascades are regulated primarily by complex interactions among numerous multidomain proteases and cofactors of these systems. It is now clear that most multidomain protease components of these plasma effector systems arose in vertebrates: orthologs for the majority of protease components and nonprotease components of vertebrate-type blood coagulation, fibrinolyis, kinin and complement pathways are missing from the *D. melanogaster*, *C. elegans* and *S. purpuratus* genomes and the majority of components of the blood coagulation and fibrinolyitic cascades are also absent in the invertebrate chordate *C. intestinalis*.

References

Abi-Rached, L., Gilles, A., Shiina, T. et al. (2002) Evidence of *en bloc* duplication in vertebrate genomes. *Nature Genetics* **31**, 100–5.
Adams, M.D., Celniker, S.E., Holt, R.A. et al. (2000) The genome sequence of *Drosophila melanogaster*. *Science* **287**, 2185–95.
Armbrust, E.V., Berges, J.A., Bowler, C. et al. (2004) The genome of the diatom *Thalassiosira pseudonana*: ecology, evolution, and metabolism. *Science* **306**, 79–86.

Andersson, S.G. & Kurland, C.G. (1999) Origins of mitochondria and hydrogenosomes. *Current Opinion in Microbiology* **2**, 535–41.

Andersson, S.G., Zomorodipour, A., Andersson, J.O. et al. (1998) The genome sequence of *Rickettsia prowazekii* and the origin of mitochondria. *Nature* **396**, 133–40.

Aparicio, S., Morrison, A., Gould, A. et al. (1995) Detecting conserved regulatory elements with the model genome of the Japanese puffer fish, *Fugu rubripes*. *Proceedings of the National Academy of Sciences of the USA* **92**, 1684–8.

Aparicio, S., Chapman, J., Stupka, E. et al. (2002) Whole-genome shotgun assembly and analysis of the genome of *Fugu rubripes*. *Science* **297**, 1301–10.

Arabidopsis Genome Initiative (2000) Analysis of the genome sequence of the flowering plant *Arabidopsis thaliana*. *Nature* **408**, 796–815.

Barabasi, A.L. & Oltvai, Z.N. (2004) Network biology: understanding the cell's functional organization. *Nature Reviews in Genetics* **5**, 101–13.

Bassett, D.E. Jr, Basrai, M.A., Connelly, C. et al. (1996) Exploiting the complete yeast genome sequence. *Current Opinion in Genetics and Development* **6**, 763–6.

Bell, P.J. (2001) Viral eukaryogenesis: was the ancestor of the nucleus a complex DNA virus? *Journal of Molecular Evolution* **53**, 251–6.

Bernardi, G. (2004) *Structural and Evolutionary Genomics. Natural Selection in Genome Evolution*. Amsterdam: Elsevier.

Bevan, M., Bancroft, I., Bent, E. et al. (1998) Analysis of 1.9Mb of contiguous sequence from chromosome 4 of *Arabidopsis thaliana*. *Nature* **391**, 485–8.

Blattner, F.R., Plunkett, G. III, Bloch, C.A. et al. (1997) The complete genome sequence of *Escherichia coli* K-12. *Science* **277**, 1453–62.

Blomme, T., Vandepoele, K., De Bodt, S. et al. (2006) The gain and loss of genes during 600 million years of vertebrate evolution. *Genome Biology* **7**, R43.

Blumenthal, T. & Spieth, J. (1996) Gene structure and organization in *Caenorhabditis elegans*. *Current Opinion in Genetics and Development* **6**, 692–8.

Boffelli, D., Nobrega, M.A. & Rubin, E.M. (2004) Comparative genomics at the vertebrate extremes. *Nature Reviews in Genetics* **5**, 456–65.

Bourque, G., Zdobnov, E.M., Bork, P., Pevzner, P.A. & Tesler, G. (2005) Comparative architectures of mammalian and chicken genomes reveal highly variable rates of genomic rearrangements across different lineages. *Genome Research* **15**, 98–110.

Boxma, B., de Graaf, R.M., van der Staay, G.W. et al. (2005) An anaerobic mitochondrion that produces hydrogen. *Nature* **434**, 74–9.

Brenner, S.E., Hubbard, T., Murzin, A. et al. (1995) Gene duplications in *H. influenzae*. *Nature* **378**, 140.

Brenner, S.E., Chothia, C. & Hubbard, T.J.P. (1997) Population statistics of protein structures: lessons from structural classifications. *Current Opinion in Structural Biology* **7**, 369–76.

Brent, M.R. & Guigo, R. (2004) Recent advances in gene structure prediction. *Current Opinion in Structural Biology* **14**, 264–72.

Bult, C.J., White, O., Olsen, G.J. et al. (1996) Complete genome sequence of the methanogenic Archaeon. *Methanococcus jannaschii*. *Science* **273**, 1058–73.

Burger, G., Forget, L., Zhu, Y., Gray, M.W. & Lang, B.F. (2003) Unique mitochondrial genome architecture in unicellular relatives of animals. *Proceedings of the National Academy of Sciences of the USA* **100**, 892–7.

Burt, D.W. (2005) Chicken genome: current status and future opportunities. *Genome Research* **15**, 1692–8.

C. elegans Sequencing Consortium (1998) Genome sequence of the nematode *C. elegans*: a platform for investigating biology. *Science* **282**, 2012–18.

Carlton, J.M., Angiuoli, S.V., Suh, B.B. et al. (2002) Genome sequence and comparative analysis of the model rodent malaria parasite *Plasmodium yoelii yoelii*. *Nature* **419**, 512–19.

Cavalier-Smith, T. (1978) Nuclear volume control by nucleoskeletal DNA, selection for cell volume and cell growth rate and the solution to the DNA *C*-value paradox. *Journal of Cell Science* **34**, 247–78.

Cavalier-Smith, T. (1985) *The Evolution of Genome Size*. New York: Wiley.

Chervitz, S.A., Aravind, L., Sherlock, G. et al. (1998) Comparison of the complete protein sets of worm and yeast: orthology and divergence. *Science* **282**, 2022–8.

Christoffels, A., Koh, E.G., Chia, J.M. et al. (2004) *Fugu* genome analysis provides evidence for a whole-genome duplication early during the evolution of ray-finned fishes. *Molecular Biology and Evolution* **21**, 1146–51.

Claverie, J.M. (2001) Gene number. What if there are only 30,000 human genes? *Science* **291**, 1255–7.

Clayton, R.A., White, O., Ketchum, K.A. et al. (1997) The first genome from the third domain of life. *Nature* **387**, 459–62.

Costa, F.F. (2005) Non-coding RNAs: new players in eukaryotic biology. *Gene* **357**, 83–94.

Cowman, A.F. & Crabb, B.S. (2002) The *Plasmodium falciparum* genome – a blueprint for erythrocyte invasion. *Science* **298**, 126–8.

Curwen, V., Eyras, E., Andrews, T.D. et al. (2004) The Ensembl automatic gene annotation system. *Genome Research* **14**, 942–50.

Davies, A.M. (1994) Tracking neurotrophin. *Nature* **368**, 193–4.

Dehal, P., Satou, Y., Campbell, R.K. et al. (2002) The draft genome of *Ciona intestinalis*: insights into chordate and vertebrate origins. *Science* **298**, 2157–67.

Derelle, E., Ferraz, C., Rombauts, S. et al. (2006) Genome analysis of the smallest free-living eukaryote *Ostreococcus tauri* unveils many unique features. *Proceedings of the National Academy of Sciences of the USA* **103**, 11647–52.

Desjardins, C., Eisen, J.A. & Nene, V. (2005) New evolutionary frontiers from unusual virus genomes. *Genome Biology* **6**, 212.

Deng, M., Templeton, T.J., London, N.R. et al. (2002) *Cryptosporidium parvum* genes containing thrombospondin type 1 domains. *Infection and Immunity* **70**, 6987–95.

Doolittle, W.F. & Sapienza, C. (1980) Selfish genes, the phenotype paradigm and genome evolution. *Nature* **284**, 601–3.

Douglas, S.E. & Penn, S.L. (1999) The plastid genome of the cryptophyte alga, *Guillardia theta*: complete sequence and conserved synteny groups confirm its common ancestry with red algae. *Journal of Molecular Evolution* **48**, 236–44.

Douglas, S., Zauner, S., Fraunholz, M. et al. (2001) The highly reduced genome of an enslaved algal nucleus. *Nature* **410**, 1091–6.

Dujon, B. (1997) The yeast genome project: what did we learn? *Trends in Genetics* **12**, 263–9.

Duret, L., Mouchiroud, D. & Gautier, C. (1995) Statistical analysis of vertebrate sequences reveals that long genes are scarce in GC-rich isochores. *Journal of Molecular Evolution* **40**, 308–17.

Edgall, D.R. & Doolittle, W.F. (1997) Archaea and the origin(s) of DNA replication proteins. *Cell* **89**, 995–8.

Eichinger, L., Pachebat, J.A., Glockner, G. et al. (2005) The genome of the social amoeba *Dictyostelium discoideum*. *Nature* **435**, 43–57.

Elgar, G., Sandford, R., Aparicio, S. et al. (1996) Small is beautiful: comparative genomics with the pufferfish (*Fugu rubripes*). *Trends in Genetics* **12**, 145–50.

Fischer, D. & Eisenberg, D. (1997) Assigning folds to the proteins encoded by the genome of *Mycoplasma genitalium*. *Proceedings of the National Academy of Sciences of the USA* **94**, 11929–34.

Fleischmann, R.D., Adams, M.D., White, O. et al. (1995) Whole-genome random sequencing and assembly of *Haemophilus influenzae* Rd. *Science* **269**, 496–512.

Fraser, C.M., Gocayne, J.D., White, O. et al. (1995) The minimal gene complement of *Mycoplasma genitalium*. *Science* **270**, 397–403.

Fraser, C.M., Casjens, S., Huang, W.M. et al. (1997) Genomic sequence of a Lyme disease spirochaete. *Borrelia burgdorferi*. *Nature* **390**, 580–6.

Frishman, D. & Mewes, H.W. (1997) Protein structural classes in five complete genomes. *Nature Structural Biology* **4**, 626–8.

Formstecher, E., Aresta, S., Collura, V. et al. (2005) Protein interaction mapping: a *Drosophila* case study. *Genome Research* **15**, 376–84.

Forterre, P. (2002) The origin of DNA genomes and DNA replication proteins. *Current Opinion in Microbiology* **5**, 525–32.

Fryxell, K.J. (1996) The coevolution of gene family trees. *Trends in Genetics* **12**, 364–9.

Fullerton, S.M., Bernardo, Carvalho, A. & Clark, A.G. (2001) Local rates of recombination are positively correlated with GC content in the human genome. *Molecular Biology and Evolution* **18**, 1139–42.

Furlong, R.F. & Holland, P.W. (2002) Were vertebrates octoploid? *Philosophical Transactions of the Royal Society of London. Series B, Biological Sciences* **357**, 531–44.

Galagan, J.E., Calvo, S.E., Borkovich, K.A. et al. (2003) The genome sequence of the filamentous fungus *Neurospora crassa*. *Nature* **422**, 859–68.

Galagan, J.E., Calvo, S.E., Cuomo, C. et al. (2005) Sequencing of *Aspergillus nidulans* and comparative analysis with *A. fumigatus* and *A. oryzae*. *Nature* **438**, 1105–15.

Gandhi, T.K., Zhong, J., Mathivanan, S. et al. (2006) Analysis of the human protein interactome and comparison with yeast, worm and fly interaction datasets. *Nature Genetics* **38**, 285–93.

Giot, L., Bader, J.S., Brouwer, C. et al. (2003) A protein interaction map of *Drosophila melanogaster*. *Science* **302**, 1727–36.

Galtier, N., Piganeau, G., Mouchiroud, D. & Duret, L. (2001) GC-content evolution in mammalian genomes: the biased gene conversion hypothesis. *Genetics* **159**, 907–11.

Gardner, M.J., Hall, N., Fung, E., White, O. et al. (2002) Genome sequence of the human malaria parasite *Plasmodium falciparum*. *Nature* **419**, 498–511.

Gerstein, M. (1997) A structural census of genomes: comparing bacterial, eukaryotic, and archaeal genomes in terms of protein structure. *Journal of Molecular Biology* **274**, 562–76.

Gerstein, M. & Levitt, M. (1997) A structural census of the current population of protein sequences. *Proceedings of the National Academy of Sciences of the USA* **94**, 11911–16.

Gilson, P.R., Su, V., Slamovits, C.H. et al. (2006) Complete nucleotide sequence of the chlorarachniophyte nucleomorph: nature's smallest nucleus. *Proceeding of the National Academy of Sciences of the USA* **103**, 9566–71.

Giovannoni, S.J., Tripp, H.J., Givan, S. et al. (2005) Genome streamlining in a cosmopolitan oceanic bacterium. *Science* **309**, 1242–5.

Glass, J.I., Assad-Garcia, N., Alperovich, N. et al. (2006) Essential genes of a minimal bacterium. *Proceedings of the National Academy of Sciences of the USA* **103**, 425–30.

Glockner, G., Rosenthal, A. & Valentin, K. (2000) The structure and gene repertoire of an ancient red algal plastid genome. *Journal of Molecular Evolution* **51**, 382–90.

Gray, M.W., Lang, B.F., Cedergren, R. et al. (1998) Genome structure and gene content in protist mitochondrial DNAs. *Nucleic Acids Research* **26**, 865–78.

Gray, M.W., Burger, G. & Lang, B.F. (2001) The origin and early evolution of mitochondria. *Genome Biology* **2**, 1018.

Grayling, R.A., Sandman, K. & Reeve, J.N. (1996) DNA stability and DNA binding proteins. *Advances in Protein Chemistry* **48**, 437–67.

Guigo, R., Flicek, P., Abril, J.F. et al. (2006) EGASP: the human ENCODE Genome Annotation Assessment Project. *Genome Biology* **7**, Suppl 1, S2.1–31.

Hebsgaard, S.M., Korning, P.G., Tolstrup, N. et al. (1996) Splice site prediction in *Arabidopsis thaliana* pre-mRNA by combining local and global sequence information. *Nucleic Acids Research* **24**, 3439–52.

Himmelreich, R., Hilbert, H., Plagens, H. et al. (1996) Complete sequence analysis of the genome of the bacterium *Mycoplasma pneumoniae. Nucleic Acids Research* **24**, 4420–49.

Himmelreich, R., Plagens, H., Hilbert, H. et al. (1997) Comparative analysis of the genomes of the bacteria *Mycoplasma pneumoniae* and *Mycoplasma genitalium. Nucleic Acids Research* **25**, 701–12.

Holland, P.W. (1999) Gene duplication: past, present and future. *Seminars in Cellular and Developmental Biology* **10**, 541–7.

Holt, R.A., Subramanian, G.M., Halpern, A. et al. (2002) The genome sequence of the malaria mosquito *Anopheles gambiae. Science* **298**, 129–49.

International Chicken Genome Sequencing Consortium (2004) Sequence and comparative analysis of the chicken genome provide unique perspectives on vertebrate evolution. *Nature* **432**, 695–716.

International Human Genome Sequencing Consortium (2001) Initial sequencing and analysis of the human genome. *Nature* **409**, 860–921.

International Human Genome Sequencing Consortium (2004) Finishing the euchromatic sequence of the human genome. *Nature* **431**, 931–45.

Iyer, L.M., Koonin, E.V., Leipe, D.D. & Aravind, L. (2005) Origin and evolution of the archaeo-eukaryotic primase superfamily and related palm-domain proteins: structural insights and new members. *Nucleic Acids Research* **33**, 3875–96.

Iyer, L.M., Balaji, S., Koonin, E.V. & Aravind, L. (2006) Evolutionary genomics of nucleo-cytoplasmic large DNA viruses. *Virus Research* **117**, 156–84.

Jiang, Y. & Doolittle, R.F. (2003) The evolution of vertebrate blood coagulation as viewed from a comparison of puffer fish and sea squirt genomes. *Proceedings of the National Academy of Sciences of the USA* **100**, 7527–32.

Kasahara, M., Hayashi, M., Tanaka, K. et al. (1996) Chromosomal localization of the proteasome Z subunit gene reveals an ancient chromosomal duplication involving the major histocompatibility complex. *Proceedings of the National Academy of Sciences of the USA* **93**, 9096–101.

Katsanis, N., Fitzgibbon, J. & Fisher, E.M.C. (1996) Paralogy mapping: identification of a region in the human MHC triplicated onto human chromosomes 1 and 9 allows the prediction and isolation of novel PBX and NOTCH loci. *Genomics* **35**, 101–8.

Kellis, M., Patterson, N., Endrizzi, M., Birren, B. & Lander, E.S. (2003) Sequencing and comparison of yeast species to identify genes and regulatory elements. *Nature* **423**, 241–54.

Khelifi, A., Meunier, J., Duret, L. & Mouchiroud, D. (2006) GC content evolution of the human and mouse genomes: insights from the study of processed pseudogenes in regions of different recombination rates. *Journal of Molecular Evolution* **62**, 745–52.

Klenk, H.P., Clayton, R.A., Tomb, J.F. et al. (1997) The complete genome sequence of the hyperthermophilic, sulphate-reducing archaeon *Archaeoglobus fulgidus. Nature* **390**, 364–70.

Kong, A., Gudbjartsson, D.F., Jonsdottir, G.M. et al. (2002) A high-resolution recombination map of the human genome. *Nature Genetics* **31**, 241–7.

Koonin, E.V. (2005) Virology: Gulliver among the Lilliputians. *Current Biology* **15**, R167–9.

Koonin, E.V. & Mushegian, A.R. (1996) Complete genome sequences of cellular life forms: glimpses of theoretical evolutionary genomics. *Current Opinion in Genetics and Development* **6**, 757–62.

Kunst, F., Ogasawara, N., Moszer, I. et al. (1997) The complete genome sequence of the Gram-positive bacterium *Bacillus subtilis*. *Nature* **390**, 248–55.

Kusche-Gullberg, M., Garrison, K., MacKrell, A.J., Fessler, L.I. & Fessler, J.H. (1992) Laminin A chain: expression during *Drosophila* development and genomic sequence. *EMBO Journal* **11**, 4519–27.

Lander, E.S., Linton, L.M., Birren, B. et al. (2001) Initial sequencing and analysis of the human genome. *Nature* **409**, 860–921.

Lang, B.F., Burger, G., O'Kelly, C.J. et al. (1997) An ancestral mitochondrial DNA resembling a eubacterial genome in miniature. *Nature* **387**, 493–7.

Leipe, D.D., Aravind, L. & Koonin, E.V. (1999) Did DNA replication evolve twice independently? *Nucleic Acids Research* **27**, 3389–401.

Li, S., Armstrong, C.M., Bertin, N. et al. (2004) A map of the interactome network of the metazoan *C. elegans*. *Science* **303**, 540–3.

Li, W.H., Gu, Z., Wang, H. & Nekrutenko, A. (2001) Evolutionary analyses of the human genome. *Nature* **409**, 847–9.

Liang, F., Holt, I., Pertea, G. et al. (2000) Gene index analysis of the human genome estimates approximately 120,000 genes. *Nature Genetics* **25**, 239–40.

Loftus, B., Anderson, I., Davies, R. et al. (2005) The genome of the protist parasite *Entamoeba histolytica*. *Nature* **433**, 865–8.

Machida, M., Asai, K., Sano, M. et al. (2005) Genome sequencing and analysis of *Aspergillus oryzae*. *Nature* **438**, 1157–61.

Margulis, L. & Sagan, D. (2002) *Acquiring Genomes. A Theory of the Origin of Species*. New York: Basic Books.

Margulis, L., Chapman, M., Guerrero, R. & Hall, J. (2006) The last eukaryotic common ancestor (LECA): acquisition of cytoskeletal motility from aerotolerant spirochetes in the Proterozoic Eon. *Proceedings of the National Academy of Sciences of the USA* **103**, 13080–5.

Martin, W. (2005) Archaebacteria (Archaea) and the origin of the eukaryotic nucleus. *Current Opinion in Microbiology* **8**, 630–7.

Matsuzaki, M., Misumi, O., Shin-I, T. et al. (2004) Genome sequence of the ultrasmall unicellular red alga *Cyanidioschyzon merolae* 10D. *Nature* **428**, 653–7.

Mattick, J.S. & Makunin, I.V. (2006) Non-coding RNA. *Human Molecular Genetics* **15**, Spec No 1, R17–29.

McLysaght, A., Hokamp, K. & Wolfe, K.H. (2002) Extensive genomic duplication during early chordate evolution. *Nature Genetics* **31**, 200–4.

McVey, J.H., Nomura, S., Kelly, P. et al. (1988) Characterization of the mouse SPARC/Osteonectin gene. *Journal of Biological Chemistry* **263**, 11111–16.

Melton, D.W. (1994) Gene targeting in the mouse. *BioEssays* **16**, 633–8.

Mouchiroud, D., D'Onofrio, G., Aissani, B. et al. (1991) The distribution of genes in the human genome. *Gene* **100**, 181–7.

Morey, C. & Avner, P. (2004) Employment opportunities for non-coding RNAs. *FEBS Letters* **567**, 27–34.

Mouse Genome Sequencing Consortium (2002) Initial sequencing and comparative analysis of the mouse genome. *Nature* **420**, 520–62.

Meunier, J. & Duret, L. (2004) Recombination drives the evolution of GC-content in the human genome. *Molecular Biology and Evolution* **21**, 984–90.

Nakabachi, A., Yamashita, A., Toh, H. et al. (2006) The 160-kilobase genome of the bacterial endosymbiont *Carsonella*. *Science* **314**, 267.

Nobrega, M.A., Ovcharenko, I., Afzal, V. & Rubin, E.M. (2003) Scanning human gene deserts for long-range enhancers. *Science* **302**, 413.

O'Leary, J.M., Hamilton, J.M., Deane, C.M. et al. (2004) Solution structure and dynamics of a prototypical chordin-like cysteine-rich repeat (von Willebrand Factor type C module) from collagen IIA. *Journal of Biological Chemistry* **279**, 53857–66.

Ohno, S. (1972) So much 'junk' DNA in our genome. *Brookhaven Symposia in Biology* **23**, 366–70.

Ohno, S. (1996) The notion of the Cambrian pananimalia genome. *Proceedings of the National Academy of Sciences of the USA* **93**, 8475–8.

Okamoto, N. & Inouye, I. (2005) A secondary symbiosis in progress? *Science* **310**, 287.

Oliver, S.G. (1997) From DNA sequence to biological function. *Nature* **379**, 597–600.

Olsen, G.J. & Woese, C.R. (1996) Lessons from an Archaeal genome: what are we learning from *Methanococcus jannaschii*? *Trends in Genetics* **12**, 377–9.

Oohashi, T., Ueki, Y., Sugimoto, M. et al. (1995) Isolation and structure of the COL4A6 gene encoding the human α6 (IV) collagen chain and comparison with other type IV collagen genes. *Journal of Biological Chemistry* **270**, 26863–7.

Orgel, L.E. & Crick, F.H.C. (1980) Selfish DNA: the ultimate parasite. *Nature* **284**, 604–7.

Park, D., Lee, S., Bolser, D. et al. (2005) Comparative interactomics analysis of protein family interaction networks using PSIMAP (protein structural interactome map). *Bioinformatics* **21**, 3234–40.

Panopoulou, G., Hennig, S., Groth, D. et al. (2003) New evidence for genome-wide duplications at the origin of vertebrates using an amphioxus gene set and completed animal genomes. *Genome Research* **13**, 1056–66.

Patthy, L. (1994) Exons and introns. *Current Opinion in Structural Biology* **4**, 383–92.

Patthy, L. (2003) Modular assembly of genes and the evolution of new functions. *Genetica* **118**, 217–31.

Patthy, L., Trexler, M., Váli, Zs., Bányai, L. & Váradi, A. (1984) Kringles: modules specialized for protein binding. Homology of the gelatin-binding region of fibronectin with the kringle structures of proteases. *FEBS Letters* **171**, 131–6.

Pereira, S.L., Grayling, R.A., Lurz, R. et al. (1997) Archaeal nucleosomes. *Proceedings of the National Academy of Sciences of the USA* **94**, 12633–7.

Popovici, C., Leveugle, M., Birnbaum, D. & Coulier, F. (2001) Coparalogy: physical and functional clusterings in the human genome. *Biochemical Biophysical Research Communications* **288**, 362–70.

Pradel, G., Hayton, K., Aravind, L. et al. (2004) A multidomain adhesion protein family expressed in *Plasmodium falciparum* is essential for transmission to the mosquito. *Journal of Experimental Medicine* **199**, 1533–44.

Raoult, D., Audic, S., Robert, C. et al. (2004) The 1.2-megabase genome sequence of Mimivirus. *Science* **306**, 1344–50.

Raven, J.A. & Allen, J.F. (2003) Genomics and chloroplast evolution: what did cyanobacteria do for plants? *Genome Biology* **4**, 209.

Reeve, J.N., Sandman, K. & Daniels, C.J. (1997) Archaeal histones, nucleosomes, and transcription initiation. *Cell* **89**, 999–1002.

Ribeiro, S. & Golding, G.B. (1998) The mosaic nature of the eukaryotic nucleus. *Molecular Biology and Evolution* **15**, 779–88.

Rivera, M.C., Jain, R., Moore, J.E. & Lake, J.A. (1998) Genomic evidence for two functionally distinct gene classes. *Proceedings of the National Academy of Sciences of the USA* **95**, 6239–44.

Riley, M. (1993) Functions of the gene products of *Escherichia coli*. *Microbiological Reviews* **57**, 862–952.

Rual, J.F., Venkatesan, K., Hao, T. et al. (2005) Towards a proteome-scale map of the human protein–protein interaction network. *Nature* **437**, 1173–8.

Rubin, G.M., Yandell, M.D., Wortman, J.R. et al. (2000) Comparative genomics of the eukaryotes. *Science* **287**, 2204–15.

Ruvkun, G. & Hobert, O. (1998) The taxonomy of developmental control in *Caenorhabditis elegans*. *Science* **282**, 2033–41.

Saccone, S., Federico, C. & Bernardi, G. (2002) Localization of the gene-richest and the gene-poorest isochores in the interphase nuclei of mammals and birds. *Gene* **300**, 169–78.

Saga, Y., Yagi, T., Ikawa, Y. et al. (1992) Mice develop normally without tenascin. *Genes and Development* **6**, 1821–31.

Schwarzbauer, J.E. & Spencer, C.S. (1993) The *Caenorhabditis elegans* homologue of the extracellular calcium binding protein SPARC/Osteonectin affects nematode body morphology and mobility. *Molecular Biology of the Cell* **4**, 941–52.

Sea Urchin Genome Sequencing Consortium (2006) The genome of the sea urchin *Strongylocentrotus purpuratus*. *Science* **314**, 941–52.

Seoighe, C. & Wolfe, K.H. (1998) Extent of genomic rearrangement after genome duplication in yeast. *Proceedings of the National Academy of Sciences of the USA* **95**, 4447–52.

Shabalina, S.A. & Spiridonov, N.A. (2004) The mammalian transcriptome and the function of non-coding DNA sequences. *Genome Biology* **5**, 105.

Sibley, M.H., Johnson, J.J., Mello, C.C. et al. (1993) Genetic identification, sequence and alternative splicing of the *Caenorhabditis elegans* alpha 2 (IV) collagen gene. *Journal of Cell Biology* **123**, 255–64.

Sidow, A. (1996) Genome duplications in the evolution of early vertebrates. *Current Opinion in Genetics and Development* **6**, 715–22.

Simmen, M.W., Leitgeb, S., Clark, V.H. et al. (1998) Gene number in an invertebrate chordate, *Ciona intestinalis*. *Proceedings of the National Academy of Sciences of the USA* **95**, 4437–40.

Starich, M.R., Sandman, K., Reeve, J.N. & Summers, M.F. (1996) NMR structure of HMfB from the hyperthermophile, *Methanothermus fervidus*, confirms that this archaeal protein is a histone. *Journal of Molecular Biology* **255**, 187–203.

Stein, L.D., Bao, Z., Blasiar, D. et al. (2003) The genome sequence of *Caenorhabditis briggsae*: a platform for comparative genomics. *PLoS Biology* **1**, E45.

Stelzl, U., Worm, U., Lalowski, M. et al. (2005) A human protein–protein interaction network: a resource for annotating the proteome. *Cell* **122**, 957–68.

Sugaya, K., Fukagawa, T., Matsumoto, K. et al. (1994) Three genes in the human MHC class III region near the junction with the class II: gene for receptor of advanced glycosylation end products, PBX2 homeobox gene and a Notch homolog, human counterpart of mouse mammary tumor gene int-3. *Genomics* **23**, 408–19.

Sugaya, K., Sasanuma, S.I., Nohata, J. et al. (1997) Gene organization of human NOTCH4 and (CTG)n polymorphism in this human counterpart gene of mouse proto-oncogene Int3. *Gene* **189**, 235–44.

Takai, D. & Jones, P.A. (2002) Comprehensive analysis of CpG islands in human chromosomes 21 and 22. *Proceedings of the National Academy of Sciences of the USA* **99**, 3740–5.

Tamames, J., Ouzounis, C., Sander, C. et al. (1996) Genomes with distinct function composition. *FEBS Letters* **389**, 96–101.

Tatusov, R.L., Koonin, E.V. & Lipman, D.J. (1997) A genomic perspective on protein families. *Science* **278**, 631–7.

Templeton, T.J., Iyer, L.M., Anantharaman, V. et al. (2004) Comparative analysis of apicomplexa and genomic diversity in eukaryotes. *Genome Research* **14**, 1686–95.

Thatcher, J.W., Shaw, J.M. & Dickinson, W.J. (1998) Marginal fitness contributions of nonessential genes in yeast. *Proceedings of the National Academy of Sciences of the USA* **95**, 253–7.

Thomas, J.W., Touchman, J.W., Blakesley, R.W. et al. (2003) Comparative analyses of multi-species sequences from targeted genomic regions. *Nature* **424**, 788–93.

Tordai, H., Nagy, A., Farkas, K., Banyai, L. & Patthy, L. (2005) Modules, multidomain proteins and organismic complexity. *FEBS Journal* **272**, 5064–78.

Ullman, C.G. & Perkins, S.J. (1997) The Factor I and follistatin domain families: the return of a prodigal son. *Biochemistry Journal* **326**, 939–41.

Van de Peer, Y. (2004) Computational approaches to unveiling ancient genome duplications. *Nature Reviews in Genetics* **5**, 752–63.

van Ham, R.C., Kamerbeek, J., Palacios, C. et al. (2003) Reductive genome evolution in *Buchnera aphidicola*. *Proceedings of the National Academy of Sciences of the USA* **100**, 581–6.

Venter, J.C., Adams, M.D., Myers, E.W. et al. (2001) The sequence of the human genome. *Science* **291**, 1304–51.

Walker, J.C. (1994) Structure and function of the receptor-like protein kinases of higher plants. *Plant Molecular Biology* **26**, 1599–609.

Watanabe, Y., Fujiyama, A., Ichiba, Y. et al. (2002) Chromosome-wide assessment of replication timing for human chromosomes 11q and 21q: disease-related genes in timing-switch regions. *Human Molecular Genetics* **11**, 13–21.

Waters, E., Hohn, M.J., Ahel, I. et al. (2003) The genome of *Nanoarchaeum equitans*: insights into early archaeal evolution and derived parasitism. *Proceedings of the National Academy of Sciences of the USA* **100**, 12984–8.

Wheeler, D.L., Church, D.M., Edgar, R. et al. (2004) Database resources of the National Center for Biotechnology Information: update. *Nucleic Acids Research* **32** (Database issue), D35–40.

Woese, C.R., Kandler, O. & Wheelis, M.L. (1990) Towards a natural system of organisms: proposal for the domains Archaea, Bacteria, and Eucarya. *Proceedings of the National Academy of Sciences of the USA* **87**, 4576–9.

Wood, V., Gwilliam, R., Rajandream, M.A. et al. (2002) The genome sequence of *Schizosaccharomyces pombe*. *Nature* **415**, 871–80.

Woodfine, K., Fiegler, H., Beare, D.M. et al. (2004) Replication timing of the human genome. *Human Molecular Genetics* **13**, 191–202.

Wuchty, S. & Almaas, E. (2005) Evolutionary cores of domain co-occurrence networks. *BMC Evolutionary Biology* **5**, 24.

Ye, Y. & Godzik, A. (2004) Comparative analysis of protein domain organization. *Genome Research* **14**, 343–53.

Yeo, G.S.H., Elgar, G., Sandford, R. et al. (1997) Cloning and sequencing of complement component C9 and its linkage to DOC-2 in the pufferfish *Fugu rubripes*. *Gene* **200**, 203–11.

Yoon, H.S., Hackett, J.D., Van Dolah, F.M. et al. (2005) Tertiary endosymbiosis driven genome evolution in dinoflagellate algae. *Molecular Biology and Evolution* **22**, 1299–308.

Zdobnov, E.M., von Mering, C., Letunic, I. et al. (2002) Comparative genome and proteome analysis of *Anopheles gambiae* and *Drosophila melanogaster*. *Science* **298**, 149–59.

Zhang, X., Vuoltecnanho, R. & Tryggvason, K. (1996) Structure of the human laminin a2-chain gene (LAMA2) which is affected in congenital muscular dystrophy. *Journal of Biological Chemistry* **271**, 27664–9.

Zimmer, C. (2006) Did DNA come from viruses? *Science* **312**, 870–2.

Zuckerkandl, E. (1976) Gene control in eukaryotes and the *C*-value paradox: 'excess' DNA as an impediment to transcription of coding sequences. *Journal of Molecular Evolution* **9**, 73–104.

Useful internet resources

Genome size database

DOGS (Database Of Genome Sizes) (http://www.cbs.dtu.dk/databases/DOGS/).

Repetitive, simple sequence, low complexity regions of DNA

SIMPLE is a program for detecting simple sequence regions in DNA (and protein sequences). It detects both tandem repeats and simple sequences (http://www.biochem.ucl.ac.uk/bsm/SIMPLE/index.html).

RepeatMasker is a program that screens DNA sequences for interspersed repeats and low complexity DNA sequences (http://www.repeatmasker.org/).

The **CENSOR** web server allows users to have query sequences aligned against a reference collection of human, rodent or plant repeats (http://iubio.bio.indiana.edu/soft/iubionew/molbio/dna/analysis/Censor/Censor_Server.html).

The **Tandem Repeats Finder (TRDB)** is a public database of tandem repeats that allows users to run their own sequences (http://tandem.bu.edu/trf/trf.html).

Segmental duplications

The **Human Genome Segmental Duplication Database** and the **Non-Human Segmental Duplication Database** contain information about segmental duplications (with sequence identity of at least 90% and sequence length of at least 5 kb) in the human, chimpanzee, mouse and rat genomes (http://projects.tcag.ca/humandup/) (http://projects.tcag.ca/xenodup).

Mining genome synteny or genome duplications

DAGchainer finds sets of genes that appear in the same order along two genome contigs (ftp://ftp.tigr.org/pub/software/DAGchainer/).

Paralogy database

Paradb is an object-oriented database created to predict and map paralogous regions in vertebrate genomes (http://abi.marseille.inserm.fr/paradb/).

Isochore tools

IsoFinder is a web tool for computational prediction of isochores in genome sequences (http://bioinfo2.ugr.es/IsoF/isofinder.html).

ISOCHORE plots isochores in large DNA sequences (http://bioweb.pasteur.fr/seqanal/interfaces/isochore.html).

CpGPlot/CpGReport/Isochore plots CpG-rich areas (cpgplot), reports all CpG-rich regions (cpgreport) and plots GC content over a sequence (isochore) (http://www.ebi.ac.uk/emboss/cpgplot/).

Identification of CpG islands

The **CpG island searcher** screens for CpG islands that meet the criteria selected in submitted DNA sequences (http://www.cpgislands.com/).

Genome databases – general

GOLD (Genomes OnLine Database) is a resource for comprehensive access to information regarding complete and ongoing genome projects around the world (http://www.genomesonline.org/).

Genomes at the EBI provides access to completed genomes (http://www.ebi.ac.uk/genomes/).

Genomes Organelle provides access to completed genomes of organelles (mithochondria, chloroplasts, nucleopmorphs) (http://www.ebi.ac.uk/genomes/organelle.html.

Genomes Virus provides access to completed genomes of viruses (http://www.ebi.ac.uk/genomes/virus.html).

Genomes Phage provides access to completed phage genomes (http://www.ebi.ac.uk/genomes/phage.html).

Entrez Genome provides access to completed genomes (http://www.ncbi.nlm.nih.gov/entrez/query.fcgi?db=Genome).

GOBASE is a taxonomically broad organelle genome database that organizes and integrates diverse data related to mitochondria and chloroplasts (http://www.bch.umontreal.ca/gobase/).

Genome databases – specific

FlyBase is a database of genetic and molecular data for *Drosophila* (http://flybase.bio.indiana.edu/).

AnoBase is a database containing genomic/biological information on anopheline mosquitoes, with an emphasis on *A. gambiae*, the world's most important malaria vector (http://www.anobase.org/).

WormBase is the repository of mapping, sequencing and phenotypic information for *C. elegans* (and some other nematodes) (http://www.wormbase.org/).

RGD (Rat Genome Database) integrates data generated from ongoing rat genetic and genomic research efforts (http://rgd.mcw.edu).

MGI (Mouse Genome Informatics) provides integrated access to data on the genetics, genomics and biology of the laboratory mouse (http://www.informatics.jax.org).

Comparative genomics tools

PipMaker computes alignments of similar regions in two DNA sequences. The resulting alignments are summarized with a 'percent identity plot', or 'pip' for short. MultiPipMaker allows the user to see relationships among more than two sequences. The program can identify evolutionarily conserved regions between genomes (http://bio.cse.psu.edu/pipmaker/).

VISTA is a comprehensive suite of programs and databases for comparative analysis of genomic sequences. Regulatory VISTA (rVISTA) combines transcription factor binding sites database search with a comparative sequence analysis. GenomeVISTA compares the query sequences with whole-genome assemblies, it will automatically find the ortholog and obtain the alignment (http://genome.lbl.gov/vista/index.shtml).

DoOP (Database of Orthologous Promoters) contains orthologous clusters of promoters from *H. sapiens*, *A. thaliana* and other organisms (http://doop.abc.hu/).

The **COG database** contains clusters of orthologous groups of proteins delineated by comparing protein sequences encoded in complete genomes, representing major phylogenetic lineages. Each COG consists of individual proteins or groups of paralogs from at least three lineages and thus corresponds to an ancient conserved domain (http://www.ncbi.nlm.nih.gov/COG/).

Sybil is a web-based software package for the visualization and analysis of comparative genomics data (http://sybil.sourceforge.net/).

Gene-finding tools

ORF Finder (Open Reading Frame Finder) is a graphical analysis tool that finds all open reading frames of a selectable minimum size in a user's sequence or in a sequence already in the database (http://www.ncbi.nlm.nih.gov/gorf/gorf.html).

Splice site prediction

GeneSplicer is a fast, flexible system for detecting splice sites in the genomic DNA of various eukaryotes (http://www.tigr.org/tdb/GeneSplicer/gene_spl.html).

SplicePredictor provides a method to identify potential splice sites in (plant) pre-mRNA (http://bioinformatics.iastate.edu/cgi-bin/sp.cgi).

The **NetPlantGene** server is a service producing neural network predictions of splice sites in *A. thaliana* DNA (http://www.cbs.dtu.dk/services/NetPGene/).

The **NetGene2 server** is a service producing neural network predictions of splice sites in human, *C. elegans* and *A. thaliana* DNA (http://www.cbs.dtu.dk/services/NetGene2/).

HSPL provides predictions of splice sites in human DNA sequences (http://softberry.com/berry.phtml?topic=spl&group=help&subgroup=gfind).

Splice Site Prediction by Neural Network (http://www.fruitfly.org/seq_tools/splice.).

GENIO/splice provides splice site and exon prediction in human genomic DNA (http://biogenio.com/splice/).

SpliceView provides splice prediction by using consensus sequences (http://l25.itba.mi.cnr.it/~webgene/wwwspliceview.html).

Searching genomic DNA with DNA or protein query

BLAT Search Genome – BLAT on DNA is designed to quickly find sequences of 95% and greater similarity of length 40 bases or more. BLAT on proteins finds sequences of 80% and greater similarity of length 20 amino acids or more (http://genome.ucsc.edu/cgi-bin/hgBlat?command=start).

Gene-finding programs

AUGUSTUS is a program that predicts genes in eukaryotic genomic sequences (http://augustus.gobics.de/).

FGENESH HMM-based gene structure prediction (http://www.softberry.com/berry.phtml).

JIGSAW predicts genes using multiple sources of evidence. It is designed to use the output from gene finders, splice site prediction programs and sequence alignments to predict gene models (http://cbcb.umd.edu/software/jigsaw/).

Genotator is a workbench for sequence annotation and mrowsing. The Genotator runs several gene finders, homology searches (using blast) and signal searches. Genotator thus automates the tedious process of running a dozen different sequence analysis programs with a dozen different input and output formats (http://www.fruitfly.org/~nomi/genotator/).

WEBGENE is an integrated computing system for protein-coding gene prediction (http://www.itba.mi.cnr.it/webgene/).

MetaGene Server is a powerful tool designed to provide exhaustive analysis of predicted gene features in sequence. It allows the researcher to submit sequences to seven gene prediction engines simultaneously to obtain a comprehensive report on sequence features (http://rgd.mcw.edu/METAGENE/).

GENEMARK.hmm predicts genes in eukaryotes
(http://exon.gatech.edu/GeneMark/eukhmm.cgi).

GeneZilla is a program for computational prediction of protein-coding genes in
eukaryotic DNA. It is based on the Generalized Hidden Markov Model (GHMM)
framework (http://www.genezilla.org/).

GENEID predicts genes in anonymous genomic sequences
(http://www1.imim.es/geneid.html).

GENESCAN predicts the locations and exon–intron structures of genes in genomic
sequences from a variety of organisms (http://genes.mit.edu/GENSCAN.html).

GrailEXP (Grail Experimental Gene Discovery Suite) is a software package that predicts
exons, genes, promoters, poly(A)s, CpG islands, EST similarities and repetitive
elements within DNA sequences (http://grail.lsd.ornl.gov/grailexp/).

FGENES uses pattern-based human gene structure prediction (http://www.softberry.
com/berry.phtml?topic=fgenes&group=programs&subgroup=gfind).

XPOUND uses a probabilistic model for detecting coding regions in DNA sequences
(http://bioweb.pasteur.fr/seqanal/interfaces/xpound-simple.html).

Gene Finder contains software tools designed to predict putative internal protein
coding exons in genomic DNA sequences (http://rulai.cshl.org/tools/genefinder/).

HMMgene predicts vertebrate and *C. elegans* genes
(http://www.cbs.dtu.dk/services/HMMgene).

Glimmer is a system for finding genes in microbial DNA, especially the genomes
of bacteria and archaea (http://www.tigr.org/~salzberg/glimmer.html).

PAIRAGON is a pair-HMM based cDNA-to-genome alignment program
(http://mblab.wustl.edu/software/pairagon/).

ACEVIEW offers a comprehensive and nonredundant cDNA-supported annotation
of human and nematode genes. The program coaligns the million mRNAs and
ESTs available from GenBank, dbEST and RefSeq on the genome sequence,
quality filters the cDNAs and clusters them into alternative transcripts and
genes (http://www.ncbi.nlm.nih.gov/IEB/Research/Acembly/).

ENSEMBL is a software system that produces and maintains automatic annotation
on selected eukaryotic genomes (http://www.ensembl.org/index.html).

EXOGEAN is a program for annotating gene structures in eukaryotic genomic
DNA (http://www.biologie.ens.fr/lgmldog/spip.php?rubrique4&lang=en).

ExonHunter is a eukaryotic gene finder that can use multiple sources of evidence
to improve prediction accuracy (http://www.bioinformatics.uwaterloo.ca/
supplements/05eh/).

ECGene provides genome annotation for alternative splicing
(http://genome.ewha.ac.kr/ECgene/).

AAT (Analysis and Annotation Tool) finds genes in genomic sequences
(http://genome.cs.mtu.edu/aat/aat.html).

EuGène is a gene finder for eukaryotic organisms (http://www.inra.fr/bia/T/EuGene/).

GeneBuilder is a gene structure prediction system (http://l25.itba.mi.cnr.it/
~webgene/genebuilder.html).

PROCRUSTES provides gene recognition via spliced alignment (http://www-hto.
usc.edu/software/procrustes/).

SIM4 is a similarity-based tool for aligning an expressed DNA sequence
(EST, cDNA, mRNA) with a genomic sequence for the gene
(http://globin.cse.psu.edu/html/docs/sim4.html).

The **ACESCAN** website identifies alternative splicing events conserved in
human–mouse. It provides ACEScan scores for orthologous human–mouse
exons (http://genes.mit.edu/acescan/).

NSCAN combines biological-signal modelling in the target genome sequence along with information from a multiple-genome alignment to generate *de novo* gene predictions (http://genome.ucsc.edu/cgi-bin/hgTrackUi?hgsid=74828811&c=chr7&g=nscanGene).

SGP2 is a program to predict genes by comparing anonymous genomic sequences from different species (http://genome.imim.es/software/sgp2/sgp2.html).

TWINSCAN predicts which regions of a genome are transcribed into pre-messenger RNA, how they are spliced, and which portions of the spliced transcript are translated into protein. TWINSCAN uses the patterns of evolutionary conservation in a set of genomes (http://mblab.wustl.edu/query.html).

The **SLAM** server compares pairs of syntenic sequences for gene annotation and alignment (http://baboon.math.berkeley.edu/~syntenic/slam.html).

Proteomes, expression profiles

The **Integr8** web portal provides easy access to integrated information about deciphered genomes and their corresponding proteomes (http://www.ebi.ac.uk/integr8/EBI-Integr8-HomePage.do).

MIPS analysis and annotation of proteins from whole genomes (http://mips.gsf.de/).

BODYMAP is a human and mouse gene expression database (http://bodymap.ims.u-tokyo.ac.jp/).

GEO (Gene Expression Omnibus) is a gene expression/molecular abundance repository (http://www.ncbi.nlm.nih.gov/geo/).

Gene Expression Atlas of human and mouse genes (http://expression.gnf.org/cgi-bin/index.cgi).

Protein sequence cluster databases

The **CluSTr database** offers an automatic classification of UniProt Knowledgebase into groups of related proteins, based on analysis of all pairwise comparisons between protein sequences. The database provides links to InterPro, which integrates information on protein families, domains and functional sites from PROSITE, PRINTS, Pfam, ProDom, SMART, TIGRFAMs, Gene3D, SUPERFAMILY, PIR Superfamily and PANTHER (http://www.ebi.ac.uk/clustr/).

Domain combination networks of proteomes

CADO (Comparative Analysis of Protein Domain Organization) is a web server that allows users to visualize the protein domain organization in a given organism and compare domain organizations among different organisms. In the language of CADO, the organization of protein domains in a given organism is shown as a domain graph in which protein domains are represented as vertices and domain combinations, i.e. instances of two domains found in one protein, are represented as edges (http://bioinformatics.burnham.org/DomainGraph/).

Protein-protein interaction databases and tools

DIP (Database of Interacting Proteins) catalogues experimentally determined interactions between proteins. It combines information from a variety of sources to create a single, consistent set of protein–protein interactions (http://dip.doe-mbi.ucla.edu/).

MIPS (Mammalian Protein-Protein Interaction Database) is a collection of manually curated high-quality protein–protein interaction data collected from the scientific literature by expert curators (http://mips.gsf.de/proj/ppi/).

InterDom is a database of putative interacting protein domains derived from multiple sources, ranging from domain fusions (Rosetta stone), protein interactions (DIP and BIND), protein complexes (PDB), to scientific literature (MEDLINE) (http://interdom.lit.org.sg/#).

MINT focuses on experimentally verified protein interactions mined from the scientific literature by expert curators (http://cbm.bio.uniroma2.it/mint/).

MPact provides a common access point to interaction resources at MIPS (http://mips.gsf.de/genre/proj/mpact).

IntAct provides a freely available, open source database system and analysis tools for protein interaction data (http://www.ebi.ac.uk/intact/site/).

BIND (Biomolecular Interaction Network Database) (http://www.bind.ca/Action).

STRING (Search Tool for the Retrieval of Interacting Genes/Proteins) (http://string.embl.de/).

OPHID (Online Predicted Human Interaction Database) is designed to be both a resource for the laboratory scientist to explore known and predicted protein–protein interactions, and to facilitate bioinformatics initiatives exploring protein interaction networks (http://ophid.utoronto.ca/ophid/).

HPRD (Human Protein Reference Database) represents a centralized platform to visually depict and integrate information pertaining to domain architecture, post-translational modifications, interaction networks and disease association for each protein in the human proteome (http://www.hprd.org/).

HPID (Human Protein Interaction Database) allows the user to use the protein IDs in ENSEMBL, HPRD and UniProt/Swiss-Prot ID to search protein interactions of interest (http://wilab.inha.ac.kr/hpid/).

Homomint is a web available tool extending protein–protein interactions experimentally verified in models of organisms, to the orthologous proteins in *H. sapiens*. Similar to other approaches, the orthology groups in HomoMINT are obtained by the 'reciprocal best hit method' as implemented in the Inparanoid algorithm (http://mint.bio.uniroma2.it/HomoMINT/Welcome.do).

DOMINO is a relational database that aims at annotating all the available information about domain peptide interactions (http://mint.bio.uniroma2.it/domino/search/searchWelcome.do).

PIN Database (Proteins Interacting in the Nucleus) (http://pin.mskcc.org/).

InterPreTS (Interaction Prediction through Tertiary Structure) is a web-based tool to predict protein–protein interactions using 3D structure information (http://www.russell.embl-heidelberg.de/interprets/).

The **DIP database** catalogues experimentally determined interactions between proteins. It combines information from a variety of sources to create a single, consistent set of protein–protein interactions (http://dip.doe-mbi.ucla.edu/).

BIND (Biomolecular Interaction Network Database) (http://www.binddb.org/BIND).

Networks, pathways – genetic network, metabolic network, molecular network

aMAZE is a database for the representation of information on networks of cellular processes: genetic regulation, biochemical pathways, signal transductions (http://www.scmbb.ulb.ac.be/amaze/).

KEGG (Kyoto Encyclopedia of Genes and Genomes) is a resource for computational prediction of higher level complexity of cellular processes and organism behaviours from genomic and molecular information (http://www.genome.ad.jp/kegg/).

KEGG PATHWAY Database is a collection of manually drawn pathway maps representing our knowledge on the molecular interaction and reaction networks for metabolism, genetic information processing, environmental information processing, cellular processes etc. (http://www.genome.ad.jp/kegg/pathway.html).

The **GeneNet** system is designed for formalized description and automated visualization of gene networks (http://wwwmgs.bionet.nsc.ru/mgs/gnw/genenet/).

Glossary

3′ flanking region

A region of genomic DNA downstream of the 3′ end of the gene. The 3′ flanking region often contains enhancers or other sites to which proteins may bind.

3′ untranslated region (3′ UTR)

The region of the mRNA between the stop codon and the poly(A) tail. The 3′ untranslated region may affect the translation efficiency of the mRNA or the stability of the mRNA and it also contains signals required for the addition of the poly(A) tail to the message.

5′ flanking region

A region of genomic DNA upstream of the 5′ end of the gene. The 5′-flanking region contains the promoter, and may also contain enhancers or other protein-binding sites.

5′-Methylcytosine

5′-Methylcytosine is a methylated form of cytosine formed by the action of DNA methyltransferases. In bacteria, 5′-methylcytosine is often used as a marker to protect DNA from restriction enzymes. In plants, fungi and animals, 5′-methylcytosine predominantly occurs at CpG dinucleotides.

5′ untranslated region (5′ UTR)

The region of the mRNA upstream of the translation initiation site. This region may contain sequences that affect the translation efficiency or stability of the mRNA.

α-helix

The most common regular secondary structure in globular proteins. It is characterized by the helical structure of the protein's backbone, a complete turn of the helix taking 3.6 residues. The structure of the helix is maintained by hydrogen bonds from the backbone carbonyl oxygen of each residue, i, in the helix to the backbone –NH of residue i + 4. Certain types of amino acid residues (e.g. alanine, arginine, leucine) favour formation of α-helices, whereas others occur less frequently in α-helices. For example, proline breaks the hydrogen-bonding pattern required for α-helix stability. *See also* **Secondary structure of proteins**.

Ab initio protein modelling. *See* **Ab initio protein structure prediction**.

Ab initio protein structure prediction

In ab initio protein structure predictions calculations are made without reference to a known structure homologous to the target to be predicted. In other words, these methods attempt to predict protein structure essentially from principles of physics and chemistry.

Acceptor splice site

The binding site of the spliceosome on the 3′ side of an intron and the 5′ side of an exon. *See also* **Donor splice site**.

Adaptation

Differential survival of phenotypes with different fitness values leads to adaptation to a given set of environmental conditions. Genetic change within a population may improve its ability to survive and reproduce in a new environment.

Adaptive evolution. *See* Adaptation.

Adenine (A)

A purine base. In DNA it pairs with thymine, in RNA it pairs with uracil. *See also* **Base pair**, **Nucleotide**.

Affine gap penalty. *See* Gap penalty.

Alanine

Alanine is a small nonpolar amino acid found in proteins. In protein sequences it is written as Ala or A.

Algae

This term designates various simple photosynthetic eukaryotic organisms belonging to diverse taxonomic groups.

Alignment

Matching the amino acids or nucleotides of a pair of (or multiple) sequences in such a way that their similarity is maximized. Alignment of homologous sequences requires that the nucleotides (or amino acids) from 'equivalent positions' (i.e. of common ancestry) are brought into vertical register. The procedures that attempt to align sites of common ancestry are referred to as sequence alignment procedures.

Alignment score

The correct alignment of two sequences or multiple sequences is the one in which only sites of common ancestry are aligned, and all sites of common ancestry are aligned. The correct alignment is most likely to be found as the one in which matches are maximized and mismatches and gaps are minimized. This may be achieved by using some type of scoring system that rewards similarity and penalizes dissimilarity and gaps. The alignment score is calculated as the sum of the scores assigned to the different types of matches, mismatches, and gaps. *See also* **Alignment**.

Allele
One of several alternative forms of a gene occupying a given locus on a chromosome. In diploid organisms a single allele for each locus is inherited from each parent. Within a population there may be many different alleles of a gene.

Alternative splicing
Different ways of combining a gene's exons.

Alu repeat
The Alu repeat family is the most common family of short interspersed elements (SINEs) present in numerous copies in the genomes of humans and other primates. Alu repeats are ~300 base pairs in length and are present in over 1 000 000 copies in the human genome. They are most commonly found in introns, 3′ untranslated regions of genes and intergenic genomic regions. *See also* **SINE**.

Amide bond (peptide bond)
The covalent bond joining two amino acids, between the carboxylic group of one and the amino group of the other to form a peptide bond.

Amino acid
The building block of proteins. Proteins are constructed from 20 types of α-amino acids which have the form $RCH(NH_2)COOH$ where R is either hydrogen or an organic group. The different amino acids are thus distinguished by the side-chain substituent R.

Amino acid exchange matrix (PAM matrix)
An amino acid exchange matrix is a 20×20 matrix which contains probabilities for each possible mutation between 20 amino acids. The matrix is symmetric, so that a mutation from amino acid A to B is assigned an identical probability as the mutation B to A. The matrix diagonal contains the probability for self-conservation.

Amino acid sequence
The relative order of amino acids in the polypeptide chain. The amino acid sequence is written in the amino-terminal $\rightarrow$ carboxyl-terminal direction, reflecting the direction of translation. The sequence of amino acids in a

protein is determined by the nucleotide sequence of the mRNA and the genetic code. The properties of a protein are determined primarily by its amino acid sequence.

Amino acid substitution matrix. *See* **Amino acid exchange matrix**.

Amino terminus
Refers to the NH_2-terminal end of a polypeptide chain.

Amphibians
The class of vertebrates containing frogs, toads, newts and salamanders.

Amphipathic
A molecule having both a hydrophilic and a hydrophobic end.

Analogous
Proteins that are similar in structure or function but show no signs of common ancestry. *See also* **Convergence**.

Ancestral genome
The genome of the cenancestor, the last universal common ancestor (LUCA) of all extant living organisms.

Aneuploidy
Aneuploidy is a chromosomal state with abnormal numbers of specific chromosomes.

Antibiotics
Substances that inhibit the growth of microorganisms.

Antibiotic resistance
A heritable trait in microorganisms that enables them to survive in the presence of an antibiotic.

Anticipation
Each generation of offspring has increased severity of a genetic disorder.

Aphids
Small insects of the family Aphididae that feed by sucking sap from plants.

Apicomplexa
A phylum of Protista. They are single-celled parasites of animals, including the organisms causing malaria. Apicomplexa are characterized by the presence of unique subcellular organelle, the apical complex.

Arginine

Arginine is a positively charged amino acid found in proteins. In protein sequences it is written as Arg or R.

Asparagine

Asparagine is a polar amino acid found in proteins. In protein sequences it is written as Asn or N.

Aspartic acid

Aspartic acid is a polar (often negatively charged) amino acid found in proteins. In protein sequences it is written as Asp or D.

ATG or AUG

The codon for methionine; the translation initiation codon. In eukaryotic DNA, the sequence is ATG, in RNA it is AUG; usually, the first AUG in the mRNA is the point at which translation starts. *See also* **Translation start site**.

Autosome

A non-sex chromosome. Autosomal chromosomes are present in identical numbers in both sexes. *See also* **Sex chromosome**.

β-strand

A segment of a polypeptide chain of a protein that is hydrogen bonded to other β-strands to form a β-sheet. The backbone in a β-strand is almost fully extended (in contrast to the α-helix structure, in which the backbone is tightly coiled). *See also* **Secondary structure of proteins**.

Backbone

When amino acids are linked in a polypeptide chain, the atoms that constitute the continuous link of the chain are referred to as the backbone atoms. Each amino acid contributes its –N–Cα–C– atoms to this backbone.

Back mutation (revertant)

Reverses the effect of a mutation that had altered a gene. Restores the wild-type phenotype.

Bacteriophage

A virus that infects bacteria.

Base

The purine or pyrimidine component of a nucleotide. This term is frequently used to refer to a nucleotide residue within a nucleic acid chain. In DNA molecules the purine bases are adenine (A) and guanine (G), the pyrimidine bases are cytosine (C) and thymine (T). In RNA, the pyrimidine base uracil (U) is used instead of thymine. *See also* **Nucleotide**.

Base composition
The percentage of G + C or A + T bases in a genome. *See also* **Isochore**.

Base pair
One pair of complementary nucleotides within a duplex strand of a nucleic acid. According to the Watson–Crick rules, the canonical base pairs consist of one pyrimidine and one purine: i.e. C–G, A–T (DNA) or A–U (RNA). In DNA two strands are held together in the shape of a double helix by the bonds between base pairs. Noncanonical base pairs (e.g. G–U) are common in RNA secondary structure.

Bilateria
Bilateria are a major group of animals. A distinguishing feature of this group is that their body has bilateral symmetry and develops from three different germ layers, called the endoderm, mesoderm and ectoderm. Deuterostomes and protostomes are two major lineages of Bilateria. *See also* **Deuterostome**, **Protostome**.

Birth-and-death evolution
The number of genes in multigene families changes over time as new genes are added by gene duplication whereas other genes are lost by gene loss or inactivation.

BLOSUM matrix
Whereas Dayhoff amino acid substitution matrices are based on substitution rates derived from global alignments of protein sequences that are at least 85% identical, the BLOSUM (BLOcks amino acid SUbstitution Matrix) matrices are derived from local alignments of conserved blocks of aligned amino acid sequence segments of more distantly related proteins. In the BLOSUM 62 matrix, for example, the alignment from which scores were derived was created using sequences sharing no more than 62% identity. *See also* **Amino acid substitution matrix**.

Bootstrap analysis
A commonly used approach for calculating reliability of inferred phylogenies. *See also* **Phylogenetic tree**.

Box
Refers to a short nucleic acid consensus sequence or motif that is universal within kingdoms of organisms. Examples of DNA boxes are the Pribnow box (TATAAT) for RNA polymerase, the Hogness box (TATA) that has a similar function in eukaryotic organisms.

bp
Abbreviation for base pair. Double-stranded DNA is usually measured in bp rather than nucleotides (nt).

Branch-length estimation
The process of estimating branch lengths in a phylogenetic tree. *See also* **Phylogenetic tree**.

Branch site
During splicing of group II introns and spliceosomal introns the 2'-OH of an adenylate (A) residue, located upstream from the 3' end of the intron, attacks the 5' end of the intron resulting in the cleavage of the phosphodiester bond between the upstream exon and the intron and the formation of a 2',5'-phosphodiester bond between this A residue and the 5'-terminal phosphate of the intron. Since this adenylate residue remains joined to two other nucleotides by normal 3',5'-phosphodiester bonds, a branch is generated at this site and a lariat intermediate is formed.

Cα
The central carbon atom, Cα, common to all amino acids to which is attached an amino group (NH_2), a carboxyl group (COOH), a hydrogen atom and a specific side chain. Different types of amino acids are distinguished by their specific side chain.

Cap
In eukaryotes the 5' end of the mRNAs has a structure called a 'cap' consisting of a 7-methylguanosine in 5'-5' triphosphate linkage with the first nucleotide of the mRNA. The 'cap' structure is added post-transcriptionally to the primary transcript and it plays a role in determining the site of translation initiation. *See also* **mRNA**, **Translation start site**.

Carboxyl terminus
Refers to the COOH end of a polypeptide chain.

Catalytic triad
A group of three amino acid residues in an enzyme structure responsible for its catalytic activity. The three residues may be far apart in the amino acid sequence of the protein but in the folded structure they are in close proximity in the active site to perform the enzyme's catalytic function. The best-known catalytic triad is the His–Asp–Ser triad of serine proteinases.

CCAAT box
A sequence found in the 5' flanking region of certain genes. The transcription factor, CCAAT-binding protein (CBP), binds to this site. *See also* **Box**.

cDNA

complementary DNA. A DNA sequence obtained by reverse transcription of an mRNA sequence. The term 'cDNA' is usually used to describe double-stranded DNA which originated from a single-stranded RNA molecule, even though only one strand of this DNA is complementary to the mRNA. Note that a full-length cDNA corresponds to a full-length mRNA sequence therefore it contains the coding region, the 5'-untranslated region, the 3' untranslated region and poly(A) tail, but the introns and promoter sequences of the corresponding protein-coding gene are absent from such cDNAs.

Cell

The basic structural and functional unit of living organisms. Cells may exist as independent units of life, as in bacteria, archaea and unicellular eukaryotes or may form colonies or tissues as in plants and animals.

Cenancestor. *See* **Ancestral genome**.

Centimorgan

cM, the unit of map distance between two genes on the same chromosome based upon recombination frequency within the interval. One centimorgan is equal to a 1% chance that a marker at one genetic locus will be separated from a marker at a second locus due to crossing over in a single generation. The correlation between centimorgan and physical distance in base pairs is not uniform within or between organisms. In human, one centimorgan is equivalent, on average, to one million base pairs. *See also* **Recombination**.

Centromere

The chromosomal region containing the kinetochore, where the spindle fibres attach during meiosis and mitosis. A centromere is essential for separation of the chromatids to daughter cells during cell division. Human centromeres are composed of a large block of a complex family of repetitive DNA, called alphoid sequences. The basic alphoid sequences of ~170 bp are similar among chromosomes. *See also* **Meiosis, Mitosis**.

Cephalochordate

Chordate marine animals that lack a true vertebral column.

Chiasma

Points where chromatids exchange genetic material during chromosomal crossover during meiosis. *See also* **Recombination**.

Chloroplast

The photosynthesizing organelle of eukaryotes such as algae and plants. The chloroplast of eukaryotes has originated by endosymbiosis of a cyanobacterium and an ancestral unicellular eukaryote. *See also* **Endosymbiotic theory**.

Chloroplast chromosome
Circular DNA found in the photosynthesizing organelle.

Chordate
A member of the animal phylum Chordata, which includes the urochordates (e.g. *Ciona intestinalis*), cephalochordates (e.g. amphioxus) and vertebrates. These animals have a rod-like notochord that forms the basis of the internal skeleton.

Chromatin
Chromatin is the complex of DNA and protein inside the nuclei of eukaryotic cells. *See also* **Euchromatin**, **Heterochromatin**.

Chromosomal deletion
Breakage and loss of a segment of a chromosome.

Chromosomal inversion
Breakage within a chromosome at two positions, followed by rejoining of the ends after the internal segment reverses orientation.

Chromosomal translocation
Breakage of two nonhomologous chromosomes with exchange and rejoining of broken fragments. A reciprocal translocation occurs when there is no loss of genetic material and each product contains a single centromere.

Chromosome
The self-replicating genetic structure of cells. In prokaryotes, one or more circular chromosomes generally contain the complete set of genes of the organism. Eukaryotic genomes consist of a number of linear chromosomes whose DNA is associated with different kinds of proteins. The number, size and gene content of each chromosome are species specific and invariant among individuals of a species. Structural features describing eukaryotic chromosomes include the centromere, the telomere and the long and short arms. *See also* **Replication**.

Clade
A monophyletic group in which all members descended from a single ancestor and the group includes all descendants of that ancestor.

Cladistic classification
The members of a group in a cladistic (or phylogenetic) classification share a more recent common ancestor with one another than with the members of any other group. *See also* **Phenetic classification**.

Cladogram
Cladograms are phylogenetic trees that represent the order in which different groups branched off from their common ancestor; the most closely related species are on adjacent branches. *See also* **Phylogenetic tree**.

Clock. *See* **Molecular clock**.

Code. *See* **Genetic code**.

Coding region
The term denotes only the regions in genomic DNA that end up translated into amino acid sequences. The coding regions are the coding fraction of coding exons and do not include untranslated exons or the untranslated segments of the coding exons. *See also* **Open reading frame (ORF)**.

Coding sequence
The portion of a gene or an mRNA which actually codes for a protein. Introns are not coding sequences, nor are the 5′ or 3′ untranslated regions of mRNA. The coding sequence in a cDNA or mature mRNA includes everything from the AUG (or ATG) initiation codon to the stop codon. *See also* **Open reading frame (ORF)**.

Codon
The triplet (three-base unit) of the genetic code that specifies the incorporation of a specific amino acid into a protein during mRNA translation. The translated region of mRNA consists of a series of triplets that are read in a 5′ to 3′ direction and specify the amino acid sequence of the protein encoded by that mRNA. The triplet of bases that is complementary to a codon is called an anticodon; the triplet in the tRNA that reads the codon is called the anticodon. The genetic code is degenerate with most of the 20 amino acids represented by several codons. Initiation and termination of translation are specified by initiator and terminator codons. Codon usage shows significant variation between different groups of organisms. *See also* **Genetic code**.

Codon usage bias
The tendency for an organism or virus to use certain synonymous codons more frequently than others to encode a particular amino acid. In the standard genetic code there are 61 codons encoding 20 amino acids. Codons that encode the same amino acid are called synonymous codons. Usage of synonymous codons is nonrandom and the preferences in codon usage are called codon usage bias. Codon usage bias reflects mostly the background base composition of the genome: an organism that has a relatively low G+C content will be less likely to have a G or C at synonymous positions than an A or T. Codon usage bias may also correlate with the relative abundance of alternative tRNA isoacceptors.

Coevolution

Coevolution is a reciprocally induced evolutionary change in a biological entity (species, organ, cell, organelle, gene, protein, biosynthetic compound) in response to a change in another with which it interacts. See for example coevolution of predator and its prey, parasite and its host. In the case of molecular coevolution, this term describes the coevolution of molecules involved in protein–protein and DNA–protein interactions, respectively. Coevolution is reflected in cladograms of the interacting entities: the cladograms of coevolving entities are said to be congruent.

Coevolution of protein residues

The three-dimensional structure of a protein is maintained by several types of noncovalent interactions (short-range repulsions, electrostatic forces, van der Waals interactions, hydrogen bonds, hydrophobic interactions) and some covalent interactions (disulphide bonds) between the side chain or peptide backbone atoms of the protein. There is a strong tendency for the correlated evolution of interacting residues of proteins. For example, in the case of extracellular proteins, pairs of cysteines forming disulphide bonds are usually replaced or acquired 'simultaneously' (on an evolutionary time scale). Such correlated evolution of pairs of cysteines involved in disulphide bonds frequently permits the correct prediction of the disulphide bond pattern of homologous proteins. The identification of protein sites undergoing correlated evolution is of major interest for structure prediction since these pairs tend to be adjacent in the three-dimensional structure of proteins. *See also* **Coevolution**.

Coil (random coil)

Irregular, unstructured regions of a protein's backbone, as distinct from the regular regions (α-helices, β-sheets, turns, ordered loops) characterized by specific patterns of main-chain hydrogen bonds. *See also* **Loop**.

Coiled coil

Coiled coils are elongated helical structures that contain a highly specific heptad repeat that conforms to a specific sequence pattern. The first and fourth positions within this pattern are hydrophobic while the remaining positions show preferences for polar residues. Coiled coils form intimately associated bundles of long α-helices.

Common ancestor

The most recent ancestral form from which two different entities evolved.

Comparative genomics

A research area dealing with the global comparison of genome information in terms of their structure and gene content. The information from whole-genome comparisons may include conservation of sequences across species, identification of orthologous and paralogous genes, conservation of noncoding regulatory regions, conservation of local gene order (synteny) etc. *See also* **Genomics**.

Comparative modelling. *See* **Homology modelling**.

Complexity
The numbers of components and interactions in a system.

Concerted evolution (molecular drive)
Tandem members of a multigene family may become homogenized over evolutionary time through unequal crossing over or gene conversion. This results in greater sequence similarity within a species than between species when comparing members of gene families. *See also* **Gene conversion**.

Consensus sequence
A consensus sequence is a sequence of nucleotides or amino acids that best represents a set of multiply aligned sequences. A consensus sequence is usually determined by selecting – at each individual position – the nucleotide or amino acid residue that is found most frequently in that position. Consensus sequences are useful in defining common features of homologous sequences, recognition sites etc. *See also* **Sequence logo**.

Consensus tree
A single tree summarizing phylogenetic relationships from multiple individual phylogenies.

Conserved synteny. *See* **Synteny**.

Control elements
DNA sequences that interact with regulatory proteins and through these interactions regulate the expression of genes. *See also* **Transcription factor**, **Enhancer**, **Silencer**.

Convergence
The process by which a similar character (structure, function, etc.) evolves independently in different entities species, biological macromolecules, etc. Similar characters that evolved independently are said to be analogous. Conversely, similar characters that reflect common ancestry are said to be homologous. *See also* **Analogous**, **Homologous**.

Crossing over
The process during meiosis in which the chromatids of homologous chromosomes of a diploid pair exchange genetic material. This process can

result in a reciprocal exchange of alleles between chromosomes. *See also* **Recombination**.

Conservative substitution
A nucleotide mutation which alters the amino acid sequence of the protein, but which causes the substitution of one amino acid with another that has a side chain with similar charge/polarity characteristics. *See also* **Non-conservative substitution**.

CpG island
A segment of genomic DNA with base composition >60% GC and CpG dinucleotide frequency 0.6 observed/expected as opposed to other regions of the mammalian genome where the average base composition is ~38% GC and the average CpG frequency is 0.2–0.25 observed/expected. CpG islands are found more often in GC-rich (gene-rich) bands, are most often unmethylated, located at or near the 5′ ends of genes and associated with promoters. *See also* **Base composition**, **Isochore**.

C-value. *See* **Genome size**.

C-value paradox
The paradox that the genome size (*C*-value) of a species does not correlate with its organismic complexity.

Cysteine
Cysteine is a nonpolar amino acid found in proteins. In protein sequences it is written as Cys or C.

Cytosine (C)
A pyrimidine base, one member of the base pair GC (guanine and cytosine) in DNA. *See also* **Base pair**.

Darwinian evolution
Evolution by natural selection acting on random variation.

Darwinism
Darwin's theory that species originated by evolution from other species and that evolution is mainly driven by natural selection.

Dayhoff amino acid substitution matrix (PAM matrix, percent accepted mutation matrix)
Amino acid substitution matrices describe the probabilities and patterns displayed by nonsynonymous mutations of nucleotide sequences during evolution. Amino acid substitution probabilities in proteins were analysed

by Dayhoff, who tabulated nonsynonymous mutations observed in several different groups of global alignments of closely related protein sequences that are at least 85% identical. Mutation data matrices, or percent accepted mutation (PAM) matrices, commonly known as Dayhoff matrices, were derived from the observed patterns of nonsynonymous substitutions, and matrices for greater evolutionary distances were extrapolated from those for lesser ones. Dayhoff amino acid substitution matrices are used in the alignment of amino acid sequences to calculate alignment scores and in sequence similarity searches. *See also* **BLOSUM matrices**.

Degeneracy
Refers to a feature of the genetic code that different – synomymous – codons can specify the same amino acid in proteins.

Deletion
Loss of a segment of the DNA – may refer to loss of a single base pair or of very large regions of a chromosome.

Deoxyribose
A five-carbon sugar that serves as a component of DNA. *See also* **Ribose**.

Deuterostome
A taxon of animals that includes echinoderms (e.g. sea urchins, starfishes), invertebrate chordates (e.g. *Ciona*), vertebrates etc. A distinguishing feature of deuterostome development is that the first opening (the blastopore) becomes the anus, whereas in protostomes it becomes the mouth. *See also* **Protostome**.

Diatom
Single-celled algae. Their two-part shells have intricate patterns.

Diploid
A full set of genetic material consisting of paired chromosomes, one from each parental set. Most animal cells except the gametes have a diploid set of chromosomes. *See also* **Haploid**.

Direct repeats
Identical or related sequences present in two or more copies in the same orientation in the same molecule of DNA.

Disulphide bridge
Two cysteine residues may form a covalent S–S bond in the oxidative extra-cytoplasmic space. Disulphide bridges usually occur only in extracellular proteins or extracellular parts of membrane proteins. *See also* **Cysteine**.

DNA

Deoxyribonucleic acid, the molecule that encodes genetic information in cellular forms of life. The four nucleotides in DNA contain the bases adenine (A), guanine (G), cytosine (C) and thymine (T). DNA is a double-stranded molecule held together by weak bonds between base pairs of nucleotides: A pairs with T; G pairs with C.

DNA–protein coevolution

Coevolution is a reciprocally induced evolutionary change in a biological entity (species, organ, cell, organelle, gene, protein, biosynthetic compound) in response to an inherited change in another with which it interacts. Protein–DNA interactions essential for some biological function (e.g. binding of a hormone receptor to the appropriate hormone response element, binding of a transcription factor to a transcription factor binding site) are ensured by specific contacts between a DNA sequence motif and the DNA-binding protein. During evolution gene regulatory networks have expanded through the expansion of the families of nuclear hormone receptors, transcription factors and the coevolution of the DNA-binding proteins with the DNA sequence motifs to which they bind. Coevolution of the novel hormone receptor (with novel ligand specificity) and the target DNA binding sites has led to an increased complexity of gene regulatory networks. *See also* **Coevolution**.

DNA replication

The use of one strand of DNA as a template for the synthesis of a new, complementary DNA strand. Bacteria typically have a single replication origin and a single terminus, making the genome a single replicon. Eukaryotic genomes have numerous origins, termini and replicons.

DNA sequence

The relative order of bases, adenine (A), cytosine (C), guanine (G) and thymine (T) in DNA. The DNA sequence is written in the $5' \rightarrow 3'$ direction, reflecting the direction of replication.

Domain. *See* **Protein domain**.

Domain family

A collection of single-domain and/or multidomain proteins that share a common domain. Domain families are thus different from protein families: the latter share similarity along the entire length of their sequences.

Dominant allele

If heterozygosity exists (i.e. two different alleles are found) at a given locus and the two alleles encode different phenotypes then the phenotype that is manifested is that of the dominant allele. *See also* **Recessive allele**.

Donor splice site

The binding site of the spliceosome on the 5′ end of an intron and the 3′ end of the upstream exon is called a donor splice site. *See also* **Acceptor splice site**.

Double helix

Two linear strands of DNA assume a double helix when complementary nucleotides on opposing strands interact through canonical base pairs. *See also* **Base pair**.

Downstream/upstream

This orientation reflects the direction of both the synthesis of mRNA and its translation from the 5′ end to the 3′ end. In the case of genomic DNA a sequence element is said to be downstream if it is located distal to a specific point in the direction of transcription, upstream if it is in the opposite direction. For example, the transcription start site is downstream of the promoter region, whereas the promoter region is upstream of the transcription start site. It should be noted that downstream/upstream is meaningful only in conjunction with a given gene. In the case of mRNA a sequence element is said to be downstream if it is located distal to a given point in the direction of translation. For example, the coding region is downstream from the initiation codon, towards the 3′ end of an mRNA molecule, whereas the coding region is upstream of the stop codon.

Drift

Synonym of genetic drift.

Duplication

Duplication of a segment of the DNA – can refer to a few base pairs or of very large regions of a chromosome. The duplicate sequence may appear next to the original or be copied elsewhere into the genome.

Echinoderm

Echinoderms are a group of marine invertebrate deuterostomes that includes starfish, sea cucumbers and sea urchins.

Endosymbiotic theory

This theory assumes that the key organelles of eukaryotic cells (mitochondria, chloroplasts) originated from prokaryotic organisms engulfed by ancestral eukaryotic cells as endosymbionts.

Enhancer

Transcriptional control elements of eukaryotic DNA that act at some distance to enhance the activity of a specific promoter sequence. They may be upstream or downstream and can exert their effects independently of orientation relative to the promoter.

EST (expressed sequence tag)
ESTs are short (usually approximately 300–500 base pairs) sequence fragments selected at random from a cDNA library.

Euchromatin
Region of chromosome that encodes actively expressed genes.

Eukaryote
Unicellular or multicellular organisms with membrane-bound, structurally discrete nucleus. *See also* **Prokaryote**.

***E*-value (expectancy value)**
The expectancy value (*E*-value) is a statistical value for estimating the significance of alignments performed in a homology search of a query sequence against a sequence database. It describes the number of hits one can expect to see by chance when searching a database of a particular size. It estimates the number of sequences randomly expected in the sequence database to provide an alignment score at least that of the alignment considered.

Evolutionary clock. *See* **Molecular clock**.

Evolutionary distance
An evolutionary distance measure is a numerical representation of the dissimilarity between two species, sequences or populations. Evolutionary distances are often computed by comparing molecular sequences. A set of evolutionary distances may be used to construct a phylogenetic tree. *See also* **Phylogenetic tree**.

Evolutionary footprinting
One can define the portions of a genomic sequence that are important by comparing that sequence with its orthologs from other species. Regions of high conservation are likely to be regions that are functional in all the test species.

Exon
Those portions of a genomic DNA sequence which are represented in the final, mature mRNA. Eukaryotic genes are split into exons and intervening sequences (introns). The primary RNA transcript contains regions corresponding to both exons and introns. The introns are removed by RNA splicing reaction, they are not present in mature mRNA. The coding region of a gene is contained within the exons, but exons also include the 5′ and 3′

untranslated regions of the mRNA. Exons or parts of exons of protein-coding genes that are translated are collectively referred to as protein-coding regions. *See also* **Coding region, Intron**.

Exon shuffling

The process whereby exons of genes are duplicated or deleted or exons of different genes are joined through recombination in introns. Exon shuffling has played an important role in the evolution of new gene functions, particularly when exons that are transferred encode a domain that constitutes a structural and functional unit. Shuffling of exons or exon sets encoding complete protein domains leads to module shuffling and to the formation of multidomain proteins with an altered domain organization. *See also* **Exon, Intron, Intron phase, Module shuffling, Multidomain protein, Protein module**.

Exonic splicing enhancers (ESEs), exonic splicing silencers (ESSs)

Cis-acting sequence elements in exons that promote or suppress splice site recognition. ESEs and ESSs also play important roles in the regulation of alternative splicing: silencers promote the skipping of an alternative exon, enhancers promote the inclusion of the alternative exon. Splicing enhancer and silencer sequences are usually short (5–10 nucleotide) sequences.

Expressed sequence tag (EST)

A short sequence fragment of a cDNA clone selected at random from a cDNA library.

Expression

Manifestation of a structural gene's function. In the case of protein-coding genes this term refers to the entire process that includes transcription, translation and post-translational modification. In the case of genes that encode RNA (for example, a tRNA gene) expression includes production and post-transcriptional modifications of the RNA.

Expression profile. *See* **Gene expression profile**.

Fitness

The success of an allele in a population in surviving and reproducing as measured by the allele's contribution to the next generation and subsequent generations.

Fixation

An allele has achieved fixation when its frequency has reached 100% in the population.

Fold

The fold of a protein describes the topology of the three-dimensional domain structure that the polypeptide chain adopts as it folds. The topology refers to both the secondary structures in the domain structure and the arrangements of those positions in three-dimensional coordinate space.

Fold recognition

Fold recognition methods attempt to predict the tertiary structure of a protein from its amino acid sequence by selecting the fold from a fold library that best matches that sequence.

Frame-shift mutation

In translated regions of protein-coding genes deletion or insertion of a number nucleotides that is not a multiple of three shortens or lengthens a codon and this results in a shift from one reading frame to another reading frame. The amino acid sequence of the protein downstream of the mutation is completely altered and is usually much shorter due to premature termination.

Function prediction – homology-based methods

Homology-based methods of function prediction usually transfer the function from a gene/protein (where functions are known) to its homologs.

Function prediction – nonhomology-based methods

Nonhomology-based comparative genomic methods of function prediction rely on the fact that genes encoding physically interacting partners tend to be proximate on the genome (genomic proximity methods), tend to evolve in a correlated manner (phylogenetic profiling methods) and to be fused as a single sequence in some organisms (domain fusion methods).

Functional genomics

A research area dealing with the determination of gene function on a genome-scale. Functional genomics uses high-throughput approaches to determine the function of every gene and every protein in an organism to understand the complex networks of metabolic pathways and regulatory control mechanisms. *See also* **Genomics**.

Functional signature

A sequence pattern that is uniquely associated with a given function in a protein.

Fungi

A group of eukaryotic organisms comprising the kingdom Fungi that includes yeasts, moulds and mushrooms. They can exist either as single cells or make up a multicellular body called a mycelium.

Gamete
Haploid male or female reproductive cells (sperm or ovum) that combine at fertilization to form the zygote.

Gap (indel)
A gap is a space introduced into an alignment in one sequence relative to another. The gaps represent sites where the affected sequences are thought to have accumulated an insertion or deletion (indel) during divergent evolution.

Gap penalty
To prevent the accumulation of too many gaps in an alignment, introduction of a gap is usually penalized with a negative score. Terminal gaps at the ends of alignments are usually treated differently to internal gaps reflecting the fact that the N-terminal and C-terminal regions of proteins are quite tolerant to extensions/deletions.

GC composition. *See* **Base composition**.

GC richness. *See* **Base composition**.

Gene
A segment of a genome (usually DNA but may be RNA in some viruses) which performs some essential biological function. In the case of structural genes the biological function is exerted through the production of RNA (RNA genes) or protein (protein-coding genes).

Gene amplification
Repeated copying of a piece of DNA resulting in multiple copies of a gene. Gene amplification is common in tumour cells.

Gene cluster
A group of related genes found in close physical proximity in genomic DNA. For example, ribosomal RNA genes and genes for histones are clustered in the human genome.

Gene conversion
The alteration of all or part of a gene by a homologous donor DNA that is itself not altered in the process.

Gene disruption
The interruption of a specific coding sequence by insertion of foreign DNA into the coding region.

Gene distribution

Gene density within chromosomes and chromosome isochores is strongly nonuniform, with G+C poor regions being relatively gene poor, whereas G+C rich regions are gene rich. See also **Isochore**.

Gene duplication

The process whereby a new copy of a gene is formed. The most common mechanism of gene duplication is unequal crossing over that leads to tandem duplication of genes. Successive duplications caused by unequal crossing over lead to gene clusters.

Gene elongation

An increase in the length of a gene. The most common mechanism of gene elongation is through intragenic duplication.

Gene expression

The process by which the information of structural genes (RNA genes and protein-coding genes) is converted into the structures that carry out specific functions of the organism or virus. Expression of RNA genes leads to the production of RNA (e.g. rRNA, tRNA), expression of protein-coding genes gives rise to mRNA which is translated into protein.

Gene expression profile

Differences in the abundance of the transcript(s) of a given gene in a number of different cell types, organs, tissues, developmental, physiological or disease conditions.

Gene family

A group of related genes that have descended from the same ancestral gene. In contrast with domain families, members of gene or protein families share significant similarity along the entire length of their sequences. Members of a gene family may be clustered at one or several chromosome locations or may be widely dispersed. Sequence divergence among members of a gene family may result in functional divergence, expression pattern divergence or conversion to pseudogenes.

Gene knockout. *See* **Gene disruption**.

Gene loss and gene inactivation

Processes by which genes may be lost from genomes or become non-functional. Gene loss and inactivation may occur by deletion of a gene, by mutations that prevent its transcription or by mutations that prevent its translation. Inactivated genes (pseudogenes) are abundant in genomes. *See also* **Pseudogene**.

Gene product
The final macromolecule product encoded by an expressed structural gene. It is RNA in the case of RNA genes whereas it is a protein in the case of protein-coding genes.

Gene sharing
A single gene encodes a protein that has more than one function. For example, in the course of animal evolution, soluble proteins having a wide variety of original functions have also been recruited to serve the role of eye crystallins.

Gene size
The amount of genomic DNA in base pairs containing all exons and introns of a gene.

Genetic code (standard genetic code, universal genetic code)
The code by which cells translate protein-coding DNA sequences into amino acid sequences. Bases in protein-coding regions are arranged into non-overlapping codons consisting of a group of three consecutive bases (triplets). The genetic code is almost universal in nuclear genomes.

Genetic drift
Changes in the frequencies of alleles in a population that occur by chance rather than because of natural selection.

Genetic polymorphism
Difference in DNA sequence among individuals, groups or populations within the same species.

Genetic redundancy
If members of a gene family perform the same function deletion of one gene may have no effect on the organism's viability. For example, the human genome contains hundreds of rRNA genes in five clusters. The number of rRNA genes in each cluster varies among individuals and entire clusters can be deleted with no obvious negative consequences to the individual.

Genome
The complete set of genetic information defining a particular organism or virus.

Genome sequence
The complete sequence of the total DNA content of a single cell of an organism or a virus. The genome sequence of a diploid eukaryotic organism covers a complete haploid set of the nuclear genome (including both sex

chromosomes) and a single copy of any non-nuclear genomes (mitochondrial and plastid genomes). In the case of bacteria the genome sequence includes the sequence of the main chromosome(s) and those of plasmids.

Genome size (*C*-value)
The number of base pairs of DNA in a haploid genome of a given species. *See also* **C-value paradox**.

Genomics
Global, genome-wide approach to the structural and functional analysis of all genes, all transcripts and all proteins encoded by the genomes of organisms. The genome-wide approach requires the use of high-throughput automated experimental technologies and computational tools. Genomics technologies include transcriptomics, proteomics, interactomics, functional genomics, comparative genomics.

Genotype
The genetic constitution of an organism as distinguished from its physical appearance and functional properties (its phenotype). *See also* **Phenotype**.

Germ cell
Cells in a diploid organism can be divided into diploid somatic cells and haploid germ cells. Germ cells are the reproductive cells that produce the gametes (sperm and egg cells).

Germline
Genetic information from one generation to the next is transferred through the germline. As a consequence, only germline mutations are inherited.

Gigabase (Gb)
Unit of length for DNA, equal to 1 billion base pairs.

Global alignment
An alignment of two or more sequences over their full lengths, as opposed to local alignment. *See also* **Local alignment**.

Globular proteins
Proteins characterized by a compact isospheric three-dimensional conformation of their polypeptide chain are called globular. They form the majority of naturally occurring proteins in solution and are distinct from structural proteins such as the fibrous proteins.

Glutamic acid
Glutamic acid is a polar (usually negatively charged) amino acid found in proteins. In protein sequences it is written as Glu or E.

Glutamine
Glutamine is a polar amino acid found in proteins. In protein sequences it is written as Gln or Q.

Glycine
Glycine is a small nonpolar amino acid found in proteins. In protein sequences it is written as Gly or G.

Glycoprotein
A glycosylated protein.

Glycosylation
The covalent addition of sugar moities to N or O atoms present in the side chains of certain amino acids of certain proteins.

Golgi apparatus
A membranous structure that is in continuity with the endoplasmic reticulum of eukaryotic cells. The Golgi apparatus plays an important role in the post-translational processing and transport of secreted proteins.

Group I intron. *See* **Intron**.

Group II intron. *See* **Intron**.

Guanine (G)
A purine base, one member of the base pair GC (guanine and cytosine) in DNA. *See also* **Base pair**.

Hairpin
A duplex region formed by base pairing between adjacent inverted complementary sequences within a single strand of RNA or DNA. In the case of hairpins the complementary regions are separated by a short region that is not part of the base paired region. *See also* **Stem loop**.

Haploid
A single set of chromosomes present in the gametes (egg and sperm cells of animals and egg and pollen cells of plants). The haploid set of chromosomes has only one copy of each locus in the genome. *See also* **Diploid**.

Hemizygous
Having only one copy of a chromosome. For example, males have only one copy of the Y chromosome.

Heterochromatin
Regions of chromosomes that are practically devoid of genes. Heterochromatin has low levels of histone acetylation, is highly condensed and contains repetitive DNA elements. *See also* **Euchromatin**.

Heterozygosity
The proportion of individuals in a population that are heterozygotes.

Heterozygote
An individual having two different alleles at a genetic locus. *See also* **Homozygote**.

Heterozygous
Having different alleles at a given locus on homologous chromosomes of a diploid organism. *See also* **Homozygous**.

Histidine
Histidine is a polar (sometimes positively charged) amino acid found in proteins. In protein sequences it is written as His or H.

Homeobox
Homeoboxes are relatively short, very similar or identical sequences of DNA, characteristic of homeotic genes that play a central role in controlling body development and are shared by almost all eukaryotic species. Homeoboxes encode a protein 'homeodomain', a protein domain that binds to DNA.

Homoallelic
Synonym of homozygous.

Homologous
Biological entities (organs, structures, macromolecules) whose similarity is attributable to descent from a common ancestor are said to be homologous. For example, two genes are homologous if they are descended from a common ancestral gene either through speciation or gene duplication. *See also* **Analogous**.

Homologous recombination
Exchange of fragments between two related but different DNA molecules resulting in a new chimeric molecule. An essential requirement is the existence of a region of homology of the recombination partners.

Homologous superfamily
A homologous superfamily consists of a group of proteins related by divergent evolution from a common ancestral protein. In contrast to a protein family, the superfamily contains more distant relatives that may have no detectable sequence similarity.

Homology
Similarity of biological entities (organs, structures, macromolecules) attributable to descent from a common ancestor.

Homology-based function prediction. *See* **Function prediction – homology-based methods**.

Homology modelling
The prediction of a three-dimensional protein structure from a protein's amino acid sequence based on an alignment to a homologous amino acid sequence of a protein with an experimentally solved three-dimensional structure.

Homology search. *See* **Sequence similarity search**.

Homozygote
An individual having two copies of the same allele at a genetic locus.

Homozygous
An individual having two copies of the same allele at a genetic locus is said to be homozygous for that locus. *See also* **Heterozygous**.

Horizontal gene transfer
Horizontal (or lateral) gene transfer occurs when a gene is transferred from the genome of one species to that of another species. A gene acquired through horizontal gene transfer that is homologous with a gene of the recipient species is called a xenolog. *See also* **Xenolog**.

Housekeeping genes
Genes encoding proteins essential for general cell functions that are transcribed and expressed in all cell types.

Hybrid
Biological entities that descended from genetically different parents.

Hydrogen bond
An interaction between two nonhydrogen atoms, one of which (the proton donor) has a covalently bonded hydrogen atom. The other nonhydrogen atom (the proton acceptor) has electrons not involved directly in covalent bonds that can interact favourably with the hydrogen atom. *See also* **Secondary structure**.

Hydrogenosome
An organelle found in some anaerobic flagellates, ciliates and fungi that produces hydrogen and ATP. Hydrogenosomes have evolved from mitochondria. *See also* **Mitochondrion**, **Organelle**.

Hydropathy (hydrophobicity)
Low affinity for water.

Hydrophilicity
High affinity for water. The term hydrophilicity is applied to polar chemical groups that have favourable energies of solvation in water. *See also* **Hydropathy**.

Hydrophobicity. *See* **Hydropathy**.

Indel
Indel denotes evolutionary insertion and deletion events. If indels have occurred during divergence of related genes/proteins the alignment of their sequences requires the insertion of 'gap' characters in order to bring corresponding regions into the correct register. *See also* **Alignment**, **Deletion**, **Gap**, **Insertion**.

Initiation codon
Also called a 'start codon'. The codon at which translation of a polypeptide chain is initiated and which determines the reading frame. In the case of eukaryotes this is usually the first AUG triplet in the mRNA molecule from the 5′ end, where the ribosome binds to the cap and begins to scan the mRNA in a 3′ direction. *See also* **Translation**, **Translation start site**, **Open reading frame (ORF)**.

Initiator sequence
Two different sequence elements are involved in specifying the transcriptional initiation site of polymerase II transcribed genes in eukaryotes. These are the TATA box (located 30–35 bp upstream from the transcription start site) and the initiator. The initiator and the TATA box bind the multiprotein complex TFIID, which is involved in specifying the initiation site. *See also* **Transcription**.

Insertion
Insertion of one or several base pairs of DNA into genomic DNA. The size of DNA inserted can be a single base pair or entire genes and even chromosomal segments. *See also* **Indel**.

Interactome
The networks of interactions between all macromolecules of the cell, including protein–protein, protein–DNA and protein–carbohydrate interactions. *See also* **Genomics**.

Interactomics
Determination of interactions between all macromolecules of the cell, including protein–protein, protein–DNA and protein–carbohydrate interactions.

Intergenic sequence

Regions in genomic DNA between genes. In the case of neighbouring genes arranged in head-to-tail orientation (i.e. genes transcribed from the same strand of DNA) this means the region between the promoter of one gene and the 3′ end of its upstream neighbour. In the case of neighbouring genes arranged in head-to-head orientation (i.e. genes transcribed from opposite strands) the intergenic distance is the distance between their promoters. Intergenic distances are highly variable in size. In G+C-rich, gene-rich regions, they tend to be short, whereas in A+T-rich, gene-poor regions intergenic distances can be very large.

Intron

Sequence within a gene that separates two exons. Regions corresponding to introns are present in the primary transcript but are removed by splicing during RNA processing and are not included in the mature mRNA, rRNA or tRNA. There are three major types of introns. Group I and group II introns are self-splicing introns that have a conserved tertiary structure. Group I introns are found in some RNA transcripts of protozoa, fungal mitochondria, bacteriophage T4 and bacteria. Group II introns are found in fungal mito-chondria, higher plant mitochondria and plastids. Introns of nuclear genes of eukaryotes are spliced by the spliceosome (a complex of proteins and small nuclear RNAs), therefore they are usually called spliceosomal introns. In the majority of nuclear genes, the sequence of introns begins with GT and ends with AG (the GT–AG rule). *See also* **Exon**, **Splicing**, **Spliceosome**.

Intron phase

The phase of an intron (located in the translated region of a protein-coding gene) refers to its position relative to the reading frame. There are three types of intron phases: phase 0, the intron lies between two codons; phase 1, the intron occurs between the first and second bases of a codon; phase 2, the intron occurs between the second and third bases of a codon. An exon in which both upstream and downstream introns have the same phase is called a symmetrical exon. If an exon from one gene is inserted into an intron of another gene (as in cases of exon shuffling), the exon must be flanked by introns of the same phase as the recipient intron, otherwise the reading frame is disrupted. Only symmetrical exons satisfy this requirement. *See also* **Symmetrical exon**.

Intronic splicing enhancers (ISEs), intronic splicing silencers (ISSs)

Cis-acting sequence elements in introns that promote or suppress splice-site recognition. Intronic splicing enhancers (ISEs) and intronic splicing silencers (ISSs) also play important roles in the regulation of alternative splicing: silencers promote the skipping of an alternative exon; enhancers promote the inclu-sion of the alternative exon. Splicing enhancer and silencer sequences are usually short (5–10 nucleotide) sequences.

Inverted repeats

Two copies of the same or related sequence of DNA repeated in opposite orientation on the same molecule. *See also* **Direct repeats**, **tandem repeats**.

Inversion

A type of mutation where a segment of DNA sequence has been placed in the reverse orientation.

Isochore

Large DNA segments (>300 kb) within a genome that are homogeneous in base composition. Genomes of warm-blooded vertebrates are mosaics of isochores belonging to four classes: the L isochores are A+T rich (~35% to 42% GC) and the H1, H2 and H3 isochores are increasingly G+C rich (42% to 46%, 47% to 52% and >52%, respectively).

Isoleucine

A nonpolar amino acid found in proteins. In protein sequences it is written as Ile or I.

Kilobase (kb)

One kilobase equals 1000 base pairs of DNA.

Kinetochore

Region of the centromere to which spindle fibres attach during mitosis and meiosis. The kinetochore is required for segregation of chromatids to daughter cells. *See also* **Centromere**.

Knock-out

The excision or inactivation of a gene within an intact organism, usually carried out by a method involving homologous recombination. *See also* **Gene disruption**.

Kozak sequence

Consensus sequence around the site of initiation of eukaryotic mRNA translation. The Kozak sequence is involved in recognition of translation start site by the ribosome. The optimal Kozak consensus sequence has been identified as CCA/GCCAUGG. The small 40S ribosomal subunit, carrying the methionine tRNA and translation initiation factors binds to the mRNA at the capped 5′ end and then scans the mRNA until it encounters the first AUG (initiator) codon. The Kozak sequence modulates the ability of the AUG codon to halt the scanning 40S subunit.

K selection

Natural selection favours organisms with long life cycles that live in stable environments. According to the r/K selection theory selective pressures drive evolution in one of two stereotyped directions: r or K selection. Typically, r-selected species produce many offspring, whereas K-selected species have fewer offspring that have a better chance of survival. *See also* **r selection**.

L1 element
Human long interspersed repetitive element (LINE) capable of transposition and present in approximately 100 000 copies in the human genome. Full-length L1 sequences are 6.4 kb long, but the majority of copies are truncated at the 5′ end. *See also* **LINE**, **SINE**.

Lariat intermediate. *See* **Branch site**.

Last universal common ancestor. *See* **Ancestral genome**.

Lateral gene transfer. *See* **Horizontal gene transfer**.

Leucine
Leucine is a nonpolar amino acid found in proteins. In protein sequences it is written as Leu or L.

LINE (long interspersed nuclear element)
A class of large transposable elements common in the human genome. Long interspersed elements differ from other retrotransposons in that they do not have long terminal direct repeats (LTRs). There are two major LINE transposable elements in the human genome, L1 and L2. The approximately half a million repeats take up more than 15% of the human genome. *See also* **L1 element**, **SINE**.

Lineage
A sequence of species, populations, cells or genes that descend from a common ancestor.

Local alignment
Alignment of the most similar consecutive segments of two or more sequences, as opposed to global alignment that aligns complete sequences. *See also* **Global alignment**.

Log-odds score. *See* **amino acid exchange matrix**.

Loop
Secondary structures such as α-helices and β-strands or sheets are connected to each other by structural segments like turns, coils or loops. Whereas turns are short ordered structures, coils are longer disordered structures, loops are intermediate in this respect inasmuch as they often have at least some definite

structure. Loops and coils are most frequently found on the surface of globular proteins and insertions/deletions (indels) usually are tolerated in loop or coil regions. *See also* **Coil**, **Indel**, **Secondary structure**, **Turn**.

Low-complexity region

A low-complexity region within a protein or nucleotide sequence is a sequence region with a biased amino acid or nucleotide composition that differs significantly from the general composition observed in either the sequence or a database of sequences.

LUCA. *See* **Ancestral genome**.

Lysine

Lysine is a positively charged amino acid found in proteins. In protein sequences it is written as Lys or K.

Main chain

The set of all backbone atoms in a polypeptide chain.

Mammals

The group of warm-blooded vertebrates, evolved from a common ancestor that share properties such as mammary glands, hair or fur.

Maximum-likelihood method of phylogeny reconstruction

A method for selecting a phylogenetic tree relating a set of sequence data based on a probabilistic model of evolutionary change.

Maximum-parsimony method of phylogeny reconstruction

The maximum-parsimony method is based on the parsimony principle: it prefers a phylogenetic tree that requires the fewest total number of evolutionary changes to explain the observed sequences.

Megabase (Mb)

Unit of length for DNA, equal to 1 million base pairs.

Meiosis

A reductive cell division whereby the ploidy (number of chromosome sets) is reduced by half. Meiosis in diploid organisms gives rise to haploid gametes. Crossing over and recombination occur during a phase of meiosis. *See also* **Mitosis**.

Messenger RNA. *See* **mRNA**.

Metazoa

Multicellular animals whose cells are organized into tissues and organs.

Methionine
Methionine is a nonpolar amino acid found in proteins. In protein sequences it is written as Met or M.

Microsatellite
Microsatellites are simple sequence repeats in DNA: they may be homo-polymers consisting of a single type of nucleotide, dinucleotide repeats, trinucleotide repeats etc. Due to polymerase slip, during DNA replication these repeat sequences may become altered; copies of the repeat unit can be created or removed. Consequently, the exact number of repeat units may differ between individuals. *See also* **Simple DNA sequence**.

Minimum-evolution method of phylogeny reconstruction
This method minimizes the sum of branch lengths of a phylogenetic tree: of all possible tree topologies the minimum-evolution topology is the one that requires the smallest sum of branch lengths.

Minisatellite
Tandem repeats of intermediate-length motifs approximately 10–100 bp in length. Similar to microsatellites in that they are variable among individuals. *See also* **Microsatellite**.

Missense mutation
A nonsynonymous mutation, i.e. it alters a codon to a codon that encodes a different amino acid. *See also* **Nonsense mutation**, **Nonsynonymous mutation**, **Synonymous mutation**.

Mitochondrial DNA
The genetic material found in mitochondria, the organelles that generate energy for the eukaryotic cell. Mitochondrial DNA encodes only a fraction of mitochondrial proteins, the majority of mitochondrial proteins are encoded in the nuclear genome. Since mitochondria are generally carried in egg cells but not in sperm, mitochondrial DNA is passed to offspring from mothers, but not fathers.

Mitochondrion
An organelle that generates energy for eukaryotic cells. *See also* **Organelle**.

Mitosis
A cell division that conserves ploidy, the daughter cells are genetically identical to each other and to the parent cell. All cell division in multi-cellular organisms occurs by mitosis except for the division called meiosis that generates the gametes. *See also* **Meiosis**.

Modular protein

Multidomain proteins containing multiple copies and/or multiple types of protein modules. *See also* **Multidomain protein**, **Protein module**.

Module shuffling

During evolution protein modules may be shuffled through either exon shuffling or exonic recombination to create multidomain proteins with various domain combinations. *See also* **Exon shuffling**, **Multidomain protein**, **Protein module**.

Molecular clock (evolutionary clock)

The theory that assumes that molecules evolve at a constant rate, therefore there is a linear relationship between the evolutionary distance and time. For example, according to this theory the difference between orthologous molecules of different species is proportional to the time since the species diverged from a common ancestor.

Molecular coevolution. *See* **Coevolution**.

Molecular drive. *See* **Concerted evolution**.

Monophyletic group

In a phylogenetic tree, a monophyletic group is a group of nodes consisting of a common ancestor and all its descendants (and only its descendants). *See also* **Phylogenetic tree**, **Taxonomic classification**.

Mosaic protein

Multidomain proteins containing multiple types of protein domains of distinct evolutionary origin. Such chimeric proteins that arose by fusion of gene segments of two or more different genes are also referred to as modular proteins, and the constituent domains are usually referred to as protein modules. *See also* **Modular protein**, **Multidomain protein**, **Protein domain**, **Protein module**.

Motif

Motifs are characteristic conserved, short regions of peptide or nucleic acid sequence.

mRNA (messenger RNA)

The mature RNA product of the transcription of protein-coding genes. In eukaryotes mRNA molecules are derived from the primary transcript through the removal of introns and the addition of a poly(A) tail and a methylated cap. In the case of eukaryotes mature mRNA is exported from the nucleus to the cytoplasm, the site of translation. Prokaryotic protein-coding genes usually lack introns and translation can take place on the nascent mRNA. Messenger RNAs usually contain a single open reading frame, although in bacteria the polycistronic message may encode more than one protein. The sequence of mRNA also includes a 5' untranslated region (5' UTR) and a 3' untranslated region (3' UTR). *See also* **Transcription, Translation**.

Multidomain protein

Proteins containing multiple domains. Many multidomain proteins contain multiple copies of a single type of protein domain, indicating that internal duplication of gene segments encoding a domain has given rise to such proteins. Some multidomain proteins contain multiple types of domains of distinct evolutionary origin. Such chimeric proteins that arose by fusion of two or more gene segments are frequently referred to as mosaic proteins or modular proteins, and the constituent domains are usually referred to as protein modules. *See also* **Modular protein, Mosaic protein, Protein domain, Protein module**.

Multiple-sequence alignment

Matching the amino acids or nucleotides of multiple sequences in such a way that their similarity is maximized. Alignment of multiple homologous sequences requires that all the nucleotides (or amino acids) from 'equivalent positions' (i.e. of common ancestry) are brought into vertical register. The procedures that attempt to align all (and only) sites of common ancestry are referred to as sequence alignment procedures.

Mutagen

A physical or chemical agent that induces a permanent change in the genetic material.

Mutagenicity

The capacity of a chemical or physical agent to cause permanent genetic alterations.

Mutant

A strain, gene, allele, protein etc. that differs from the wild type.

Mutation

A permanent alteration in the genetic material.

Mutation matrix. *See* **Amino acid exchange matrix**.

Natural selection

Natural selection is the process in which organisms that are better adapted to their environment increase in frequency relative to less well-adapted forms because individuals better suited to their environment contribute a higher proportion of progeny to the next generation. *See also* **Darwinism**.

Neighbour-joining method of phylogeny reconstruction
The neighbour-joining method for inferring phylogenetic trees is based on the minimum-evolution principle. Neighbour joining resolves the phylogeny by successively joining pairs of taxa in such a way that pairing minimizes the sum of branch lengths.

Networks of genetic, metabolic and molecular interactions
Biological molecules (DNA, RNA, proteins, polysaccharides, lipids etc.) interact with each other, affecting each other's conformations, activities etc. Interacting molecular components thus form molecular networks. There are several distinct types of interactions depending on the type of interacting molecules. For example, protein–DNA interactions are crucial for transcriptional regulation (e.g. interaction of a transcription factor and a DNA *cis*-regulatory region), protein–RNA interactions play important roles in post-transcriptional regulation, regulation of splicing, translation etc. Protein–protein interactions are crucial in the formation of multimeric complexes and their regulation through post-translational modification. Molecular interactions between biological molecules are often represented as a graph, with genes as vertices and interactions between them as edges. *See also* **Interactome**.

Neutral drift
Synonym of genetic drift.

Neutral mutation
A mutation with the same fitness as the other allele or alleles at its locus.

Neutral theory
Evolutionary theory suggesting that most evolutionary change at the molecular level is caused by the random genetic drift of mutations that are selectively neutral or nearly neutral.

Nonconservative substitution
A mutation which results in the substitution of one amino acid in a polypeptide chain with an amino acid that has drastically different physicochemical properties. *See also* **Conservative substitution**.

Nonsense codon. *See* **Stop codon**.

Nonsense-mediated mRNA decay (NMD)
An mRNA surveillance mechanism that ensures the degradation of aberrant mRNAs that contain premature termination codons as a result of mutations or errors during transcription or splicing of RNA.

Nonsense mutation
A mutation that causes a nonsense (stop or termination) codon to replace a codon representing an amino acid. *See also* **Sense codon**, **stop codon**.

Nonsynonymous mutation
Nucleotide substitution in translated regions of protein-coding genes that alters the amino acid. *See also* **Missense mutation**, **Synonymous mutation**.

N-terminus
In a polypeptide chain the first residue has a free N-amino (NH2-) group, therefore this terminus of the polypeptide is referred to as the N-terminus. During translation of mRNA, protein synthesis proceeds from the N-terminus towards the C-terminus. *See also* **Amino terminus**.

Nuclear intron. *See* **Intron**.

Nucleic acid
A macromolecule composed of nucleotide units. There are two types of nucleic acids: DNA and RNA.

Nucleolar organizer region
A part of some chromosome containing tandem repeats of ribosomal RNA genes. In the human genome, nucleolar organizer regions are located on the short arms of chromosomes 13, 14, 15, 21 and 22.

Nucleomorph
A eukaryotic endosymbiont alga whose cellular structure is reduced to only a membrane-bound nucleus and a chloroplast. Nucleomorph genomes are extremely reduced and compact compared with those of free-living algae.

Nucleotide
Basic building block of DNA and RNA. A nucleotide consists of a sugar (deoxyribose in DNA and ribose in RNA) and phosphate backbone with a nitrogenous base (adenine, guanine, thymine or cytosine in DNA; adenine, guanine, uracil or cytosine in RNA) attached.

Nucleus
A region of eukaryotic cells enclosed within a membrane that contains the nuclear DNA genome. *See also* **Eukaryote**.

Open reading frame (ORF)
The sequence of cDNA or mRNA located between the start codon (initiation codon) and the stop codon (termination codon). In prokaryotes, which have no introns, ORFs of cDNA and mRNA correspond to ORFs of genomic DNA.

In eukaryotes genes introns may interrupt the ORFs of the corresponding mRNA. *See also* **Translation**.

Operational taxonomic unit (OTU)
In a phylogenetic tree, the terminal nodes correspond to observable genes or species. The term is used to distinguish these observable nodes from internal nodes, which are assumed to have existed earlier in evolution. *See also* **Phylogenetic tree**.

Operon
A unit of genetic regulation comprising a set of genes under the coordinate regulation of an operator gene.

Optimal alignment
The optimal alignment of a pair or multiple sequences refers to the highest scoring alignment out of many possible alignments, as assessed using a scoring system consisting of a residue exchange matrix and gap penalty values. *See also* **Alignment**.

ORF. *See* **Open reading frame**.

Organelle
A distinct structure found in the cytoplasm of eukaryotic cells (e.g. mitochondrion and chloroplast).

Organismic complexity
Organismic complexity is a function of the number and types of cells, tissues and organs they are constructed from, the number of different states of these constituents as well as the number and dynamics of their interactions with other constituents and with the environment.

Ortholog
Homologous genes in different species that arose through speciation, i.e. they derive from a single ancestral gene in the last common ancestor of the respective species. We speak of one-to-one orthologs if, following speciation, the orthologs did not undergo gene duplication in either species, one-to-two orthologs if the gene was duplicated in one of the species following their divergence etc. Orthology relationships are usually determined through sequence similarity analyses and phylogenetic analysis. Sets of orthologous genes are collected in the COG (clusters of orthologous groups) data set. *See also* **Reciprocal best hits**.

Orthology
Homology of genes in different species derived from a common ancestor through speciation, i.e. they are direct evolutionary counterparts in different species.

Pairwise alignment
Matching the amino acids or nucleotides of a pair of sequences in such a way that their similarity is maximized. Alignment of homologous sequences requires that the nucleotides (or amino acids) from 'equivalent positions' (i.e. of common ancestry) are brought into vertical register. The procedures that attempt to align all sites and only sites of common ancestry are referred to as sequence alignment procedures. *See also* **Alignment**.

PAM matrix of amino acid substitutions. *See* **Amino acid exchange matrix**, **Dayhoff amino acid substitution matrix**.

PAM matrix of nucleotide substitutions
PAM matrices for scoring DNA sequence alignments incorporate the information from mutational analyses which revealed that point mutations of the transition type (A $\leftrightarrow$ G or C $\leftrightarrow$ T) are more probable than those of the transversion type (A $\leftrightarrow$ C, A $\leftrightarrow$ T, G $\leftrightarrow$ T, G $\leftrightarrow$ C). *See also* **Transition**, **Transversion**.

Paralog
Two genes are said to be paralogous (or paralogs) if they descended from a common ancestral gene through gene duplication. *See also* **Ortholog**.

Paralogy
Paralogy describes the relationship of homologous genes that arose by gene duplication.

Parasite
An organism that takes nutrients from a different species without providing any benefit to the host.

Penalty. *See* **Gap penalty**.

Peptide bond
A covalent bond between two amino acids, in which the carboxyl group of one amino acid and the amino group of another amino acid react to form an amide bond.

Percent accepted mutation matrix. *See* **Dayhoff amino acid substitution matrix**.

Phage
A virus for which the natural host is a bacterial cell.

Phenetic classification
Classification methods that cluster descendants according to similarities. *See also* **Cladistic classification**.

Phenogram
Tree that describes the observed similarity relationships among descendants.

Phenotype
The physical and functional characteristics of an organism produced by the interaction of its genotype and the environment.

Phenylalanine
Phenylalanine is a large aromatic amino acid found in proteins. In protein sequences it is written as Phe or F.

Photosynthesis
The biological process by which photosynthetic organisms make organic compounds such as carbohydrates from atmospheric carbon dioxide and water using light energy.

Phylogenetic footprinting
A technique involving the use of comparative genomics to distinguish conserved functional DNA elements (e.g. enhancers, transcription factor binding sites) from surrounding nonfunctional and nonconserved DNA sequence. *See also* **Comparative genomics**.

Phylogenetic tree (phylogeny, phylogeny reconstruction)
A tree representing the evolutionary relationships of biological entities. A phylogenetic tree contains interior and exterior (terminal) nodes. The contemporary sequences are referred to as the operational taxonomic units (OTUs) and they correspond to the exterior nodes of the evolutionary tree. The internal nodes represent ancestral sequences. The length of each branch connecting a pair of nodes may correspond to the estimated number of substitutions between two associated sequences.

Phylogenomics
Genome-scale use of phylogenetic methods.

Phylogeny
The ancestral relations among biological entities.

Phylogeny reconstruction. *See* **Phylogenetic tree**.

Plasmid
An extrachromosomal genetic element capable of autonomous replication and transmission to progeny cells. In bacteria plasmids can be passed between cells independently.

Ploidy
Refers to the number of complete sets of chromosome of an organism: haploid (1N); diploid (2N); triploid (3N); tetraploid (4N).

Point mutation
The change of a single base pair of the genetic material.

Polyadenylation
Addition of a poly(A) tail to an mRNA molecule. The site of polyadenylation is generated by cleavage of the primary RNA transcript. In higher eukaryotes, the site of polyadenylation is specified by the sequence AAUAAA located upstream from the cleavage site.

Poly(A) tail
A string of adenylic acid residues (typically 50–200 bases) added by poly(A) polymerase to the polyadenylation site at the 3′ of the primary mRNA transcript. The presence of the poly(A) tail increases the stability of the mRNA.

Polymorphism
The condition in which the DNA sequence shows variation between individuals in a population.

Polypeptide
Linear polymer of amino acids joined covalently by peptide bonds between the α-carboxylic acid group and the α-amino group of constituent amino acids. Proteins are large polypeptides.

Polyploid
An organism containing more than two sets of genes and chromosomes.

Polypyrimidine tract
Spliceosomal introns usually have a pyrimidine-rich segment in the vicinity of the 3′ end of the intron. During splicing U2 snRNP binds to the branch site and a polypyrimidine tract. *See also* **Acceptor splice site**, **Branch site**.

Population
A group of organisms that interbreed and share a gene pool.

Position-specific scoring matrix
A commonly used representation of patterns in biological sequences. Position-specific scoring matrices assume independence between positions in the pattern.

Positive Darwinian selection (positive selection)
Positive Darwinian selection is natural selection that acts to favour amino acid replacements. *See also* **Natural selection**, **Purifying selection**.

Positive selection. *See* **Positive Darwinian selection**.

Post-translational modification
Protein modifications following its initial synthesis, e.g. phosphorylation, glycosylation, sulphation, addition of fatty acid moieties.

Post-translational processing
Reactions that alter a protein's covalent structure, such as proteolytic cleavage (e.g. removal of signal peptides, pro-regions) or various types of post-translational side-chain modifications.

Power law (Zipf's law)
A frequency distribution of entities in which a small number of entities are very frequent. The best-known example of the power law is the usage of words in texts. Zipf's law states that some words are used frequently while most other words are used infrequently and the frequency distribution can be approximated to a power law function. It has been found that power law is also applicable in contexts such as the occurrence of protein families, folds, domains or protein–protein interactions, i.e. within a larger population a few members are dominant.

pre-mRNA
The primary transcript of protein-coding genes that is subsequently processed through splicing reactions, polyadenylation etc. to generate the mRNA.

Pribnow box
The Pribnow box is the sequence TATAAT that is essential for transcription initiation in prokaryotes. It is located about 10 base pairs upstream from the first transcribed nucleotide.

Primary structure
Refers to the linear sequence of amino acid residues or nucleotides within protein or nucleic acid molecules, respectively. *See also* **Quaternary structure**, **Secondary structure**, **Tertiary structure**.

Primary transcript
When a structural gene (RNA gene, protein-coding gene) is expressed, the initial product is the primary transcript. This primary transcript may be then processed (e.g. to remove introns, to polyadenylate the 3′ end).

Primate
A mammal belonging to the order primates, which includes prosimians, monkeys, apes and humans.

Prokaryote

Organisms (Bacteria and Archaea) lacking a membrane-bound, structurally discrete nucleus and other subcellular compartments. *See also* **Eukaryote**.

Proline

Proline is a small nonpolar amino acid found in proteins. In protein sequences it is written as Pro or P.

Promoter

The site to which RNA polymerase binds and initiates transcription. The promoter is located close to the transcription start site and it specifies the site of initiation of transcription. Polymerase I and II genes have their promoter sequences located immediately upstream and downstream from the transcription initiation site. RNA polymerase III promoters are located upstream from the transcription start site in some genes (snRNA genes) and mainly within the transcribed region in others (5S RNA and tRNA genes). For protein-coding genes the promoter starts with the nucleotide immediately upstream from the cap site, and includes binding sites for one or more transcription factors.

Protein

A macromolecule composed of one or more polypeptide chains.

Protein domain

A structurally independent, compact spatial unit within the three-dimensional structure of a protein. Distinct structural domains of multidomain proteins usually interact less extensively with each other than do structural elements within the domains. *See also* **Multidomain protein**.

Protein family

Groups of proteins sharing homology. *See also* **Domain family**, **Homologous superfamily**.

Protein module

A protein module is a structurally independent protein domain that has been spread by module shuffling and may occur in different multidomain proteins with different domain combinations. *See also* **Module shuffling**, **Multidomain protein**, **Protein domain**.

Protein–protein coevolution

Specific protein–protein interactions essential for some biological function (e.g. binding of a protein ligand to its receptor, binding of a protein inhibitor to its target enzyme) are ensured by a specific network of interresidue contacts between the partner proteins. During evolution, sequence changes accumulated by one of the interacting proteins are usually compensated by complementary changes in the other. Coevolution of protein–protein interaction partners is reflected by congruency of their cladograms. Thus, comparison of phylogenetic trees constructed from multiple-sequence alignments of interacting partners (e.g. ligand families and the corresponding receptor families) shows that protein ligands and their receptors usually coevolve so that each subgroup of ligands has a matching subgroup of receptors. *See also* **Coevolution**.

Protein structure
The folded three-dimensional structure of a protein. Protein structure is usually represented as a set of three-dimensional coordinates for each atom of the protein.

Proteome
The full protein complement of a genome, cell, organism or other biological entity.

Proteomics
A research area dealing with the study of the full set of proteins encoded by a genome, cell, organism or other biological entity. *See also* **Genomics**.

Protostome
A taxon of invertebrate animals that includes arthropods, nematodes, roundworms, molluscs, annelids, flatworms etc. A distinguishing feature of protostome development is that the mouth forms at the site of the blastopore and the anus forms as a second opening. *See also* **Deuterostome**.

Protozoa
A group of unicellular eukaryotic organisms belonging to various phyla. Most feed on decomposing organic matter but many of them are parasites, including the agent that causes malaria (*Plasmodium*).

Pseudogene
A sequence of DNA homologous to a gene but nonfunctional.

Pseudorevertant
A mutant which has recovered a wild-type phenotype due to a second mutation that does not involve the reversal of the initial mutation.

Psyllids
Plant lice of the family Psyllidae. Psyllids include several small cicada-like insects that feed on plant juices.

Purifying selection (negative selection)
Natural selection that acts to eliminate mutations that are harmful to the fitness of the organism.

Purines
Nitrogen-containing, double-ring, basic compounds that occur in nucleic acids. The purines in DNA and RNA are adenine (A) and guanine (G).

Pyrimidines
Nitrogen-containing, single-ring, basic compounds that occur in nucleic acids. The pyrimidines in DNA are cytosine (C) and thymine (T); the pyrimidines in RNA are cytosine (C) and uracil (U).

Quaternary structure
Three-dimensional arrangement of two or more interacting macromolecules. Complexity of macromolecular assemblies may range from simple dimers to large multiprotein complexes and to protein–RNA assemblies such as the ribosome, spliceosome etc. *See also* **Primary structure**, **Secondary structure**, **Tertiary structure**.

Random drift
Synonym of genetic drift.

Reading frame
Since codons are nucleotide triplets, during translation of mRNA its nucleotides are read three at a time. In principle, each mRNA could be read in three reading frames depending on the choice of the nucleotide where translation starts. The actual reading frame of an mRNA is determined by the initiation codon. *See also* **Initiation codon**, **Open reading frame (ORF)**.

Recessive allele
If two different alleles for a single gene are present in a diploid cell each resulting in different phenotypes in the haploid state, only one of those phenotypes will be manifested in the diploid. The allele that encodes the hidden phenotype is the recessive allele.

Reciprocal best hits
Reciprocal best hits are proteins, cDNAs from different organisms that are each other's top hits in sequence similarity searches (e.g. BLAST searches) when the complete proteomes or transcriptomes from those organisms are compared to each other. Orthologous sequences are usually reciprocal best hits. *See also* **Ortholog**, **Sequence similarity search**.

Recombination

Recombination is the rearrangement of genetic material. Crossing over of homologous chromosomes during meiosis leads to the exchange of DNA between a pair of chromosomes. *See also* **Crossing over**, **Chiasma**.

Redundancy

In genomics redundancy refers to the fact that some genes appear to be superfluous since there are additional genes that fulfil similar roles. A manifestation of redundancy is that they can be eliminated without major effect on viability of the organism. *See also* **Genetic redundancy**, **Gene disruption**.

Regulatory region

A DNA region that controls gene expression.

Repetitive DNA

Sequences of varying lengths that occur in multiple copies in the genome. These may be short repeats just a few base piars long, like CACACA. They can also range up to a few hundred base pairs. Examples of the latter include Alu repeats, LINEs, SINEs. In shorter repeats like di- and trinucleotide repeats, the number of repeating units can frequently change during evolution. *See also* **Microsatellite**, **Minisatellite**.

Repetitive sequence. *See* **Simple DNA sequence**.

Replication

The copying of a nucleic acid molecule into a new nucleic acid molecule of the same type, i.e. DNA to DNA or RNA to RNA.

Replication origin

The location or locations at which DNA replication initiates within a genome. Bacteria usually have a single replication origin while eukaryotic genomes have numerous origins.

Replication terminus

The point at which DNA replication terminates. Bacteria usually have a single replication terminus while eukaryotic genomes have numerous termini.

Replicon

The region between the replication origin and the replication terminus. In bacteria usually a single replicon corresponds to the whole genome, but in eukaryotes there are numerous replicons.

Retrogene

A gene produced through reverse transcription of an RNA and the insertion of the resulting cDNA in the genome.

Retrosequence
A genomic sequence that originates from the reverse transcription of an RNA molecule and its subsequent integration into the genome. Usually, the RNA template is the RNA transcript of a gene. Most of the retrosequences derived from protein-coding genes are derived from processed mRNAs which thus lack introns. A retrosequence that is functional is usually called a retrogene, while nonfunctional retrosequences are called retropseudogenes or processed pseudogenes.

Reverse transcriptase
A DNA polymerase that copies an RNA molecule into single-stranded complementary DNA (cDNA).

Ribonucleotide. *See* **Nucleotide**.

Ribose
A five-carbon sugar that serves as a component of RNA. *See also* **RNA (ribonucleic acid), Deoxyribose**.

Ribosomal RNA (rRNA)
Ribosomal RNAs (rRNAs) provide the structural and catalytic core of the ribosome, the ribonucleoprotein complex that is the site for protein synthesis. In prokaryotes (Eubacteria, Archaea) and organelles of prokaryotic origin (mitochondria, chloroplasts) the small ribosomal subunit contains the 16S rRNA, the large ribosomal subunit contains two rRNA species (the 5S and 23S rRNAs). In eukaryotes the rRNA of the small ribosomal subunit is the 18S rRNA, the large subunit contains three rRNA species (the 5S, 5.8S and 25S/28S rRNAs). Prokaryotic 16S, 23S and 5S rRNA genes are organized in cotranscribed operons and they usually have multiple copies of the operon. Eukaryotes usually have many copies of the rRNA genes organized in tandem repeats; in humans approximately 300–400 rDNA repeats are present in five clusters (on chromosomes 13, 14, 15, 21 and 22).

Ribosome
A cellular particle involved in the translation of mRNA. The ribosome is composed of ribosomal RNAs and ribosomal proteins. Ribosomes consist of two subunits. Prokaryotes have 70S ribosomes (consisting of a small 30S and a large 50S subunit), whereas eukaryotes have 80S ribosomes (consisting of a small 40S and a large 60S subunit).

Ribosome binding site
The site on the mRNA that directs its binding to the ribosome and helps to define the position of the translation start site. The sequence nature of these sites differs between prokaryotes and eukaryotes. The prokaryotic

ribosome-binding site is known as the Shine–Dalgarno sequence; it has the consensus AGGAGG and is located 4–7 bp upstream of the AUG translation initiation codon. The Shine–Dalgarno sequence is complementary to the sequence at the 3′ end of 16S rRNA and is involved in binding of the ribosome to mRNA. In eukaryotes the equivalent is the Kozak sequence.

Ribozyme

A catalytically active RNA.

RNA (ribonucleic acid)

RNA is one of the two major forms of nucleic acids. The nucleotides of RNA are composed of three parts: a sugar, a phosphate and one of the bases: adenine, guanine, cytosine and uracil. The major chemical difference between RNA and DNA is that the sugar is ribose in RNA whereas it is deoxyribose in DNA.

RNAi

RNA interference or RNA silencing is the mechanism by which small double-stranded RNAs can interfere with expression of any mRNA having a similar sequence. These small RNAs are known as short interfering RNAs (siRNA). The mode of action for siRNA appears to be via dissociation of its strands, hybridization to the target RNA, extension of those fragments by an RNA-dependent RNA polymerase, then fragmentation of the target. Since remnants of the target molecule act as an siRNA itself the effect of a small amount of starting siRNA is effectively amplified and can have long-lasting effects on the recipient cell. The RNAi technique is widely used to study the biological role of genes. *See also* **Gene disruption**.

RNA splicing

A series of reactions in which introns are excised from the primary transcript and exons are covalently linked to produce a mature RNA molecule. *See also* **Splicing**.

RNA structure

The spatial organization of an RNA molecule determined by interactions between RNA bases. The most important secondary structural elements of RNA are the stems maintained by consecutive canonical base pairs (i.e. G:C and A:U base pairs) between antiparallel segments of RNA. Unpaired nucleotides connecting antiparallel segments of a stem are usually referred to as hairpin loops. The overall tertiary structure of RNA is maintained by canonical and noncanonical (e.g. A:A) base pairs between bases that may be more distant in the primary sequence. *See also* **Hairpin**, **Stem loop**.

rRNA. *See* **Ribosomal RNA**.

r selection
Natural selection favours organisms with maximum reproductive and/or growth rate. According to the r/K selection theory selective pressures drive evolution in one of two stereotyped directions: r or K selection. Typically, r-selected species produce many offspring, whereas K-selected species have fewer offspring that have a better chance of survival. *See also* **K selection**.

Satellite DNA
Tandemly repeated simple sequence DNA found primarily in heterochromatin. Satellite DNA families include the human centromeric alphoid sequence.

Scoring matrix (substitution matrix)
A matrix of pairwise values defining relationships between nucleotides or amino acid residues to calculate alignment scores in sequence comparison methods. In the simplest type of scoring matrix – the unitary matrix – the pairwise value for identities is 1, whereas the value assigned to nonidentical pairs is 0. More sophisticated scoring matrices have been derived for both nucleic acid and protein sequence comparisons that reflect the observed substitution rates. The most commonly used scoring matrices of this type are the Dayhoff mutation data matrices (PAM matrices) and the blocks substitution matrices (BLOSUM matrices) for protein sequence comparisons. *See also* **Alignment, BLOSUM matrix, Dayhoff amino acid substitution matrix**.

Secondary structure
Local structure within a protein or an RNA. *See also* **Primary structure, Quaternary structure, Tertiary structure**.

Secondary structure of proteins
Local conformation of the protein's backbone. The secondary structure can be irregular or regular; regions that have an irregular conformation are said to have random-coil conformation. The regular conformations are stabilized by hydrogen bonds between main-chain atoms; the most common of these conformations are the α-helix, the β-sheet and turn. *See also* **α-helix, Backbone, β-strand, Turn**.

Secondary structure of RNAs
Local conformation of the RNA (e.g. hairpins, stem-loop structures) determined by base pairing of nucleotides which are relatively closely positioned within the sequence. *See also* **RNA structure**.

Secondary structure prediction of protein
Prediction of the regions of secondary structure in a protein (e.g. α-helix, β-strand, coil) from its amino acid sequence.

Sequence logo

A sequence logo is a graphic representation of an aligned set of sequences displaying the frequencies of nucleotides (or amino acids) at each position as the relative heights of letters, along with the degree of sequence conservation as the total height of a stack of letters.

Sequence motif. *See* **Motif**.

Sequence of nucleic acids

The order of neighbouring nucleotides in a nucleic acid, listed from the 5′ to the 3′ end.

Sequence of proteins

The order of neighbouring amino acids in a protein, listed from the N- to the C-terminus. *See also* **Primary structure**.

Sequence similarity

Related nucleic acid and protein sequences usually retain a significant degree of sequence similarity that can be studied by comparing their aligned sequences. The degree of similarity is measured by some sort of scoring matrix that rewards similarity and penalizes dissimilarity. *See also* **Scoring matrix**.

Sequence similarity search

Sequence similarity searches align the query nucleic acid or protein sequence with all sequences in a database and calculates the similarity scores. The search provides a list of sequences ranked in the order of decreasing similarity scores.

Serine

Serine is a small polar amino acid found in proteins. In protein sequences it is written as Ser or S.

Sex chromosome

Chromosome present in different numbers in the two sexes of an organism and whose presence or numbers determines the sex of the organism. For example, in mammals the X and Y chromosomes are the sex chromosomes (females are XX, males XY). Unlike autosome pairs, the different sex chromosomes do not have identical gene contents. For example, the human X chromosome is larger than the Y chromosome and contains unique, essential genes. This leads to the possibility of sex-linked inheritance of some traits. *See also* **Autosome**.

Side chain

In a protein all atoms other than the backbone atoms are termed side-chain atoms. Interactions between side chains help determine the structure and stability of the folded state.

Signal sequence
A hydrophobic amino acid sequence that directs a growing peptide chain to be secreted into the endoplasmic reticulum.

Silencer
A control element whose interaction with a regulatory protein has a negative effect on transcription of a gene.

Silent mutation. *See* **Synonymous mutation**.

Simple DNA sequence
DNA sequence composed of repeated identical or highly similar short motifs. Simple sequences may be polymorphic in length. *See also* **Microsatellite**, **Minisatellite**.

SINE (short interspersed nuclear element)
Retrotransposons less than ~500 bp in length. SINEs constitute approximately 10% of the human genome, the majority of these belonging to the Alu family. In addition to the Alu repeat, many of the other SINEs are copies of tRNAs or small nuclear RNAs.

Single-nucleotide polymorphism (SNP)
Single-nucleotide polymorphisms are DNA variations in a single base within a genomic sequence within a population. *See also* **Mutation**, **Polymorphism**.

snRNA
Small nuclear RNA – forms complexes with proteins to produce snRNPs that are involved in RNA splicing and polyadenylation. *See also* **Spliceosome**.

snRNP
Small nuclear ribonucleoprotein particles are complexes of small nuclear RNAs and proteins. These particles are involved in RNA splicing and polyadenylation. *See also* **Spliceosome**.

Somatic cell
All cells of the body of multicellular organisms except gametes and their precursors. *See also* **Gamete**.

Somatic mutation
Mutation in somatic cells. Somatic mutations are not inherited. *See also* **Germline**.

Splice site
Refers to the location of the exon–intron junctions in genes or primary transcripts.

Spliceosome

The ribonucleoprotein complex involved in splicing of nuclear protein-coding genes of eukaryotes. Spliceosomes consist of multiple types of snRNAs and proteins. *See also* **Intron, Splicing**.

Splicing

The process of removal of introns from RNAs to produce the mature RNA molecule.

SR protein

The term 'SR protein' derives from the presence of an RS domain that is rich in serine (Ser, S) and arginine (Arg, R) residues. SR proteins are conserved in evolution; they contain one or two RNA binding motifs in addition to the RS domain. SR proteins facilitate the recruitment of the components of spliceosomes and are involved in regulating and selecting splice sites in eukaryotic mRNA. Alternative splicing requires SR proteins to select which alternative splice sites should be used.

Standard genetic code. *See* **Genetic code**.

Start codon. *See* **Initiation codon**.

Stem loop

A basic local element of RNA secondary structure, in which two anti-parallel sequences form a 'stem' through complementary base pairing. A short sequence that separates the complementary sequences of the stem forms a 'loop' at one end.

Stop codon

Nonsense codon. A codon that terminates translation.

Subfunctionalization

Subfunctionalization occurs when, following duplication of an ancestral gene with a broad functional spectrum (broad molecular specificity, broad expression spectrum etc.) or multiple functions, the daughter genes diverge in a way that they specialize to a subset of the functions of the ancestral gene. *See also* **Gene sharing**.

Substitution

Replacement of one nucleotide by another nucleotide in a nucleic acid sequence or replacement of one amino acid by another amino acid in a protein. *See also* **Mutation**.

Substitution matrix. *See* **BLOSUM matrix, Dayhoff amino acid substitution matrix**.

Supersecondary structure
Commonly occurring assemblies of secondary structure elements that constitute a higher level of structure than secondary structures.

Symbiont
The smaller participant living in symbiosis with another, unrelated organism. *See also* **Symbioisis**, **Parasite**.

Symbiosis
Mutually beneficial association of two organisms. The larger member of the symbiotic relationship is usually referred to as the host, whereas the smaller member is called the symbiont. *See also* **Parasite**.

Symmetrical exon
An exon flanked by introns of the same phase at both its 5' and 3' ends. *See also* **Intron phase**, **Exon shuffling**.

Synapsis
The process during meiosis where paired homologous chromosomes are attached along their lengths by the synaptonemal complex.

Synonymous mutation (silent mutation)
Due to the degeneracy of the genetic code more than one codon may encode the same amino acid. Nucleotide substitutions occurring in translated regions of protein coding genes are called synonymous (or silent) if they cause no amino acid change.

Syntenic block
Syntenic blocks are large chromosomal segments inherited from a common ancestral chromosome without major chromosomal rearrangements, i.e. in which gene order is conserved.

Synteny
This term refers to the state of (genes) being on the same chromosome. The phrase 'conserved synteny' is usually used to indicate that gene order is conserved on homologous (or homologous parts of) chromosomes.

Tandem repeats
Identical or related DNA sequences immediately adjacent to each other and in the same orientation. *See also* **Microsatellite**, **Minisatellite**.

TATA box
Consensus sequence located 30–35 bases upstream from the transcription start site of RNA polymerase II transcribed genes.

Taxonomic classification

A hierarchically structured terminology for organisms. There are seven levels in the present classification system: (1) kingdom (Archaea, Bacteria and Eukarya); (2) phylum; (3) class; (4) order; (5) family; (6) genus; (7) species.

Telomerase

The enzyme that adds specific DNA sequence repeats to the 3′ end of the telomere regions found at the ends of linear chromosomes.

Telomere

DNA sequences found at the ends of linear chromosomes. Telomeres are composed of simple repeats (TTAGGG in all vertebrates) reiterated several hundred times. The telomere stabilizes the chromosome ends and allows DNA replication to proceed to the end of the chromosome without loss of DNA material.

Termination codon

Stop codon or nonsense codon that does not specify any amino acid and causes termination of the translation of mRNA.

Terminator

A sequence downstream from the 3′ end of an open reading frame that terminates transcription by the RNA polymerase. In bacteria terminator sequences are palindromic capable of forming hairpins.

Tertiary structure

Three-dimensional structure of proteins or nucleic acids defined by inter-actions between amino acid residues or nucleotides which are distant within the primary structure of the molecule. *See also* **Primary structure**, **Quaternary structure**, **Secondary structure**.

Tetraploid

Organisms containing four homologous sets of chromosomes are said to be tetraplod. *See also* **Ploidy**.

Tetrapod

The vertebrate lineage with four leg-like appendages. This group includes amphibians, reptiles, birds and mammals.

Threading

Protein threading methods align and assess the compatibility of the sequence of an unknown structure with known structures.

Threonine

Threonine is a polar amino acid found in proteins. In protein sequences it is written as Thr or T.

Thymine (T)
A pyrimidine base in DNA. It forms a base pair with adenine.

TIM barrel
Triose phosphate isomerase barrel, one of the most common tertiary folds. A β barrel consists of eight parallel β-strands with each pair of adjacent strands connected by a loop containing an α-helix.

Transcript
The RNA produced by transcription of DNA. *See also* **Transcription**.

Transcription
The synthesis of a single-stranded RNA molecule from a sequence of DNA, e.g. from protein-coding genes or RNA genes. It is the first step of gene expression. Transcription is initiated at the transcription start site under the control of the promoter and other regulatory elements and proceeds until termination occurs. The product of transcription is an immature RNA that may undergo a variety of processing reactions to produce a mature RNA.

Transcription factor
Proteins that act in conjunction with an RNA polymerase and modify the initiation of transcription and control gene expression. Transcription factors usually bind to DNA at specific transcription factor binding sites. *See also* **Transcription**, **Transcription factor binding site**.

Transcription factor binding site
Sites on genomic DNA (in promoter regions and other regulatory regions) to which transcription factors bind.

Transcription start site
The point (adjacent to the promoter region at the 5' end of a gene) at which RNA polymerase initiates transcription.

Transcriptional regulatory regions
Sequences involved in regulating the expression of a gene, including sequences in promoter regions, enhancers, silencers. *See also* **Enhancer**, **Promoter**, **Transcription factor**, **Transcription factor binding site**, **Transcription start site**.

Transcriptome
The full complement of transcripts in a particular cell, tissue, organ or organism.

Transcriptomics
The measurement of the expression of all genes in an organism by micro-array analysis of cDNA.

Transfer RNA (tRNA)
A class of small, tightly folded RNA molecule. tRNAs have triplet nucleotide sequences (anticodons) that are complementary to the triplet codons of mRNA. The role of tRNAs in protein synthesis is to bond with amino acids and transfer them to the ribosomes, where proteins are assembled according to the genetic code carried by mRNA. There are 20 types of isoaccepting tRNA molecules, one type recognizing each of the 20 main amino acids.

Transition
A nucleotide substitution in DNA in which one pyrimidine is replaced by the other pyrimidine (C → T), or one purine is replaced by the other purine (A → G). *See also* **Transversion**.

Translation
The process of converting the sequence information encoded in a mature mRNA into a protein sequence using the genetic code. Translation is carried out by the ribosome in association with tRNAs, translation initiation, elongation and termination factors.

Translation start site
The position of an mRNA at which protein synthesis begins. The translation start site is usually an AUG codon that codes for methionine. In prokaryotes binding of the ribosome to the start AUG is mediated by the Shine–Dalgarno motif (UAAGGAG), a sequence complementary to the 3′ of the 16S rRNA of the 30 S subunit of the ribosome. In eukaryotes, the 40 S ribosomal subunit starts at the 5′ end and scans the message in the 5′ to the 3′ direction until it arrives at the AUG codon. A short recognition sequence, the Kozak motif (with a consensus sequence of ACCAUGG) usually surrounds the initiation codon.

Translocation
A type of mutation that results from the breakage and fusion of one chromosomal segment to another chromosome.

Transposable element
A class of DNA sequences that can move from one genomic site to another.

Transposition
The movement of DNA from one location to another location on the same molecule, or a different molecule within a cell.

Transposon
A transposable genetic element capable of moving from one site to another in a DNA molecule without any requirement for sequence relatedness at the donor and acceptor sites.

Transversion
A mutation changing a purine into a pyrimidine, or vice versa (i.e. changes from A or G to C or T and changes from C or T to A or G).

Trinucleotide repeat
A tandem repeat of three nucleotides.

tRNA. *See* **Transfer RNA**.

Tryptophan
Tryptophan is a large aromatic amino acid found in proteins. In protein sequences it is written as Trp or W.

Turn
A reversal in the direction of the backbone of a protein that is stabilized by hydrogen bonds between backbone NH and CO groups and which is not part of a regular secondary structure region such as an α-helix or β-sheet. *See also* **α-helix**, **Backbone**, **β-strand**, **Secondary structure**.

Tyrosine
Tyrosine is a large aromatic amino found in proteins. In protein sequences it is written as Tyr or Y.

Unequal crossing over
A crossing over in which the two chromosomes do not exchange equal lengths of DNA. *See also* **Crossing over**.

Universal genetic code. *See* **Genetic code**.

Untranslated region (UTR, 5′ UTR, 3′ UTR)
Regions of an mRNA molecule that do not code for a protein. The 5′ UTR is the portion of an mRNA from the 5′ end to the position of the first codon used in translation. The 5′ untranslated region contains the Kozak sequence which is involved in recognition of the translation start site. The 3′ UTR is the portion of an mRNA from the position of the last codon used to the 3′ end (e.g. the poly(A) site). Both 5′ and 3′ untranslated regions may contain sequences that mediate post-transcriptional regulation by affecting RNA stability.

UPGMA
The unweighted pair-group method using arithmetic averaging is a phenetic method of tree reconstruction. It can be used to construct phylogenetic trees only if the rates of evolution are constant among the different lineages. *See also* **Phenetic classification**.

Upstream/Downstream
This orientation reflects the direction of both the synthesis of mRNA, and its translation from the 5' end to the 3' end. In the case of genomic DNA a sequence element is said to be downstream if it is located distal to a specific point in the direction of transcription, upstream if it is in the opposite direction. For example, the transcription start site is downstream of the promoter region, whereas the promoter region is upstream of the transcription start site. It should be noted that downstream/upstream is meaningful only in conjunction with a given gene. In the case of mRNA a sequence element is said to be downstream if it is located distal to a given point in the direction of translation. For example, the coding region is downstream from the initiation codon, toward the 3' end of an mRNA molecule, whereas the coding region is upstream of the stop codon.

Uracil
A pyrimidine base found in RNA but not DNA; uracil is capable of forming a base pair with adenine.

Urochordate
A chordate marine animal. Urochordates (tunicates) have an unsegmented body and a notochord.

Valine
Valine is a nonpolar amino acid found in proteins. In protein sequences it is written as Val or V.

Vertebrates
The monophyletic group of animals that have an internal skeleton made of bone or cartilage.

Virulence
The disease-producing ability of a microorganism.

Virus
A noncellular biological entity that can replicate only within a host cell. Inside the infected cell the virus uses the synthetic capability of the host to produce progeny virus.

Wild type
The genotypes or phenotypes found most frequently in nature.

Wobble
The ability of the third base in some anticodons of tRNA to bond with more than one kind of base in the complementary position in the mRNA codon.

X chromosome
One of the two sex chromosomes, X and Y. *See also* **Y chromosome**, **Sex chromosome**.

Xenolog
Homologous genes acquired through horizontal transfer of genetic material between different species are called xenologs. *See also* **Homology**, **Horizontal gene transfer**.

Y chromosome
One of the two sex chromosomes, X and Y. *See also* **X chromosome**, **Sex chromosome**.

Yeast two-hybrid screening
A molecular biology technique used to discover protein–protein interactions by testing for physical interactions between two proteins.

Zipf's law. *See* **Power law**.

Zygote
The diploid cell formed by the fertilization of male and female gametes.

Index

Page numbers in *italics* represent figures, those in **bold** represent tables.